D1666327

Future Fixed and Mobile Broadband Internet, Clouds, and IoT/AI

IEEE Press
445 Hoes Lane
Piscataway, NJ 08854

Future Fixed and Mobile Broadband Internet, Clouds, and IoT/AI

Toni Janevski
Ss. Cyril and Methodius University
Skopje, North Macedonia

WILEY

Published by John Wiley & Sons, Inc., Hoboken, New Jersey. Published simultaneously in Canada.

Library of Congress Cataloging-in-Publication Data:

Names: Janevski, Toni, author.
Title: Future fixed and mobile broadband Internet, clouds and IoT/AI / Tori
  Janevski.
Description: Hoboken, New Jersey : Wiley, [2024] | Includes index.
Identifiers: LCCN 2023057788 (print) | LCCN 2023057789 (ebook) | ISBN
  9781394187966 (hardback) | ISBN 9781394187973 (adobe pdf) | ISBN
  9781394187980 (epub)
Subjects: LCSH: Broadband communication systems. | Cloud computing. |
  Internet of things. | Artificial intelligence.
Classification: LCC TK5103.45 .J36 2024 (print) | LCC TK5103.45 (ebook) |
  DDC 621.39/81—dc23/eng/20240128
LC record available at https://lccn.loc.gov/2023057788
LC ebook record available at https://lccn.loc.gov/2023057789

Cover Design: Wiley
Cover Image: © Yuichiro Chino/Getty Images

Set in 9.5/12.5pt STIXTwoText by Straive, Chennai, India

To my great sons, Dario and Antonio, and to the most precious woman in my life, Jasmina.

Contents

About the Author

Toni Janevski, Ph.D., is Full Professor in telecommunications at the Faculty of Electrical Engineering and Information Technologies (FEEIT), Ss. Cyril and Methodius University (UKIM), Skopje, North Macedonia. During 1996–1999, he worked for T-Mobile Macedonia. Since 1999, he has been with FEEIT. During 2005–2008, he was Member of the Commission of the Agency for Electronic Communications of Macedonia. From 2008 to 2016, he was Member of the Senate of UKIM. In 2009, he established the ITU Center of Excellence (CoE) of FEEIT, and served as its head during the period 2009–2022. Since 2009, he has successfully tutored and/or coordinated many international ITU courses in the ITU Academy every year. He received the "Goce Delchev" state award for science in 2012 and "UKIM best scientists" award for 2013, both of which can be received once in a lifetime. Professor Janevski has written multiple books with the well-known publishers Wiley and Artech House, as well as over 200 scientific papers in the field of telecommunications in journals and conferences. He is elected Head of Telecommunications Institute of FEEIT for the term 2023–2026. Prof. Janevski is elected member of the ITU Group on Capacity Building Initiatives (GCBI) for two 4-year terms, 2019–2022 and 2023–2026.

1

Fixed and Mobile Broadband Evolution

The telecommunication world has demonstrated its power to digitalize the life in the past years, providing all services and applications from the physical world to the cyber (or otherwise called digital) world. When we talk about the digital world, digitalization, etc., we mean the open Internet, which exists parallel to our physical world. So the open Internet is the main "platform" for all the services we enjoy today in our work and life and society in general.

When did the growth of the Internet begin? For the world as a whole, the growth of the Internet began in the 1990s and continues at the same pace until today [1–3]. Nowadays, Internet technologies are the main network technologies in telecommunication networks, including fixed and mobile.

What was crucial for Internet/IP to become what it is today, a major networking technology in today's telecommunications world? Well, it is due to the design of Internet Protocol (IP, as a protocol) to be flexible to accommodate different underlying transport technologies (e.g. Ethernet, Wi-Fi, mobile access networks, optical transport networks, and satellite networks) and all the various applications and services running over the top (therefore called Over The Top – OTT services/applications).

1.1 Evolution of Fixed and Mobile Telecommunications

The evolution of telecommunications started with fixed access networks dedicated to telephony at the end of the 19th century, which continued through most of the 20th century. Telecommunications include all technologies that are available at the given time for transfer of different types of information, such as audio (e.g. voice, or in other words, telephony), video (e.g. television at the beginning and many new video services at the present time), and data (everything else that is not included as audio or video as media, such as various types of services, e.g. email and Web). What technologies existed in the 19th century? Well, that was electricity, so different types of information could be transmitted on distance by using electric signals (e.g. DC – direct current electric signals).

1.1.1 Initial Telecommunication Technologies

What is the first telecommunication service? Well, it's telegraphy, which is a type of data service, because telegrams were messages transmitted by electrical signals over long-distance wires. Telegraphy predates telephony, which was invented in 1876 by Alexander Graham Bell. Working

Future Fixed and Mobile Broadband Internet, Clouds, and IoT/AI, First Edition. Toni Janevski.

telegraphy began in the early 19th century, with the first commercial electric telegraph service opened in London in 1839 with a system created by Charles Wheatstone [4]. In the United States, Samuel Morse created the well-known Morse code in 1844 for use in telegraphy. At the same time in 1843, Alexander Bain patented in the United Kingdom the first image transmission system, which is considered a precursor to the fax services that were used later in the 20th century for business communications. Thus, the first telecommunication service was actually telegraphy, which belongs to data services. With the advent of telegraphy came national and international telegraph services, and international traffic raised the issues of politics, language, and economics, which are also issues nowadays that telecommunications must constantly deal with in parallel with all the technological changes over time. For example, a telegram (that is, a message transmitted through a telegraphic system) written in one language (e.g. English) needed to be translated into the recipient's language (e.g. German, Italian, Spanish, and French). It refers to the content of the message. However, the correct transmission of messages from country A to country B required the same approach to encoding and decoding messages adopted by both countries. Also, the financial part, such as who will pay for sending a telegram (sender, receiver, or partly both parties) was also an issue that needed to be resolved at the national and especially the international level.

The telegraph connected major cities in many different countries around the world in a short period of time. The first submarine cable was used in 1850 to connect France and Great Britain, while the first transatlantic cable was laid in 1858 to connect North America to Europe. All that imposed the need for international agreements between governments, i.e. administrations, and the need to process standardization in telecommunications (at that time telecommunications was basically just telegraphy). This led to the founding of the International Telegraph Union (ITU) in 1865. Later the ITU got the name International Telecommunication Union where the word "telegraphy" was changed to "telecommunication," considering that telephony appeared in 1876 and television was demonstrated for the first time in 1925 in London [4].

But before television, wireless telecommunications began with the invention of the wireless telegraph, invented in 1895 by Guglielmo Marconi in Bologna [5]. The wireless, that is radio telegraphy, was analogous to wire telegraphy used earlier in the 19th century. The Marconi Wireless Telegraph Company was later formed to provide World Wide Wireless (the initial WWW about a century before the emergence of the WWW as well-known web services).

Radio broadcasting and television were the next telecommunication services, after telegraphy as the first and telephony as the second. So, after the data service (telegraphy) and audio service (telephony) appeared radio broadcasting (as an audio-based service) and television (TV) as a multimedia service, consisting of video with accompanying audio. However, the spread of radio broadcasting was in the first half of the 20th century and TV broadcasting (initially also based on radio) in the second half.

Telecommunications originally used so-called analog signals obtained by modulating electrical signals in copper cables or radio signals (for radio/wireless communication) with audio, video, or information data and transmitting such a signal from a transmitter at one end to a receiver at the other. However, it required separate networks for different types of signals, such as separate telephone networks, separate radio broadcast networks, separate TV broadcast networks/equipment, and separate data networks. With the transition of legacy telecom world to Internet technologies it became possible for all services and applications to be provided over the same broadband IP-based networks and services (as shown in Figure 1.1). That provided the possibility to have one broadband network (with fixed and mobile access) for all existing and future services.

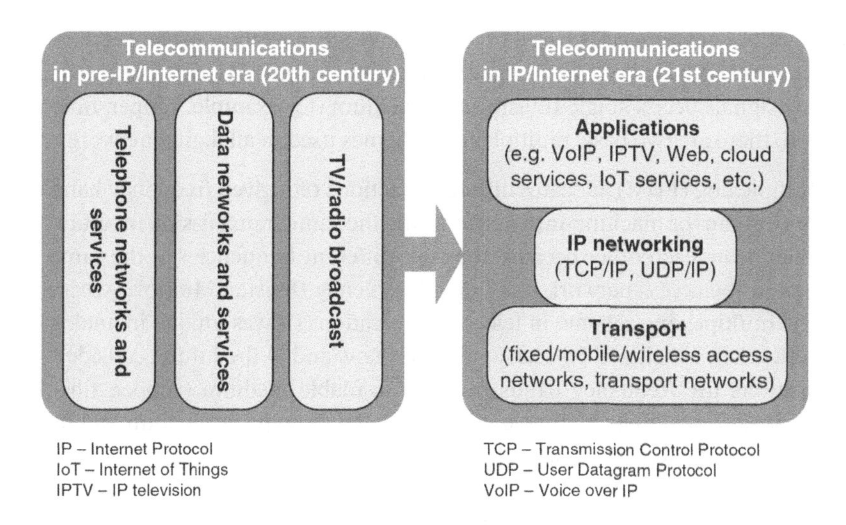

Figure 1.1 Convergence of legacy telecommunication to IP-based networks and services.

1.1.2 Digital Telecommunication World

The driver for the convergence of different types of information into one network was digitalization of signals and systems, driven by the introduction of computer science and informatics into the telecommunications world gradually from the 1960s. By digitizing naturally analog signals, all signals are represented as a series of digits. In the world of telecommunications, it is an unwritten rule that the simplest solution that gets the job done is often the best solution. Thus, although different digital systems can be defined (with a digital base of 2 digits, 3 digits, 4 digits, and so on), the usual approach to encoding information is with the binary system, which consists of two digits, one (1) and zero (0). The simplicity lies in the fact that when a given signal representing a binary 1 or a binary 0 (with noise added to the transmission path) is received, a single threshold is required to decide at the receiver's end whether the digit sent in the given time interval was a binary "1" or "0."

With the transformation of telecom networks from analog to digital since the 1970s and 1980s, it became possible to use the same network for different types of information or media (i.e. audio, video, multimedia, and data). However, different types of media and different services also required different capacity (expressed in bits per second) and different performances by the networks (in terms of end-to-end delay and losses). At that time, it was evident that video requires the most bandwidth (i.e. bitrates), which directly was related to television and its transition from analog to digital. On the other side, voice services (without accompanying video) require much less bitrate than video (e.g. TV).

1.1.2.1 Circuit Switching

Originally, digital telecommunications networks followed the same approach as analog networks before them, that is, they used an allocation of a fixed (i.e. dedicated) amount of bandwidth per flow (for example, a voice call in a given direction) called circuit switching. In fixed digital telephony, the dedicated circuit switching channel was 64 kbit/s in each direction over the telecommunications networks it traversed. The allocation of dedicated bandwidth (e.g. a time slot on a given frequency in a wired or wireless medium) was based on the high synchronization in the network required to multiplex different input streams into larger aggregated streams. Multiplexing is the technique of placing many signals over a single transmission medium (for example, copper, fiber optics, or radio). In general, there are two basic multiplexing schemes used in all digital networks:

- Frequency Division Multiplexing (FDM) allocates different fractions of a given frequency band to different connections (human- or machine-initiated) sharing the same transmission medium. However, they are isolated from each other because they use different frequencies at the same time. When FDM is used in the access network it is called Frequency Division Multiple Access (FDMA). This is the first multiplexing scheme in telecommunications (it was unique in analog telecommunications networks), which will be used at all times now and in the future, considering that telecommunications use frequency bands in every available medium (copper, fiber optics, and radio). Of course, the frequency range used depends on the type of medium and its characteristics in terms of signal attenuation, as well as whether the signals are electrical, optical, or radio signals over copper, optical, and radio links, respectively.
- Time Division Multiplexing (TDM) uses different time intervals, called time slots, which are assigned to different links at the same time using the same frequency in the case of copper cables or radio transmission, or the same wavelength in the case of fiber as the transmission medium. All digital telecommunication transmission systems use TDM. When TDM is used in access networks, it is called Time Division Multiple Access (TDMA). The use of TDM/TDMA requires a certain level of synchronization in the networks.

For multiplexing purposes, circuit-switched transport networks were based on Synchronous Digital Hierarchy (SDH) [3], which uses a centrally positioned primary reference clock with the highest accuracy, and its reference clock is distributed to all nodes in the SDH network via synchronization paths. The American version of SDH was called Synchronous Optical Network (SONET). SDH was based on time slots with a bitrate of 64 kbit/s, originally designed to carry digital voice traffic, although later (from the 1990s onward) the same SDH/SONET transport networks were also used to carry data traffic (e.g. Internet traffic).

What was the main disadvantage of circuit switching? It was the deterministic allocation of resources, such as the mentioned 64 kbit/s in digital telephony. With the aim to provide efficient allocation of network resources, such as frequencies and time slots (as the noted examples of FDM and TDM), circuit switching was inefficient due to several reasons including (but not limited to) slotted resources (e.g. multiples of 64 kbit/s, based primarily on resources needed for digital telephony in the 20th century [3]) as well as use of fixed amount of resources even in cases when there is no traffic to transfer over the given channel (lower flexibility). Therefore, the next stage in the evolution of telecommunications was packet switching, which aimed to provide efficiency and flexibility in resource allocation (frequencies and time slots) in telecom networks.

Figure 1.2 shows mobile and fixed networks based on circuit switching. As for mobile networks, the first digital circuit switching networks belong to the so-called 2G (second generation mobile systems), of which Global System for Mobile Communications (GSM) is the most famous representative that appeared in the 1990s (2G era), standardized by the European Telecommunications Standardization Institute (ETSI). Also, the next generation of mobile networks called 3G had its Circuit-Switching (CS) part, at least the 3G mobile systems that evolved from GSM mobile networks, and were standardized by 3G Partnership Project (3GPP), led by ETSI [6]. The main service in 2G and 3G mobile networks based on circuit switching was voice (i.e. mobile telephony). On the other side, all fixed telephone networks in the 20th century (until the 1990s) were based on circuit

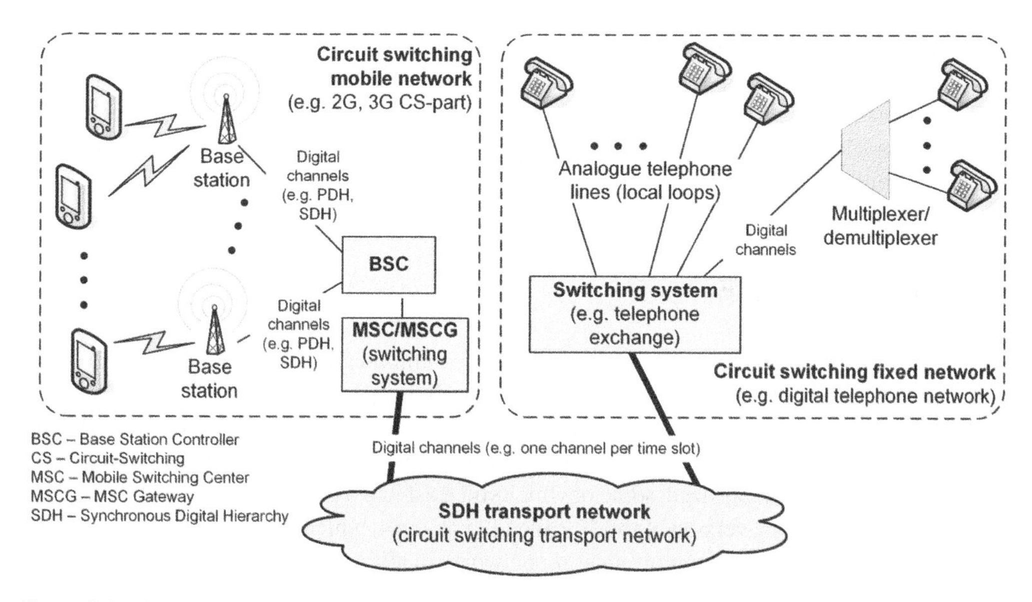

Figure 1.2 Fixed and mobile networks based on circuit switching.

switching (also shown in Figure 1.2). So, the telecom world in the 20th century was based on circuit switching until the 1990s (i.e. before the spread of the Internet/IP technologies).

1.1.2.2 Packet Switching

Packet switching is based on the transmission of information through the so-called packages. What is a packet in telecommunication terms?

Well, a packet refers to a small portion of data that is transmitted based on information contained in so-called packet headers, which carry the necessary control and addressing information for the packet. So, in general, packets are similar to traditional postal packets, with packet "content" (that is, the data, such as digital form of audio, video, and other data, in the packet payload) and an address for the packet's destination (in the header on the packet).

Packet switching is possible only in digital networks, while circuit switching is possible in both analog and digital telecom networks, because the concept of packet header and payload is only possible when the data is in digital form, mainly in the form of ones and zeros (i.e. binary form).

There have been and still are a variety of packet switching technologies. As usual in the world of telecommunications, any new technology is first implemented in the control part of the network, which mainly refers to signaling as a standardized exchange of various control information. For example, in telephony, the important control information is the telephone number for which the dialed digits are transmitted by signaling. Thus, the world's first packet switching system implemented in telecommunication networks was Signaling System 7 (SS7), which was standardized by the ITU for signaling in telecommunication networks in the last two decades of the 20th century.

On the other hand, for packet switching technology to transfer user data there were two main candidates, Internet technologies developed in the United States and Asynchronous Transfer Mode (ATM) developed and supported by European countries. However, ATM, although a packet technology, still largely implemented a circuit switching philosophy, which requires signaling connections to be established before any data transmission. On the other hand, the Internet was created on a best-effort basis, with no guarantees that a given connection will be established or a packet will reach the destination, but the IP network does its best to deliver the IP packet to the destination. The best-effort approach to Internet networking won the packet switching "battle" until the late 1990s (with ATM as the main competition), and then the path was traced to establish Internet technologies as the new telecommunications paradigm in the 21st century for all types of services, including legacy, existing and future ones.

1.2 Internet Evolution

The Internet was the winning technology in the telecommunications world's transition from circuit-switched networks in the 20th century to packet-switched networks in the 21st century. The evolution of the Internet in the world of telecommunications began in the 1990s, although the initial work on Internet protocols began in the 1970s, and especially in the 1980s when the main Internet protocols were standardized, some of which still exist half a century later initial standardization (for example, the first version of the IP, called IP version 4, which was standardized in 1981, was then implemented in all telecommunication networks in all countries around the world in the decades after its introduction [6]).

All the technologies that make the Internet functional and work are standardized by the Internet Engineering Task Force (IETF). For example, the noted IP version 4 (IP) is standardized by RFC 791 [7]. The RFC series was started in 1969 by Steve Crocker and was actually an approach used to take working notes of the ARPANET program (the predecessor of the current Internet). RFC editors' operations were funded by the US government's Defense Advanced Research Projects Agency (DARPA) until 1998, and since then they have been governed by the Internet Society.

What were the main reasons for the Internet's success as a packet switching technology for the telecommunications world? Well, the Internet was created on several principles that made it a global success, of which the following can be considered the most important:

- There is separation of Internet applications and services from the underlying transport technologies (e.g. mobile or fixed access networks, transport networks) via the common networking protocols for all nodes in the network (including end user nodes and network nodes such as switches and routers). For example, Ethernet (the IEEE 802.3 family of standards, started in the 1980s) has been initially created for Internet protocol stack to be implemented over it.
- All network nodes and end user/machine devices have the main Internet protocol stack based on transport layer protocols, which were primarily User Datagram Protocol (UDP) [8] and Transmission Control Protocol (TCP) [9] over the IP, which currently exists in its two versions, IP version 4 (standardized with RFC 791 in September 1981 [7]) and IP version 6 (originally appeared in December 1995 with RFC 1883, was later updated in 1998 with RFC 2460, and finally standardized in 2017 with RFC 8200 [10]).
- Best effort principle of Internet (and generally IP networks) which provided lower costs for Internet network equipment when compared with the costs of the traditional telecom networks in the circuit switching era (i.e. until the 1990s). In fact, in native Internet networking approach there were no performance guarantees end to end, so signaling (which was mandatory in the telephone networks, which were the main circuit-switched telecom networks in the 20th century) became obsolete for many services (e.g. Web services).
- The emergence of Hyper Text Transfer Protocol (HTTP) in the early 1990s and with it the World Wide Web (WWW). In fact, the HTTP as the communication protocol for WWW appeared as an invention by Tim Berners Lee in 1989–1990; however, the crucial moment was not patenting the

HTTP by CERN (where the HTTP inventor has worked in that period), but generally providing to the Internet community (i.e. IETF, to become an open standard) in early 1993. Later appeared standardized HTTP 1.0 (defined by RFC 1945 in May 1996 [11]) and then the long-lasting standard HTTP 1.1 (defined initially by RFC 2068 in 1997 [12], and later updated with RFC 2616 in June 1999 [13]). Until the end of the 1990s, the Web together with email (which was initially standardized in 1970s) was a worldwide success, driving the commercialization of the Internet (moving from being a US federal project to a global network managed by representatives from the players from the Internet ecosystem).

1.2.1 Comparison of Internet and Legacy Telecommunications

A comparison of the traditional telecommunications layering protocol and the IP model is given in Figure 1.3. Originally, the IP model in the early days (in the 1970s) was based on three layers: an interface layer at the bottom, a Network Control Protocol (NCP) in the middle, and an application layer at the top. In 1981, NCP split into TCP (or UDP) over IP, so they became the four-protocol layer model as the native Internet model from the 1980s. However, the network interface layer is typically split into the physical layer and data-link layer by all Standards Development Organizations (SDOs), so with such classification the basic Internet protocol layering model has five layers.

Layering does not forbid multiple protocols to be implemented within a given layer, which in jargon is called layer splitting. Thus, as an example, for audio/video streaming, Real Time Protocol (RTP) is used over UDP, where both protocols (RTP and UDP) belong to Layer 4 — the transport layer. The underlying transport technologies define the network interface, which includes OSI protocol layers 1 and 2. In that manner, these two layers at the bottom of the protocol stack are specific to each transport or access technology, and they are defined in specifications (standards) by the SDO that develops that technology. In that manner, for example, 4G LTE (Long Term Evolution) and 5G NR (New Radio) mobile networks used worldwide are standardized by 3GPP, so OSI layers 1 and 2 for these technologies are standardized in the 3GPP specifications. Another

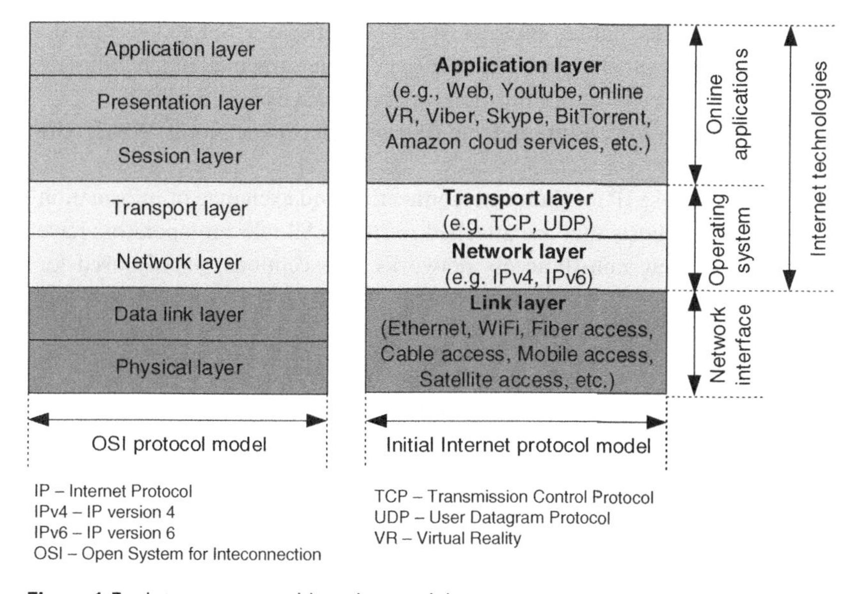

Figure 1.3 Internet protocol layering model.

example, Ethernet (IEEE 802.3 family of standards) and Wi-Fi (IEEE 802.11 family of standards) are standardized by the Institute of Electrical and Electronics Engineers (IEEE) which only standardize the lowest two protocol layers (OSI layers 1 and 2) and assume that Internet protocols will be implemented from Layer 3 (network layer) and above.

Crucial protocols for Internet are IP on OSI layer 3 (network layer) and UDP and TCP on OSI Layer 4 (transport layer). These two protocol layers, which include TCP/IP or UDP/IP, are implemented in the Operating System (OS) of all Internet hosts, including all terminals (user devices such as computers and smartphones, Internet of Things – IoT, devices, and servers) and all network nodes in Internet/IP networks are called routers (because they route IP packets on the path from source to destination). The applications on Internet are implemented on the top protocol layers which include the session, presentation, and application layers under the OSI layering model, and only the application layer under the Internet layering model. Such applications include standardized application protocols such as HTTP and Hyper Text Markup Language (HTML) for Web services, Simple Mail Transfer Protocol (SMTP) and Post Office Protocol 3 (POP3) for email services, as well as numerous proprietary (i.e. not standardized) services, which may be found in application online stores (e.g. Google Play and Apple Store) and other websites on the open Internet (e.g. YouTube, Viber, WhatsApp, Skype, Facebook, Instagram, Twitter, and TikTok).

In fact, the Internet provides access to communications and access and exchange of information between people or machines (i.e. computers), similar to the electrical distribution network for electrical appliances (e.g. washing machines, vacuum cleaners, and mobile phone chargers). For example, power plugs and sockets are the same for different electrical appliances. In the case of the Internet, Internet access is actually a "socket" for all kinds of telecommunication services that allow access to certain information (e.g. certain web content, video content, and various data) or interactive exchange of information (e.g. Voice over IP – VoIP and messaging).

As the winner technology in the packet switching "battle" in the 1980s and 1990s due to noted advantages, from the beginning of the 21st century, Internet technologies gradually penetrated into the legacy telecom networks run by national telecom operators in each country worldwide, supported by convergence of standardization work of different SDOs, including ITU, IETF, 3GPP, IEEE, and others on the national, regional, and global scales. However, IP-based telecom networks differ from the open (or, in other words, public) Internet. Why? Well, because not all telecommunications services are implemented over the open Internet (e.g. carrier-grade telephony and TV services) even though the networks of telecommunications operators have become all-IP.

Figure 1.4 shows the relation between open Internet and IP-based telecom networks. What is the Internet and what are IP networks?

IP networks are all networks that use IP for their interconnection and exchange of information (regardless of its type) between end hosts that are attached to them. All telecom operators have transited to all-IP networks (with few non-IP access networks as exceptions, mainly used for

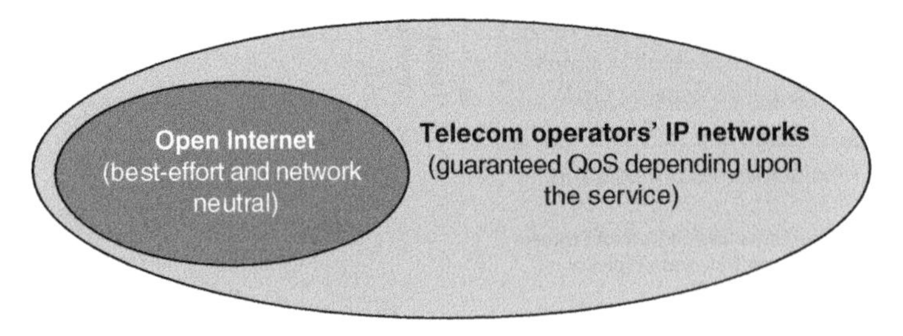

Figure 1.4 Internet vs. IP networks.

connecting low power IoT devices), which are used for open Internet access as well as for provision of legacy telecom services over IP networks with guaranteed Quality of Service (QoS), which are not part of the open Internet network. Such telecommunications services include legacy services such as telephony, TV, and leased lines (for services to businesses), which were transferred from circuit-switched to fully packet-switched telecom networks (here, packet-switched networks are in fact the IP networks), as well as any new services that require guarantees of network performances (e.g. throughput in bits per second, end-to-end delay, and delay variation).

So, the open Internet consists of interconnected IP networks. On the other side, not all IP-based networks are part of the open Internet. As already stated, IP networks carrying carrier-grade telephony, carrier-grade TV, and carrier-grade business services are not part of the open Internet, although such networks are fully based on Internet technologies (e.g. TCP/IP and UDP/IP, as those used in the open Internet). In that way, open Internet can be seen as a subset of telecom operators' IP networks.

But are the open Internet and carrier-grade services (provided by telecom operators with guaranteed QoS and standardized signaling) deployed using physically separated IP networks?

Well, they are not deployed using separate networks; instead, the same fixed and mobile access networks are used for delivery of both open Internet access (or, in other words, Internet Access Service – IAS) and carrier-grade services (i.e. services provided by telecom operators with guaranteed QoS); however, the different telecom services are logically separated from each other (e.g. Internet, telephony, TV, and business services). The same approach of logical separation (i.e. slicing) of deployed network resources and capacity is expected to continue with increasing pace in the future, considering many new emerging services (e.g. IoT services with critical quality and security requirements, such as use cases in industry, transportation, and healthcare).

1.2.2 The Broadband Evolution of Open Internet and IP Networks

Open Internet is characterized by many heterogeneous services that can be provided over it. One of the main features of the open Internet is its network neutrality, which may be considered as a built-in feature from the beginning. In short, network neutrality refers to the principle that all Internet traffic should be treated equally [14]. So, according to the network neutrality rules, the traffic from a niche website should be treated in the same manner as the traffic from global big online services providers such as Google, Amazon, Facebook, and others. However, in 1998, Google was also a niche website, and it has grown on the "wings" of network neutrality of open Internet, which was applicable almost everywhere even without explicit regulation or awareness about it at the beginning of the public Internet (in the 1990s and 2000s).

Network neutrality provided access to the global market for all services and application on the Internet, something that was not possible before (e.g. telephony was the main telecommunication service before the appearance of the open Internet, and it was heavily regulated in all countries). So, the open internet, with its natively built-in network (i.e. net) neutrality, opened up unimaginable opportunities for innovations without asking for permission (e.g. from governments and regulators) that changed the way we lived and worked in the decades since. This has resulted in many new services emerging on the open Internet and shaping the new telecommunications horizon of the 21st century. All such services provided over the open Internet are referred (more like jargon than a strong definition) as OTT services and applications. So, one may say that all services provided through the open Internet access are OTT services, which includes online Web services (e.g. all public websites), online video streaming and video on demand (e.g. YouTube and Netflix), online cloud services (e.g. Google Docs and Amazon cloud), online social networking (e.g.

Facebook, Twitter, Instagram, and TikTok), online gaming (e.g. Steam, Epic games store, and PlayStation gaming platforms), and all other services and applications (including also those with niche market shares) provided through the open Internet access. What is the business logic for telecom operators for provision of open Internet access based on network neutrality and best-effort approach?

Well, with the transition of telecom operators to IP networks from the 2000s onward, Internet access services have emerged as one of the main service offerings from telecom operators. In the last century, until the 1980s, almost all countries had one national telecom operator, managed by the state, aimed at providing telephony and telegraphy as the main services. The advent of the Internet triggered telecom operators to become Internet Service Providers (ISPs), as the Internet itself has used existing telecommunications infrastructure since its inception, first with dial-up modems in digital telecom networks in the 1990s and early 2000s. Later the telecommunications network transitioned to IP-based networking, which was naturally perfect for providing IAS, over either fixed or mobile networks. Telecom operators charge for their IAS services by time usage (that was in the era of dial-up modem access in the early years of the commercial Internet, the 1990s and early 2000s), volume-based, or by a flat fee (in the era of IP-based telecommunications networks). Thus, telecom operators receive revenues by providing access to the Internet as a whole, while online service providers (i.e. OTT providers) usually offer access to certain services for free (e.g. by obtaining revenues from commercial ads on websites) or for a fee (e.g. with appropriate authentication and authorization of the customers). However, without OTT services end users will not need Internet access services from telecom operators. So, it can be said that telecom operators and OTT service providers are related to each other like gas and a car.

The most demanding services in the evolution of the Internet in the 2000s were video services, which required higher bitrates (i.e. higher throughputs) compared to other services consumed by human end users (e.g. VoIP, web, and email). Video bitrates are highly dependent on video resolution and video coders/decoders [15]. Providing video services has multiple dependencies. On one hand, video services require fixed and mobile access networks as well as transport networks to support individual bitrates for video services ranging from hundreds of kbit/s (for lower resolutions, e.g. 480p) up to multiple Mbit/s for High Definition (HD) and tens of Mbit/s for higher video resolutions (e.g. 4K), as shown in Table 1.1. However, one should note that the video compression ratio of the codec (that is, coder at transmitter side – e.g. video server, and decoder at the receiver side – e.g. video player) is dependent upon the processing capabilities of end user devices, such as personal computers, laptops, and others. Higher compression ratio of the codec (which results in lower bitrates needed for a given video stream) on average requires higher processing power of the receiving device to perform rendering in real time with appropriate initial buffering of the video content, needed for smooth reproduction of the video due to delay variations of IP packets carrying

Table 1.1 Required bitrates for video with different resolutions.

Video resolution and frame rate	Approximation of the required bitrate (varies due to codec and frames per second – fps rate)
Below SD (Standard Definition): 180–270p	90 kbit/s to 1 Mbit/s
SD (Standard Definition): 360–540p	150 kbit/s to 4 Mbit/s
HD (High Definition): 720–1080p	500 kbit/s to 5 Mbit/s
Above HD (High Definition), Ultra HD: 1440–2160p	1.5 Mbit/s to 45 Mbit/s

the video data. So, the required bitrates for a given video stream are inversely proportional to the compression ratio of the video, which on the other side is proportional with the processing power of end devices which play the video content.

What is needed for video streaming and video on demand as well as many other different services with various requirements to be deployed on the same physical network infrastructure, either fixed or mobile/wireless one?

Well, the network needs to provide high data rates, or in other words bitrates. Having enough high bitrates (i.e. capacity) to individual end devices, which can provide services with satisfactory quality, is denoted as broadband. Broadband can have various adjectives such as high and ultra when needed to denote speeds that are higher than existing broadband at a given point in time. However, in general, the term "broadband Internet" in telecommunications networks is relative. Why?

Well, because what is broadband today may be considered as narrowband in the future (in a decade or two). For example, in the 1990s and early 2000s individual bitrates of several hundreds of kbit/s were considered as broadband. However, only a decade or two after that time, in the 2010s and 2020s, such bitrates of few hundreds of kbit/s are considered as narrowband. In that manner, bitrates in range of Mbit/s or tens of Mbit/s considered as broadband in the 2010s and early 2020s can be considered as narrowband in the 2030s or 2040s.

The initial development of broadband in the last two decades of the 20th century was intended for transfer of HD video contents, such as HD television. However, the real broadband development was triggered with the spread of Internet, which was suitable for all different types of information, including videos of different resolutions, easily searchable through the WWW, with video content hosted on websites. However, it is impossible to say whether broadband has driven higher resolution video content or whether video content has driven the development of broadband technologies. But it is clear that their development, broadband and video, has been synchronous since the mid-2000s, since the largest video-sharing platform YouTube appeared in 2005.

Figure 1.5 shows fixed Internet access speeds from 1995 onward [16]. The prediction of the average growth of speeds in the future until 2060 is made with the assumption of 20% average annual

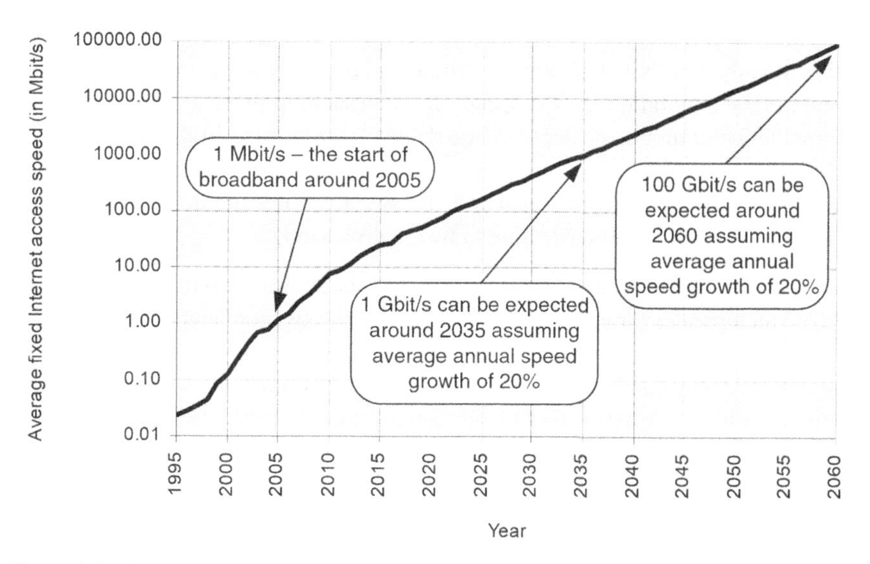

Figure 1.5 Average speeds for fixed Internet access, current and future predictions, 1995–2060.

growth. Thus, with such an assumption, the average speed of fixed Internet access can be expected to reach 1 Gbit/s around 2035 and 100 Gbit/s around 2060 (as shown in Figure 1.5). This is also consistent with Moore's Law [17], which states that processor capacity roughly doubles every two years by doubling the number of transistors on a chip of a given size, with advances in their production over time.

Video sharing platforms such as YouTube were primarily aimed at residential end users, meaning that they emerged at a time when fixed broadband technologies were available for residential users, such as Asymmetric Digital Subscriber Line (ADSL) and cable networks (originally deployed in the late 20th century to broadcast television over fixed access networks). Of course, ADSL and xDSL (digital subscriber line) technologies in general were reusing the existing copper infrastructure consisting of twisted pair lines that were deployed for telephone networks during the 20th century. Twisted pair local loop and coaxial cable access infrastructure (in countries where it already existed) facilitated the rapid deployment of broadband in the 2000s, given that most telecommunications network infrastructure costs come from the deployment of access networks.

However, the highest capacity of all three main media in telecommunication networks has the fiber. It has better performances than copper; therefore, all newly deployed fixed networks are fiber-based (e.g. no one deploys copper, i.e. metallic access, although it continues to be used until fiber is deployed on a given location). Also, fiber has better performances (in terms of capacity and packet losses) than any radio interface, including wireless (e.g. Wi-Fi), mobile (e.g. mobile networks such as 4G and 5G), and satellite broadband access. So, in the long run, all fixed networks will become optical networks. Fiber began to penetrate telecommunications networks at the end of the 20th century, and in the 21st century, especially from the 2010s onward, fiber is the main and only fixed broadband access technology that is deployed globally in all countries. The capacity of the fiber is in the range of tens of Gbit/s in access networks and it is increasing toward the future, while single fiber can carry multiple Tbit/s (Terabits per second, where 1 Tbit/s = 1000 Gbit/s) by using wavelength multiplexing (that is, the data is transferred simultaneously over multiple wavelengths in a single fiber). Of course, fiber used in transport networks needs more capacity to carry the aggregated traffic to/from access networks, and therefore uses more wavelengths than fiber deployed in access networks.

In parallel with fixed broadband, mobile broadband also started with 3G in the 2000s, continued with 4G in the 2010s and 5G in the 2020s, and will continue with 6G in the 2030s, and assuming the pace of innovation already in mobile technologies, there will probably be 7G in the 2040s, and so on. With each new mobile generation, access speeds are approaching those available on fixed networks, and the capabilities of mobile devices are becoming closer to those of desktop computers and laptops, so mobile devices have also become heavy broadband users, especially in the era of 5G (2020s). On the other side, the mobile broadband access contributes to higher affordability and more frequent access to broadband Internet due to two main reasons:

- Every mobile user has their own mobile device, a smartphone, at all times, so the availability of mobile broadband is much greater for end users due to the ubiquitous mobile networks in land-based locations.
- Many countries with developing telecommunications markets had almost no copper (i.e. metallic) access infrastructure due to the absence of higher penetration of telephone networks in the 20th century, resulting in mobile broadband networks in many countries being the main mode of broadband access to the Internet since the 2010s (when mobile broadband became available in most parts of the world) [18].

In both fixed and mobile broadband networks, the main service from the telecom operator's point of view is Internet access services (also called data or mobile data in mobile networks). The other services for which broadband is used by telecommunications operators are the provision of TV (i.e. IPTV, bearing in mind that TV is also provided using an IP network, with the advent of all telecom networks being fully IP networks) and the provision of business services over Virtual Private Networks (VPNs) to business customers (e.g. companies, organizations, institutions, and academia) as a replacement for leased lines (used in circuit-switched telecommunications networks in the 20th century for business services, i.e. key customers).

On the other hand, broadband Internet access is primarily used by end users to access content and applications/services provided by OTT service providers. Then, from the perspective of the OTT service provider, the main use cases in the 2020s for broadband Internet access (Figure 1.6) are the provision of video services (streaming video, video on demand, including social media videos) that contribute to about 71–80% of all internet traffic (according to 2022 statistics and 2028 predictions [19]), then about 7–9% of social networks (excluding video, which is included in the percentage of video), 3% software updates, 1% Web browsing, 1% audio traffic, and other that includes miscellaneous Internet traffic not included in the noted types. So, it can be concluded that in the 2020s the dominant type of internet traffic is video. Considering that TV services provided by telecommunication operators, which are not part of Internet traffic, are also a type of video traffic, it seems that fixed and mobile broadband networks in the 2020s are built primarily for accessing video content. Why?

Since human users consume most of the world's information using their eyes, either as 2D (on screen) or 3D (in real life), video results in a large volume of data to be transmitted. For example, video with resolution 1080p (that is 1920 × 1080 pixels), with 30 fps (frames per second) and 8 bits per color, when uncompressed requires 2.23 Gbit/s, [20], 4K uncompressed video requires 8.91 Gbit/s [20], and so on. If we double the video frame rate to 60 fps that will result in proportionally higher uncompressed video data rate, e.g. 4.46 Gbit/s for 1080p with 60 fps [20], and so on. The same proportional increase applies to increasing the number of color bits. The bottom line is that digital video when uncompressed requires very high data rates that cannot be accommodated by

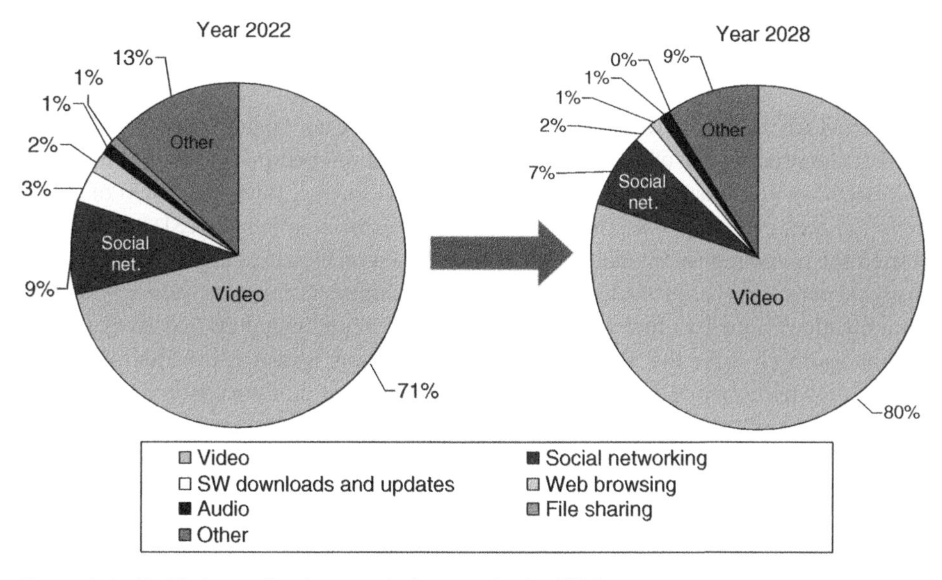

Figure 1.6 Traffic by application type in Internet in the 2020s.

any of the existing and near-future broadband speeds, so digital video is always compressed at source encoding, transmitted as such, and decoded (uncompressed) on the receiver side. The used codecs and compression of the video are also influenced by average processing power of the consumer devices at the given point in time (the reader may recall the noted Moore's law). Audio media (without any video), on the other hand, is a one-dimensional media that requires throughput in the range of kbit/s (tens or hundreds, depending on the quality of the audio streams); therefore, its percentage in the total Internet traffic continuously decreases, because the bandwidth (i.e. throughput) demands for audio are not increasing drastically, while on the other side broadband speeds and required speeds of video streams are increasing in parallel. Broadband is a driver for higher video resolutions and fps, which results in more bandwidth required transporting the video stream or, vice versa, higher quality video contents are the drivers for higher speeds for consumer devices.

What about the future evolution of the fixed and mobile broadband vs. Internet applications? Well, one can expect much more demanding applications and services, which are based on the use of Virtual Reality (VR) and Augmented Reality (AR) or generally X-Reality (XR), which demand much higher speeds than 2D video contents. While VR is not convenient for outdoor use as video (e.g. via smartphones), AR may create possibilities for many different use cases, which will require very low delays and very high throughputs from fixed or mobile access networks.

As with Internet video and other OTT services on the one hand and IPTV as a service offering from telecom operators on the other hand, also new bandwidth-intensive services such as VR/AR/XR can be provided through open Internet or through specialized or private logically separated networks of the telecommunications operator that are not part of the open Internet, although they may use, in most scenarios, the same physical broadband infrastructure. So, the broadband evolution continues with even more demanding services than video, although video is expected to progress further with higher resolutions (e.g. 8K and 16K) and higher frame rates (e.g. 60 and 120 fps), which further drive the need for more speed in fixed and mobile broadband networks.

1.3 Convergence of Telecom and Internet Worlds

Telecom and Internet worlds were developing independently from each other in the 1960s–1980s. While ARPANET, the predecessor of the Internet, was being developed in the US, the telecom world was still circuit switch–based with telephony as the main service. With the advent of computer science in the period 1960s–1980s, the telecom world (the telecom industry, SDOs, telecom operators) started the transition from analogue networks (based on electrical engineering for the hardware) to digital networks (based on electronics for the hardware). Computers were introduced into telecommunications networks first in the signaling part, hence signaling first went to the packet switching approach with the already noted SS7 signaling system [3] standardized by the ITU and used by all telecommunications operators until the end of the 20th century. Due to the already noted advantages of Internet technologies compared to other packet switching technologies such as the European ATM, and especially with the explosion of the Web after 1993, by the end of the 1990s it was clear in all SDOs that packet switching in telecom networks will be based on IP and Internet technologies. Since then, the convergence of the telecommunications and Internet worlds has begun, which includes the convergence of protocols, architectures, as well as the convergence of services and digital markets.

1.3.1 Protocols Convergence

Initially, telecom networks and the Internet used different protocols until the early 1990s. In order to observe the protocol convergence between telecommunications networks and the Internet, let us compare their main approaches and points of convergence. Telecom networks are organized in three so-called planes. They are:

- User plane: Refers to all protocols and nodes (in which they are implemented) used to carry user traffic (e.g. audio, video, and various data) that is traffic generated or consumed by end user devices (including consumer as well as IoT devices).
- Control plane: Refers to all protocols and nodes that are used to carry control information, which includes signaling protocols. Examples of protocols in control plane in legacy telecom networks are SS7 signaling protocols, which are defined on all 7 OSI protocol layers [3], used in telecom networks in the 1980s, 1990s, and 2000s.
- Management plane: Refers to traffic that is used for configuration and monitoring of all protocol layers in all devices in the telecom operator's network (for example, including operating system updates for network nodes, alarming for network and system failures).

All telecommunication networks for all services have the implementation of protocols and architectures in all three planes — user, control, and management planes. On the other hand, Internet access is based on the client–server paradigm, in which the server listens to requests from the client (for example, a web server listens to requests from web clients that are mainly web browsers), so there is no control plane traffic that would be required to establish a connection between the client and the server. However, the native Internet also has control traffic, which involves the exchange of routing information between network nodes called routers, the use of configuration protocols such as Dynamic Host Configuration Protocol (DHCP) for automatic IP address configuration (and Internet access in general), and Domain Name System (DNS) for translating domain names into IP addresses and so on. So, the Internet also has control traffic, but it is originally with different purposes (due to different addressing and naming schemes in Internet and legacy telecom networks) and different protocols. Signaling is not mandatory for every Internet connection (between a given client–server pair, once Internet access is established) unlike legacy telephony telecommunications. Also, the Internet has standardized protocols for managing different protocols on different hosts (including routers, switches, servers), such as Simple Network Management Protocol (SNMP) [2].

The key point for the convergence of the internet/IP and telecom worlds was to have a common user plane and control plane protocols. Common Layer 3 and Layer 4 networking protocols, i.e. TCP/IP and UDP/IP, came into use in telecommunication networking with the transition from circuit-switched to packet-switched telecommunication networks. With the move to an all-IP network, traffic in all three main planes (user, control, and management) became IP traffic, carried over TCP/IP or UDP/IP. The main differences between legacy telecommunications networks in the pre-IP era and the native Internet (before entering into the telecommunications world) were the following:

- Telecom networks are based on providing services with guaranteed QoS for each service [21], unlike the open Internet which is based on best effort and is network neutral.
- Telecom networks use mandatory signaling for their services, dedicated and specific to each service, while in the Internet network signaling is not mandatory and the exchange of control information is usually included in the same protocols used for data transmission (e.g. UDP, TCP, and HTTP) by using fields in the packet headers.

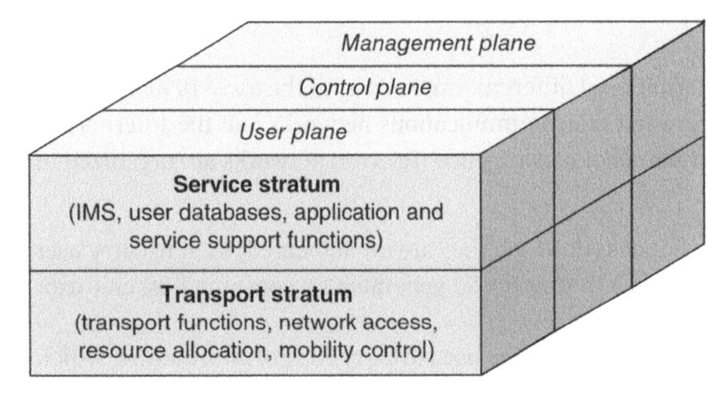

Figure 1.7 NGN transport and service stratum.

The main protocol convergence was done with synergy of the main SDOs (ITU, IETF, 3GPP, IEEE) regarding the standardization of signaling protocols that will run over the IP protocol stack considering the planned transition to all-IP telecom networks around year 2000. While the noted SS7 was also packet switching technology, as IP is, the transition to IP required synergy regarding the signaling protocols.

The main umbrella framework in this regard established by the ITU is called Next Generation Networks (NGNs) [3]. The ITU's development of NGN started in 2003 [3], with general overview of NGN specified in 2004 [22]. The NGN was primarily targeted to standardization of the legacy telecommunication services (telephony and TV) by using standardized Internet technologies while keeping backward compatibility of these two services toward legacy end user equipment. However, NGN also introduced the framework for the IoT as well as further development of NGN specifications toward future networks [23], including cloud computing and further development of the IoTs. To incorporate Internet technologies in the telecom worlds, NGN is created by incorporation of Internet native principle of separation of services on the top from underlying transport technologies at the bottom. Therefore, NGN defined two main stratums, transport stratum and service stratum (as shown in Figure 1.7).

Before NGN standardization, the IETF standardized Session Initiation Protocol (SIP), initially in 1999 and finally in 2002 [24]. In addition to SIP, another standardized IETF control protocol is the Diameter protocol, which is intended for use in AAA (Authentication, Authorization, and Accounting) in all-IP access networks including fixed and mobile networks. It was standardized by the IETF in 2003 and later updated in 2012 [25]. Both SIP and Diameter became the main protocols in IP Multimedia Subsystem (IMS), which was originally developed by the ETSI technical committee called Telecommunications and Internet Converged Services and Protocols for Advanced Networking (TISPAN), but its standardization was later transferred to 3GPP, which standardized the common IMS in the late 2000s (with 3GPP Release 8) [2]. Originally, IMS was primarily intended to be used for transition of legacy telephony to VoIP as Public Switched Telephone Network (PSTN) replacement. Although the IMS standardization is implemented by 3GPP, it is access-independent and can be used for both fixed and mobile telecommunication networks. So, it can be noted that IMS actually replaced SS7 signaling, having the SIP as the main signaling protocol.

Overall, the IP is placed in the middle of the protocol hourglass; however, its introduction in telecom networks provided possibilities to have standardized and non-standardized applications/

services deployed by telecom operators as well as OTT service providers. However, the same protocols used for open Internet are also used for so-called specialized IP networks that are logically isolated from the open Internet network. However, on protocol layers 1 and 2 (i.e. network interfaces) each technology uses its own protocols adapted to the transmission media (e.g. copper, fiber, or radio).

1.3.2 Architectural Convergence

Convergence of the architecture is crucial for the continuous development of the broadband infrastructure. Why? Because broadband networks become more economically viable if the same network can be used for a wider range of applications and thus reduce the number of separate networks required. However, the convergence is not only challenging from a technical point but includes technical, market, and regulatory issues.

Network convergence refers to offering different heterogeneous services and applications through a single network infrastructure (e.g. telephony, TV, Web, IoT services, and cloud services). The end user terminals also converge with the network convergence. That means a single device can be used for many services, such as OTT services through the open Internet access, telecom operator's services such as voice services, TV, messaging, cloud services, and IoT services. On the other side, with network convergence on IP networks, the same services can be used through different devices which may be connected to different IP networks at a time. For example, a smartphone can access OTT services such as social networks, websites, video streaming, and OTT voice services., by being connected to a mobile data service (to access the Internet as a whole) provided by a telecom operator or by using a Wi-Fi hotspot in a cafeteria, shopping center, airport, hotel, home, and office.

Native Internet architecture consists of interconnected IP networks. According to the network size (that is, number of hosts as well as distance between the hosts and network nodes), IP networks are typically classified into the following types [2]:

- Local Area Network (LAN): It connects several hosts to several hundreds of hosts via network switches which work on OSI layers 1 and 2. It is connected to the public Internet via a router, typically called a gateway router, which works on OSI layers 1 to 3. LANs are used everywhere for access to IP-based infrastructure, including home networks, corporate networks, or networks in public places (e.g. in cafeterias, airports, and hotels). The main LAN technologies in the 21st century are Ethernet (as wired LAN) and Wi-Fi (as Wireless LAN – WLAN).
- Metropolitan Area Network (MAN): It refers to a network which provides access to hosts on a territory in size of a metropolitan area (e.g. a city area) and may be used to connect multiple LANs.
- Wide Area Network (WAN): It refers to nationwide transport or regional networks which are used to connect various core networks and/or access networks (e.g. LANs and MANs). Typically, WANs are deployed through fiber links due to the highest capacity of the fiber and its longer reach. They are used for building IP regional and transit networks on national, regional, and global levels.
- Radio Access Network (RAN): This type of access network refers to access of the mobile networks (e.g. 2G, 3G, 4G, 5G, and 6G). For example, RAN in 4G mobile networks from 3GPP is based on LTE technology, while RAN in 5G mobile networks is NR.

However, with convergence of telecom networks to all-IP networks and use of Internet technologies, signaling and QoS became mandatory parts. From the telecom operator's viewpoint, networks can be divided into three main types of networks which can be mapped to LAN, MAN, WAN, and RAN network topologies (Figure 1.8):

- Access networks: They include all types of fixed and mobile access networks, including LAN, MAN, or WAN, through which end fixed and mobile devices (including IoT devices) are connected to the open Internet or specialized telecom networks.
- Regional networks: They connect different access networks which are under administrative control of a single network provider such as a telecom operator, and provide their connection to the Internet and to a set of common functionalities (e.g. control functions and management functions). They consist of core networks (where the main gateways are located) and transport networks (which are used for interconnection of different nodes in the core and in the access parts).
- Transit networks: These types of networks are used for interconnection of different regional networks on national or regional basis.

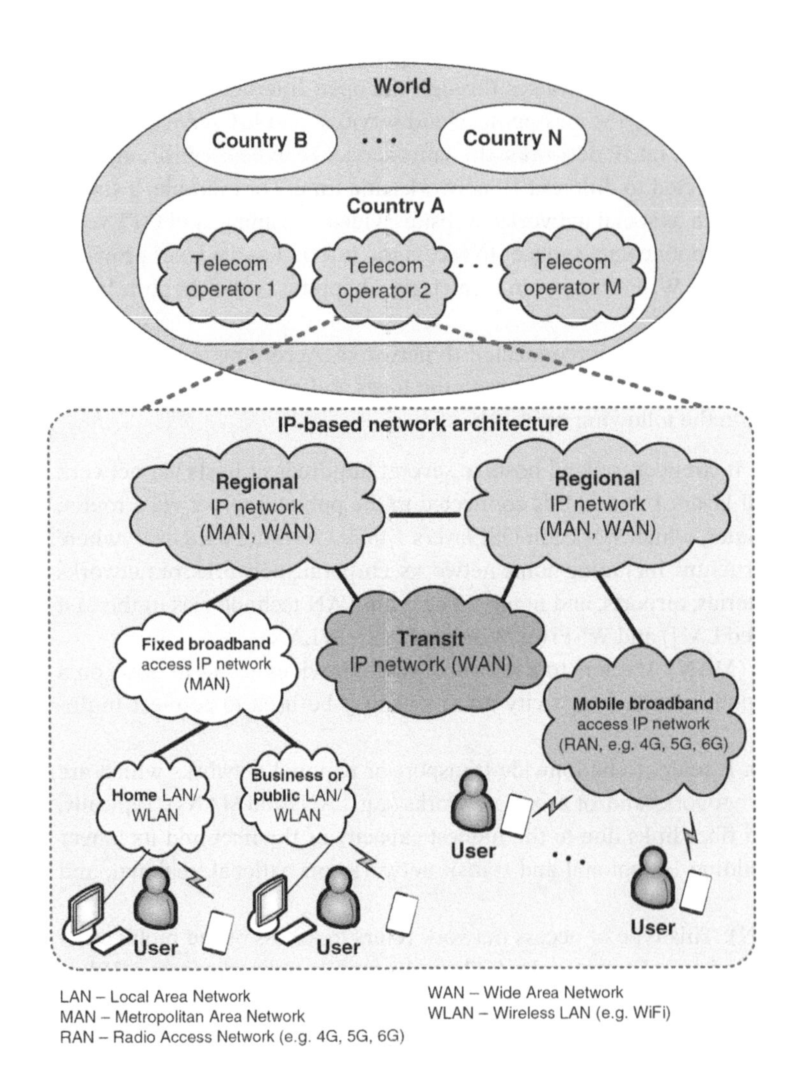

Figure 1.8 Network convergence of telecom operators toward IP and Internet technologies.

Thus, IP network types (e.g. LAN, MAN, and WAN) transitioned into telecom networks; however, the telecom approach of standardized QoS and signaling shaped the initial best-effort networks based on Internet technologies to be based on telecom standards in terms of the quality and reliability of the services, as well as their design and subsequent operation, maintenance, and administration.

With the development of broadband access, the volume of various data transferred over all types of networks increases exponentially. In such cases, manual processing of large volumes of data is becoming less feasible, hence automatic use of network and user generated data for different network tasks becomes a necessity toward the future. In that manner, for example, Big Data Driven Networking (bDDN) uses the big data generated by a network itself to improve the capability of network control and management [26].

Future networks are evolving toward automated networks in terms of their operation and maintenance, which will be to telecom operators what smartphones are to ordinary mobile users. Also, the large amount of data are possible to be used for training the so-called Machine Learning (ML) techniques (as part of the Artificial Intelligence [AI] use cases in the telecommunications field), which are aimed to enable computational systems to understand data and gain knowledge from it without necessarily being explicitly programmed [27].

In the ITU standardization roadmap, the process of convergence of telecom networks over IP-based infrastructure continued in future networks standardization, which includes Software Defined Networking/Network Function Virtualization (SDN/NFV) aspects, IMT-2020/5G systems, IoT, and Smart Sustainable Cities, and use of emerging technology areas such as AI (with its main used form ML).

1.3.3 Services Convergence and Digital Market – the Digitalization

With convergence of telecom and Internet worlds, all existing and future services are provided over the same fixed and mobile broadband infrastructure, which is "shared" between OTT services provided through Internet access and specialized (e.g. telecom operator provided) services with performance guarantees. All telecom/ICT services form the so-called digital market, which on the other side is related to digitalization, i.e. digital transformation of societies around the globe. But, what does that mean?

Telecommunications are global and because of that telecommunication networks and services do not really know the borders between countries. On the other hand, each country has jurisdiction over telecommunications that take place in its own territory. It should be emphasized that telecommunications have been characterized by rapid development since the beginning of the 21st century (if we compare it with the 20th century in which the main telecommunication services were telephony and television), which is particularly encouraged by the spread of the Internet and the Internet technologies in the telecom world from the 1990s onward. Internet technologies today are also the basic technologies that we use in telecommunications to build networks and services. All communication over the open Internet network is referred to as a cyberspace. The cyberspace is truly global (network neutrality mainly contributes to that), although it is possible for Internet traffic to be filtered by certain telecom operators or countries due to national regulations. Spreading of broadband access to Internet provided the basis for many services from the real world (e.g. paper-based services such as local or government administrations, and shopping) to move online, that is, on the open Internet. Also, open Internet access provided the possibility for all companies, including big, medium, and small enterprises, as well as all organizations and institutions, to provide information and access to their services online or selling their products on the Internet.

The movement of services from the real (physical) world to the cyber world (i.e. on the open Internet) is called digitalization, and such OTT services are called digital services. Digitalization is actually putting an emphasis on how digital services and applications (enabled through open access to the Internet through fixed and mobile broadband telecommunications networks) change and transform the experience of citizens and the way of living and working in a society. The main goal of digitalization is to improve the quality of life and well-being and achieve the so-called sustainable development by using telecom/ICT technologies.

Having one network for all digital services in all countries worldwide (that network in the Internet) provides the possibility to have a single digital market. For example, such legislation in Europe is developing since 2015 and especially progressing in 2020s toward 2030, which is also called a digital decade, marking the targeted digital transformation of the society. In that manner, the single digital market approach is based on several main pillars, which include the following [28]:

- Infrastructure: It is aimed to provide better access for consumers (i.e. residential customers of telecom operators) and businesses (i.e. business customers of telecom operators) to digital goods and services over the open Internet.
- Environment pillar: It refers to developing the appropriate playing field for digital services. The main target of digital single market is opening new opportunities by removing the differences between the online (i.e. cyber) and offline (i.e. real) worlds, particularly removing the barriers to cross-border online activity (where possible according to national telecom/ICT regulations). It also means provision of high-speed broadband access to Internet, secure and trustworthy networks and platforms, as well as setting the right regulations for important (for the society) digital services (e.g. cybersecurity, online data protection and privacy, and fairness and transparency of online platforms).
- Economy and society pillar: It aims to maximize the growth potential of the national and regional economy by using the digital single market so that every citizen and business can fully enjoy its benefits. For that purpose, there are needed digital skills of the citizens for an inclusive digital society. The aim is for the digital society to be a digital twin of the real world human society.

1.4 Legacy, Over-The-Top (OTT), and Critical Services

Regarding the technical standardization and governance of Internet technologies, IP has evolved over the years and still needs to evolve [29]. In terms of continuous evolution of Internet technologies (not only via transition from IPv4 to IPv6), the future Internet will continue to use the successful use cases of Internet technologies and continue to improve in the technologies in the IP ecosystem. The main future targets are higher speeds, lower latencies, as well as ultra higher reliability and synchronization for critical services (e.g. motion control in smart factories). The evolution of Internet is not by redesigning the IP in terms of one-protocol-fits-all, but for the virtualization of networks when needed. For example, virtualization is already present in IP networks since the beginning of the 21st century, because that is needed for delivery of services with QoS support over the IP networks.

Overall, one may expect that future telecom IP-based networks will be based on the following main types of services (Figure 1.9):

- OTT services over fixed and mobile Internet access, including Web browsing, email, social networking, video streaming, VR, OTT voice services (e.g. Skype, Viber, and WhatsApp), and so on. Also, it includes OTT massive IoT services (e.g. fleet tracking and various IoT services based on use of sensors and actuators)

- Massive IoT services – these can be provided either through the open Internet (as OTT service) or through specialized networks that generally may be IP or non-IP. In cases of non-IP massive IoT, their identifiers need to be translated to IP addresses at their gateway node toward the open Internet.
- Ultra-reliable and low-latency communications (the term is taken from 5G service types, they can also be referred to as critical IoT services) – this is the most complex type of services that require ultrahigh reliability and availability and ultralow latency (in the range of 1 ms). Of course, one should note the limitations imposed by the maximum signal speed, which is the speed of light, so that for 1 ms delay the maximum distance between the two endpoints of a communication session is less than 300 km (considering that the speed of light is 300000 km/s, i.e. 300 km/ms). These services cannot normally be provided over the open internet (e.g. based on the principles of network neutrality) but over specialized networks, which can be operated by telecommunications operators or by private entities (e.g. an Industry 4.0 utility factory, a manufacturer of cars for V2X (vehicle-to-everything) communication, city for critical smart city services, and so on).
- Legacy telecom services (voice, TV, and VPN business services) – they are not provided over the open Internet, but over specialized (i.e. managed) IP networks, with guaranteed QoS via logical isolation of different services. The QoS support is provided typically by use of QoS solutions in underlying technologies in respect to IP, such as Ethernet, Wi-Fi, and mobile technologies (e.g. 4G and 5G).

One should note that natively the success of the IP-based Internet protocol suite was and still is tied to the notion of simple (in comparison with the legacy telecom) global network connecting smart edges (e.g. computers and smartphones). The Internet and, generally, the IP model have led to the high speed of innovations with ecosystems of new applications at an unprecedented rate. With the aim for all services, including OTT as well as specialized services (that are not provided over the open Internet), to be feasible over the same broadband access and transport networks, the initially simple IP networks have an increased complexity by adding network functions needed for specialized services, such as guaranteed QoS and security for legacy telecom services, and additionally very low latency and ultrahigh reliability as well as high level of in-network

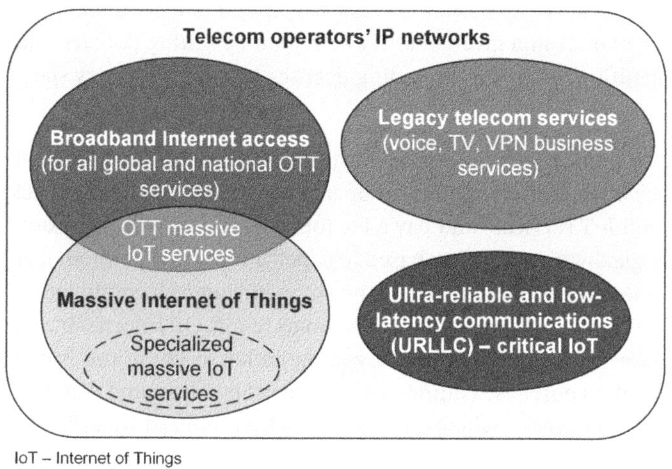

Figure 1.9 Main types of services in current and future IP networks of telecommunication operators.

synchronization for critical IoT services. So, the convergence of ICTs/telecommunications onto the IP-based networks and services resulted in merging of the legacy telecom and Internet worlds, with inclusion of the best principles from both.

1.5 Discussion

The telecommunications sector, according to the volume of traffic, the networks' capacity, variety of services, as well as impact on society and economy has seen an exponential growth in the past decades. Of course, this is due to the expansion of new innovative services based on use of the Internet (IP) technologies as well as possibilities provided by mobile broadband networks (e.g. 4G and 5G) and fixed broadband networks (e.g. Fiber To The Home – FTTH). All of that contributed to the emergence of a large number of "players" with different goals and purposes in the field of telecommunications (Figure 1.10). According to Figure 1.10, the telecom world today includes several types of entities that are interconnected, which include the following:

- End users – typically grouped into two main categories: residential customers (e.g. homes and mobile subscriptions) and business or key customers (e.g. VPNs for enterprises and IoT services such as fleet tracking, smart manufacturing, and smart city).
- Telecom operators – provide services to end users. The relation between the end users (customers) and telecom operators is defined via Service Level Agreement (SLA), which typically includes QoS parameters (e.g. provided throughputs), data caps (e.g. in GBs or unlimited), and pricing.
- Manufacturers – produce the equipment used by operators, services providers, and end users (that are either human users or machines in cases of IoT services). They produce switches, routers, radio units, servers, as well as user equipment, such as telephones, smartphones, computers, and IoT devices.
- SDOs (e.g. ITU, IETF, 3GPP, and IEEE) – make standards to ensure that there is compatibility between devices produced by different manufacturers by adopting appropriate standards or recommendations for a given equipment or service. However, there are also applications that are not based on standards, especially in OTT service space – for example, Skype, Viber, or WhatsApp are not based on standards, while Web pages use standardized protocols for HTTP communication.
- Telecom regulators (i.e. National Regulatory Agencies – NRAs), which aim to ensure the functioning of the telecommunications market in a given country or region by setting policies and rules (e.g. QoS, security, and competition) as well as providing licenses (e.g. for frequency spectrum to telecom operators).

There are relationships between different players in the telecommunications market. Thus, at one end are the end users who receive services from telecom operators (e.g. telephony, Internet access services, TV, cloud services, and IoT services) and pay a fee for that service to the operator, which can be volume-based (e.g. per gigabyte – GB), time-based (e.g. per minute and per hour), or flat fee (e.g. for daily, weekly, or monthly service usage). Also, the payment can be a combination between different types of billing/accounting options with set thresholds (e.g. X GB Internet traffic per month for x money and purchase of additional Y traffic volume, after reaching the initial threshold, for y money). Telecom operators purchase equipment from manufacturers and vendors. For the compatibility of telecom/ICT equipment which can be generally produced by different manufacturers (e.g. smartphones from different manufacturers should be able to connect to mobile networks), there are SDOs that make the standards (e.g. ITU, IETF, 3GPP, and IEEE). Manufacturers, operators, governments, universities, and regulators participate in these standardization bodies with their representatives. As in any other game, in telecommunications there must be rules that

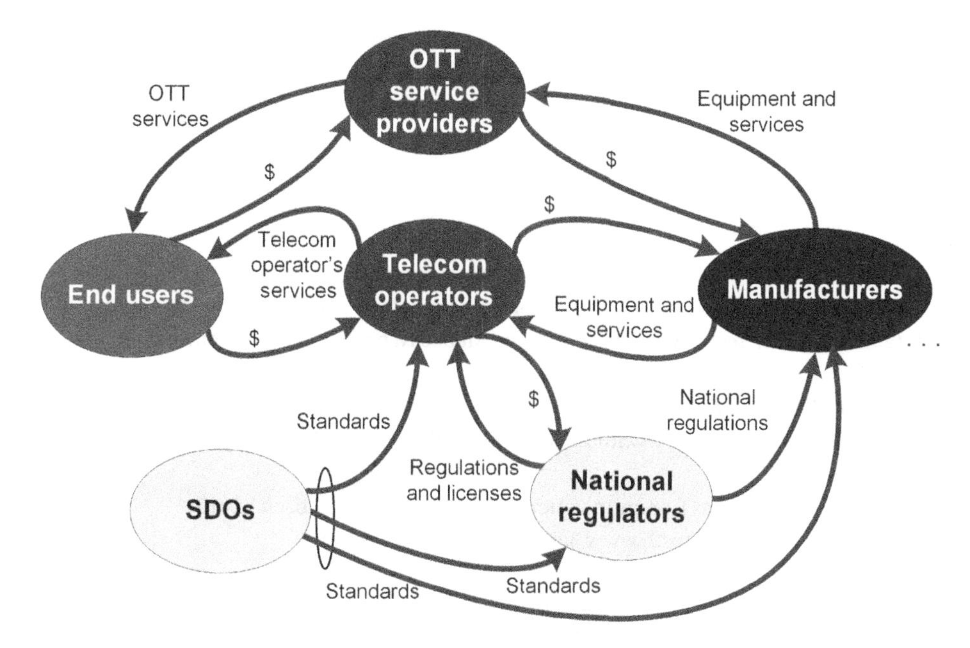

Figure 1.10 Main players in the telecom/ICT world.

must be respected by players in the telecom market, and for this purpose there are telecom regulatory bodies in all countries worldwide (they are also called NRAs).

Due to the presence of telecom (digital) infrastructure in businesses, governments, and at the individual user level, broadband policy is increasingly incorporating issues from other sectors that were not naturally part of the telecom/ICT sector in the past, and vice versa. The increasing reliance on digital infrastructure, services, and applications in almost every sector is resulting in the development of national (or regional) digital transformation strategies in all countries. In this direction, investments in non-ICT sectors such as agriculture, education, healthcare, and finance in some cases encourage direct investments in digital infrastructure to support digital services in those sectors (provided or supported by telecom operators and OTT providers). Various sectors that do not natively belong to telecom/ICT world are also called verticals. Digital transformation in various vertical sectors also creates a need for collaborative cooperation between telecom regulators on one side and non-ICT regulators on the other on different topics [30], as shown in Table 1.2.

Table 1.2 Possible cooperation between telecom regulator and non-ICT agencies.

Non-ICT regulator/agency	Potential collaboration topics with the telecom regulator
Commerce and trade	Digital taxation (for online purchases), online services
Cybersecurity	Data protection, clouds, end user equipment, IoT, AI/ML
Education	Child online protection
Energy	IoT, AI/ML, blockchain
Finance	Blockchain, cybersecurity and privacy, financial inclusion, online transactions
Transportation	Cybersecurity and privacy, IoT, AI/ML

In general, the digital transformation of all sectors and all aspects of human life based on the evolution of fixed and mobile broadband, using telecom and OTT services (i.e. digital services), is a long-term goal for the sustainable development of society and for the higher quality of life around the globe.

References

1 Janevski, T. (2019). *QoS for Fixed and Mobile Ultra-Broadband*. USA, April: John Wiley & Sons.
2 Janevski, T. (2015). *Internet Technologies for Fixed and Mobile Networks*. USA, November: Artech House.
3 Janevski, T. (2014). *NGN Architectures, Protocols and Services*. UK, April: John Wiley & Sons.
4 Overview of ITU's History, https://www.itu.int/en/history/Pages/ITUsHistory.aspx (accessed in 2023).
5 Yokogawa, "A History of Wireless Communication and Yokogawa's Approach", 2013.
6 3GPP (3G Partnership Project), https://www.3gpp.org/ (accessed in 2023).
7 RFC 791, "Internet Protocol", IETF, September 1981.
8 RFC 768, "User Datagram Protocol", IETF, August 1980.
9 RFC 793, "Transmission Control Protocol", IETF, September 1981.
10 IETF RFC 8200, "Internet Protocol, Version 6 (IPv6) Specification", July 2017.
11 IETF RFC 1945, "Hypertext Transfer Protocol – HTTP/1.0", May 1996.
12 IETF RFC 2068, "Hypertext Transfer Protocol – HTTP/1.1", January 1997.
13 IETF RFC 2616, "Hypertext Transfer Protocol – HTTP/1.1", June 1999.
14 ITU, "Trends in Telecommunication Reform 2013", April 2013.
15 ITU-T Recommendation P.1204, "Video quality assessment of streaming services over reliable transport for resolutions up to 4K", January 2020.
16 "Global average Internet speed, 1990-2050", https://www.futuretimeline.net/data-trends/2050-future-internet-speed-predictions.htm (accessed in 2023).
17 Moore, G. (1965). Cramming more components onto integrated circuits. *Electronics* 38 (8) April.
18 ITU Broadband Commission, "The State of Broadband 2022: Accelerating broadband for new realities", September 2022.
19 Ericsson, "Video traffic update", https://www.ericsson.com/en/reports-and-papers/mobility-report/dataforecasts/traffic-by-application (accessed in 2023).
20 Extron, "8K data calculator", https://www.extron.com/product/videotools.aspx (accessed in 2023).
21 ITU Recommendation Y.1291, "An architectural framework for support of Quality of Service in packet networks", May 2004.
22 ITU-T Recommendation Y.2001, "General overview of NGN", December 2004.
23 ITU-T Recommendation Y.3001, "Future networks: objectives and design goals", May 2011.
24 IETF RFC 3261, "SIP: Session Initiation Protocol", June 2002.
25 RFC 6733, "Diameter Base Protocol", October 2012.
26 ITU-T Recommendation Y.3653, "Big data-driven networking – Functional architecture", April 2021.
27 ITU-T Recommendation Y.3172, "Architectural framework for machine learning in future networks including IMT-2020", June 2019.
28 European Commission, "2030 Digital Compass: the European way for the Digital Decade", March 2021.
29 ICANN Office of the Chief Technology Officer, "New IP", October 2020.
30 ITU, Broadband Commission, "The State of Broadband: People-Centered Approaches for Universal Broadband", September 2021.

2

Internet Technologies

2.1 Open Internet Architecture

The main Internet technologies are standardized by Internet Engineering Task Force (IETF). All IP networks use Internet networking, including open Internet network as well as specialized (i.e. private) IP networks. However, the Internet architecture is a matter of network design [1]. Because IP traffic in general can be transferred over many access and transmission technologies, the network architecture can be diverse.

Open Internet architecture by definition is a global network, consisting of a huge number of individual networks intercommunicating with a common protocol – the Internet Protocol (IP), used to carry traffic to end hosts' applications through the Internet access service (either as a service provided to residential users or to businesses and organizations). However, one may distinguish between IP architecture and Internet network architecture [2].

2.1.1 Internet Protocol Architecture

The IP architecture typically uses TCP and UDP on transport layer (Layer 4) over IP (IPv4 or IPv6) on network layer (Layer 3), implemented in the Operating System (OS) of the device. In all Internet hosts (computers, smartphones, servers, and routers), the so-called socket interface is placed between the OS and the applications on the top. This refers to all hosts, including those connected to the open Internet as well as those connected to specialized IP networks which are not part of the open Internet. For example, telephony by using Voice over IP (VoIP) or television by using IPTV offered by the telecom operators are not provided via open Internet but over dedicated IP networks, which are typically based on Next Generation Network (NGN) framework from ITU [3].

The main end point of each application on a host connected to an IP network (that is, either the Internet or a private/specialized IP network) is called socket. What is a socket? The socket is an end-point of an Internet connection which connects the application on the top with the transport protocol (e.g. TCP or UDP) and network protocol (e.g. IPv4 and IPv6) that are implemented in the OS of the device, as shown in Figure 2.1. The socket binds the IP address in IP packet headers (IP address) and port number in transport protocol header (e.g. TCP and UDP) with the aim to identify the connection end point. In fact, each Internet connection on the Internet can be easily identified by using the IP addresses on each end, port numbers used on each of the two end hosts, and transport protocol (e.g. TCP or UDP). For TCP, the socket is referred to as "stream" socket because it provides a stream of bytes from the client to the server and vice versa (or between two end peers) in order and without any losses (TCP ensured lossless transmission, as we will discuss in the TCP

Future Fixed and Mobile Broadband Internet, Clouds, and IoT/AI, First Edition. Toni Janevski.

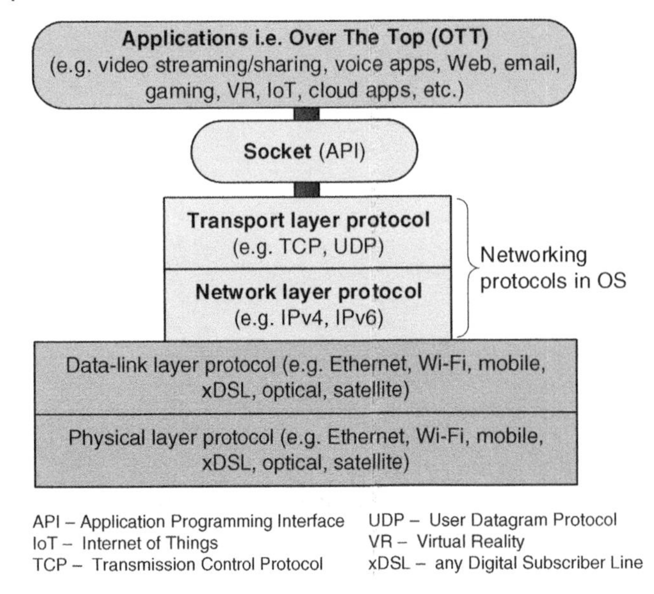

Figure 2.1 Internet protocol stack and socket interface.

subsection). On the other side, UDP socket is called a "datagram" socket considering that UDP uses transmission of messages (called datagrams) instead of byte stream.

So, the entire open architecture of the Internet is based on the standardized IP protocol stack, designed to connect any two IP networks and any two individual hosts connected to the global Internet. However, different hosts and network nodes may have different internal hardware, software, and vendor-specific design solutions in addition to standardized ones. Once two networks are interconnected, end-to-end IP communication is enabled, so that any host connected to the Internet has the ability to communicate with any other, no matter where they are located. This openness of the design has allowed the open Internet architecture to grow to a global planetary scale. Having a single protocol at the network layer, which is IP (either IP version 4 or version 6), enables uniform networking across competing, public, multi-vendor, multi-provider networks (which are IP-based).

2.1.2 Open Internet Network Architectures

The global telecom architecture consisting of large-scale public networks such as the Internet and other telecommunication networks is difficult to change in a short period of time as a whole due to the enormous amount of resources needed to build, operate, and maintain them. Therefore, the common IP absorbs and hides from the applications on the top the different protocols and implementations of underlying layers and, vice versa, the IP protocol hides the application operation on the top from the underlying transport technologies. Only communication on Layer 3, which is the network layer (in which IP is implemented), and above protocols have end-to-end communication (Figure 2.2). What does that mean? That means messages created by each protocol on one end can be "understood" only by the same protocol (on the same layer) on the other end. Protocol layers 1 and 2 (below the network layer) are limited to a link (e.g. a wire or wireless, i.e. radio link). In Figure 2.2 are shown two hosts and two network nodes between them (one switch which works up to Layer 2, and one router which works up to Layer 3), which results in three links in this example. Each of these links can be different from the others. However, the hosts or nodes on both ends of a single link should use the same Layer 1 and Layer 2 protocols (that is why Layer 2 is also called

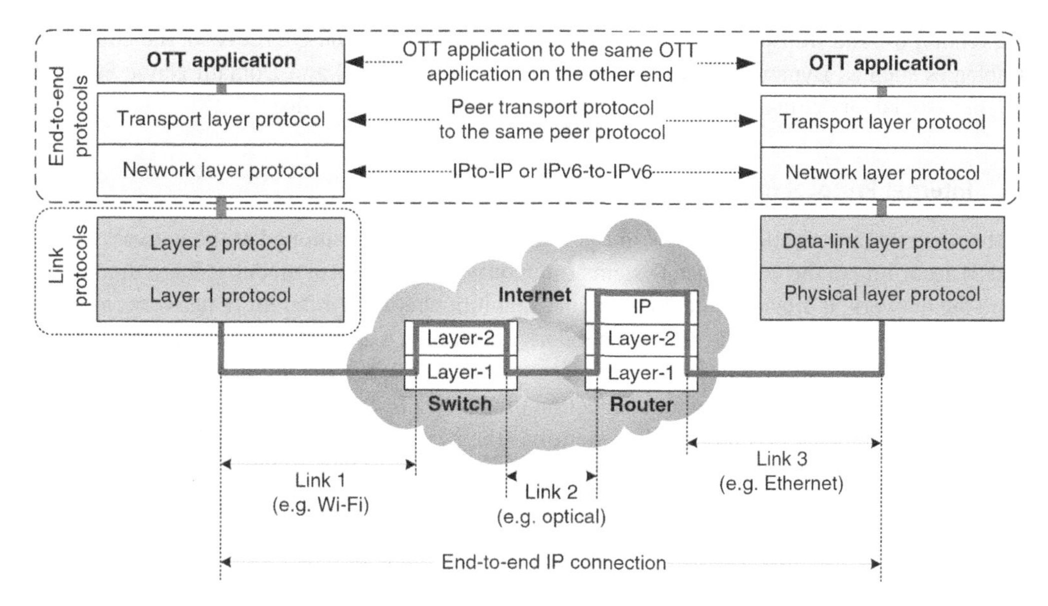

Figure 2.2 End-to-end IP communication.

link layer). For example, one link can be Wi-Fi (e.g. link 1 in the given example in Figure 2.2), another link (link 2) can be an optical transport (based on certain standards – there are multiple such standards), and link 3 can be an Ethernet link. Of course, there can be multiple links in a row of the same type, as well as other types of link that are not shown in Figure 2.2, such as mobile radio links, satellite links, and so on.

In general, there are two main networking architectures in IP networks; they are:

- Client-server – in this case, an up and running server machine is connected to the open Internet and listens to requests coming from other hosts, called clients, for a particular service. By design, the client machine is the one that sends a request to a chosen server machine, and the server is the one that listens on open sockets on predefined ports and responds to clients. So, the client-server approach is based on request-response principles.
- Peer-to-peer (P2P) – in P2P networking, the end hosts have both client and server capabilities. They act as a client (when request to a remote peer) or server (when responding to a remote peer). So, one may say that the peer-to-peer model is a symmetric one regarding the end hosts.

Well-known Internet services that use client-server communications are electronic mail (email), File Transfer Protocol (FTP), Hypertext Transfer Protocol (HTTP) – which is used for Web services, various Over-the-Top (OTT) services such as YouTube, social networking, and Amazon cloud. On the other hand, all conversational communication over IP networks uses P2P networking (e.g. telecom-grade voice and OTT voice like Skype, Viber, and WhatsApp). However, P2P services typically start with client-server mode for signaling purposes to locate the target destination host (e.g. to deliver a voice call or message).

2.2 Main Internet Technologies

Main Internet technologies are standardized protocols on Layer 3 (network layer) and Layer 4 (transport layer), and enabling technologies. That includes IP in its two existing versions—version

4 and version 6—and transport layer protocols such as TCP, UDP, and QUIC. Also, the enabling technologies such as Dynamic Host Configuration Protocol (DHCP) and Domain Name System (DNS) are crucial for "plug-and-play" functioning of Internet access service.

2.2.1 Internet Protocol (IP): IPv4 and IPv6

IP is the main communication protocol in the IP model, which is positioned at the network layer (Layer 3) according to the OSI (Open Systems Interconnection) protocol model. It has a dual role. Toward the network, it provides interconnection capabilities between different IP networks as well as providing locator capability of the given network interface of a given host (one should note that a given host can have multiple network interfaces – for example, a smartphone has a 5G interface and Wi-Fi interface, which are used for connection to different IP networks). On the other side, the IP protocol from application point of view identifies the network interface of the host to applications running on that host (that is, identifier role of the IP address).

In general, IP is designed as a connectionless protocol which means that there is no requirement for connection establishment before the data transmission. In that way each IP packet (i.e. datagram) is independently transferred over the open Internet. Each IP network which is part of the open Internet does its best effort to transmit every IP packet toward its destination (or the next network on the path); however, without any guarantees for a successful delivery (the IP packet can be lost due to a congestion at a network node or damaged due to bit errors on some links on the path). Also, IP does not guarantee that packets will arrive in the original sequence. Lossless packet delivery is left to be assured by functionalities of the upper protocols, such as transport protocols and/or applications.

The native Internet is built over IP version 4 (IPv4), which has been standardized in 1981 with RFC 791 [4], and implemented in 1983 (on 1st of January, also referred to as a flag day, when TCP/IP protocol stack was introduced in all hosts, several hundreds of them at that time). We refer to IPv4 as IP, considering that it was the only version of the IP in its appearance.

All functions of IP are defined by different fields in the IP header, while data is transferred in the IP payload. However, IP payload also carries the headers from upper protocol layers (e.g. TCP or UDP header and applications header).

The newer version of the IP, IPv6, is expected to be the foundation of existing and future broadband Internet/IP networks and services. The penetration of IPv6-based networks has grown steadily since the 2010s, as all IPv4 address space was exhausted in all five Regional Internet Registries (RIRs) during the 2010s, hence no new IPv4 address blocks to allocate (except for those IPv4 addresses that are traded regionally between organizations and corporations).

To overcome the exhaustion of IPv4 addresses (which was predicted in the mid-1990s due to the unexpected growth of the public Internet at the time), version 6 of the Internet Protocol (IPv6) was developed initially with RFC 1883 in 1995 (it was an early IPv6 specification) and RFC 2460 from 1998, which was the main IPv6 specification for a long time (almost two decades).

Finally, IPv6 entered its final standardized form with IETF RFC 8200 in July 2017 [5] which is the official IPv6 standard. This is the first IPv6 standard from the IETF, although IPv6 originally appeared in the late 1990s (in 1995 and then updated in 1998), but still was not an official IETF standard until 2017. IPv6-based networks are expected to replace IPv4-based networks, but the process of transitioning from IPv4 to IPv6 will certainly take many years. Why? Because such a process is gradual and has already started in the past decade (2010–2020), but will occur in the current decade (2020–2030), as the IPv4 address space has been drained.

So, IPv6 is the latest version of the IP, but it is not directly compatible with its predecessor, IPv4. Each IPv6 address has a length of 128 bits, which allows for the existence of a huge number of IPv6 addresses (theoretically 2^128 IPv6 addresses) compared to the number of IPv4 addresses (with 32 bits, there are theoretically 2^32 IPv4 addresses). In addition to the larger addressing space (which was the main driver for its standardization), IPv6 also brought several other important innovations compared to IPv4. These novelties include the new Flow Label and Next Header fields, as well as the exclusion of the Header Checksum and the fields related to IP packet segmentation (which turned out to be unnecessary in IPv4). A comparison of IPv4 and IPv6 headers is shown in Figure 2.3.

As a novelty compared to IPv4, IPv6 supports per-flow Quality of Service (QoS) at the network layer by using the Flow Label. A flow is a sequence of related packets sent from a given source to a given destination. This means that flow-based QoS (which is generally determined by loss, packet delay, and bandwidth given in bit/s) will be easier to implement on the Internet. In IPv4, a packet is classified to a particular flow by using 5-tuple, consisting of source and destination IP addresses, source and destination ports, and type of transport protocol (e.g. TCP or UDP). With IPv6, a packet can be classified to a particular flow by the triplet consisting of a Flow Label, a source IPv6 address, and a destination IPv6 address. Labeling a flow with the Flow Label field enables the classification of packets belonging to a specific flow. Without the Flow Label, the classifier must use port numbers which are carried in the headers of transport layer protocols such as TCP and UDP. So, traffic classifier needs to read headers of protocol Layer 4 with the aim to see the port numbers, while with Flow Label in IPv6 only Layer 3 header read is sufficient to identify the flow – if the Flow Label is actually used. However, the flow state should be established on a subset or all IPv6 nodes (routers) on the path, which should monitor all triples of all flows in use. But, similar to IPv4, IPv6

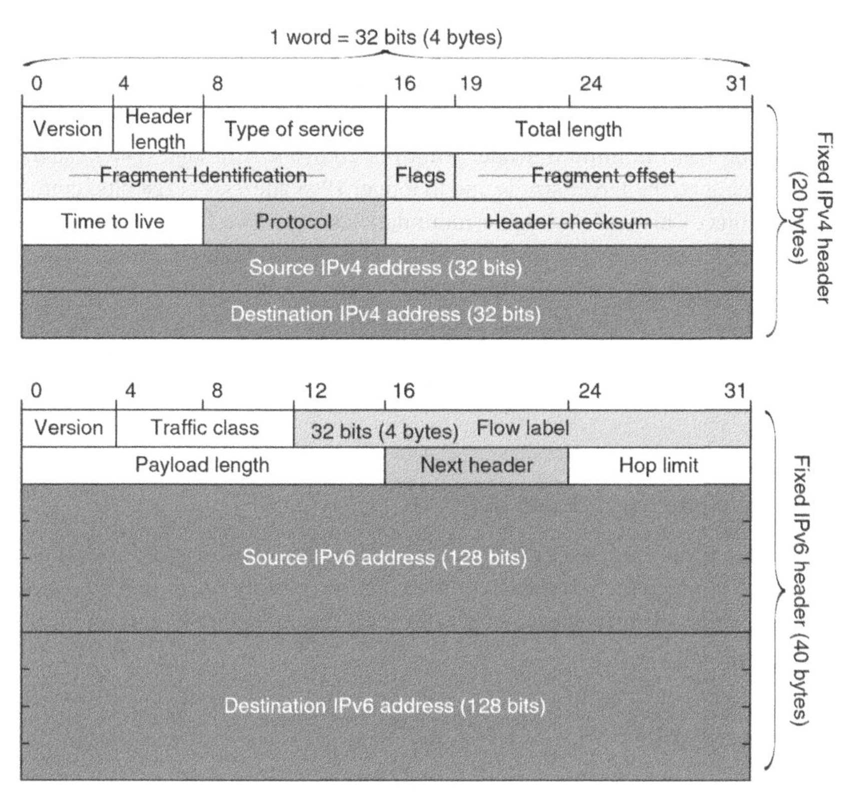

Figure 2.3 IPv6 vs. IPv4 header comparison.

does not guarantee by itself end-to-end QoS because there is no reservation of network resources (this has to be provided by other mechanisms in each of the IPv6 networks on the path).

Another important novelty of IPv6 is the Next Header field which identifies the type of header immediately following the IPv6 header, although that is the same purpose field as the field "Protocol" in the IPv4 header. Single IPv6 packet may carry zero, one, or more next headers, placed between the IPv6 header and the upper protocol header (e.g. TCP header and UDP header). These so-called extension headers may carry routing information, authentication, authorization, and accounting information, which provides better network layer functionalities in all network environments.

Significant changes are made regarding the fragmentation of data in IPv6, which is done at the source host contrary to IPv4 where it is performed in routers.

Header Checksum in IPv6 header is omitted in order to reduce the processing of IP headers in routers (as a reminder, in IPv4, each router for each packet has to calculate a new Header Checksum due to changes in the TTL field). The error control in IP header is redundant because it is provided in lower protocol layers (data link layer, i.e. Layer 2) and upper protocol layers (e.g. TCP and UDP, i.e. Layer 4). Hence, checksum is omitted in IPv6. So, IPv6 header has a fixed format that allows hardware processing for faster routing.

Also, all fragmentation fields in IPv4 header are omitted in IPv6 headers, because fragmentation on IP layer caused more troubles than benefits, so it has no practical use in the open Internet era (that is, since the mid-1990s).

Hop limit in IPv6 header is an 8-bit value that provides the same functions as the Time To Live (TTL) field in IPv4, which decreases by one for each hop (i.e. decreased by each router on the path).

Overall, IPv6 is a newer IP version that is not significantly different from the previous version, IPv4. Networks still are assigned network address blocks or network prefixes, IPv6 routers route packets hop by hop, providing connectionless delivery, and network interfaces still must have valid IPv6 addresses. However, IPv6 is seen as a simpler, scalable, and more efficient version of the IP.

The minimum IPv6 header length is 40 bytes. Hence, the IPv6 packet header is at least twice as long as the IPv4 header which has a minimum header length of 20 bytes. Although IPv6 header has fewer fields than IPv4 headers, the larger size is due to longer IPv6 addresses (128 bits) compared to IPv4 addresses (32 bits). This leads to higher redundancies with IPv6 headers when the communication requires smaller packets (i.e. small packet payload). For example, in real-time communications (e.g. VoIP) we have to use smaller packets. In such a case, higher header redundancy of IPv6 (the header is used for addressing and data control fields, not for user data, so it is called redundant with respect to user data such as voice and video) leads to inefficient utilization of available links and network capacity. And, the payload data is even smaller for massive IoT services.

2.2.2 Transport Protocols in Internet: TCP and UDP

Internet networking is based on network layer protocol (IP) and transport layer protocols (TCP, UDP). Network protocols are implemented in the OS of the hosts or network nodes and they are connected to the application using an abstraction called "socket" as the endpoint of a communication call/session in IP environments.

2.2.3 User Datagram Protocol (UDP)

The User Datagram Protocol (UDP) is a standard protocol with STD number 6. UDP was described by RFC 768 in 1980 [6].

UDP/IP is typically used for real-time data (e.g. VoIP and IPTV) and some control traffic (e.g. DNS uses UDP/IP). In practice, UDP does nothing but add the source port and destination port numbers (Figure 2.4), which are used by the ports opened at both ends of the communication link. Port numbers identify the application using the given transport protocol layer (e.g. UDP in this case). All well-known protocol numbers are defined by Internet Assigned Numbers Authority (IANA), which is part of Internet Corporation for Assigned Names and Numbers (ICANN), the non-governmental body that governs the Internet technologies globally.

UDP has no flow control mechanism. Also, UDP client and server do not establish a connection prior to sending an IP packet in either direction (client to server or vice versa). Of course, the server needs to listen to an open UDP port for incoming packets from clients. Considering that each IP packet is independently sent by using UDP/IP the term "datagram" is justified. All these features of UDP make it a very simple and lightweight protocol. However, with its simplicity UDP provides

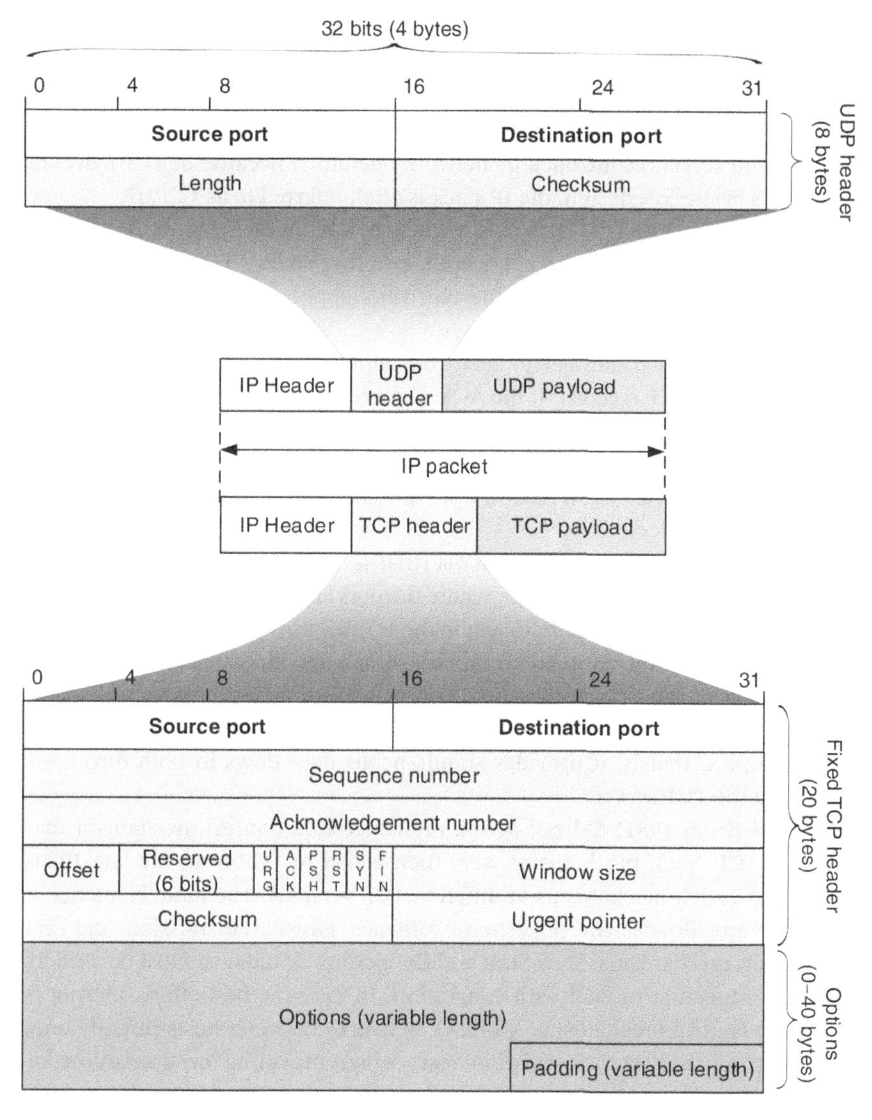

Figure 2.4 UDP versus TCP headers.

the lowest end-to-end latency (since there is no retransmission of lost packets, which takes extra time and hence adds latency) required for some time-sensitive services.

UDP is typically used in cases where reliability is not required and in cases where higher level protocols can provide flow control and error control. It is used as a transport protocol for several well-known application protocols such as DNS, Simple Network Management Protocol (SNMP), as well as real-time services over the Internet and specialized IP networks, such as VoIP and IPTV. For example, all VoIP services use UDP/IP communication for voice data, including VoIP provided by telecom operators as well as OTT VoIP (e.g. Skype, Viber, and WhatsApp).

2.2.4 Transmission Control Protocol (TCP)

The Transmission Control Protocol (TCP) is a standard protocol with IETF Standard Number 7, which was established by RFC 793 in 1981 [7]. TCP is a transport layer protocol in Internet hosts (client and server machines) for the reliable transmission of information. The most popular Internet services today are based on TCP, such as email and World Wide Web (WWW), as well as FTP and others. Unlike UDP, TCP incorporates means of Internet congestion control which are located in the end hosts. This is also a very different approach from the traditional approach in telecommunications where the end hosts are "dumb" devices and all control resides in the network (e.g. network nodes and servers) controlled by network operators. Because of TCP's overall importance in the Internet's best-effort design, the IP stack is often referred to as TCP/IP.

From an application's point of view, TCP transmits a constant stream of bytes across the network. TCP ensures that data sent by the sender's application over the sender's TCP port is received in an orderly, lossless, and error-free manner by the receiver's application over the receiver's TCP port.

The TCP sender assigns a sequence number to each transmitted byte and expects a positive acknowledgment (ACK) from the TCP receiver. If the ACK is not received within the time interval, the data is retransmitted.

TCP performs flow control, that is, when the receiving TCP sends an ACK back to the sender, it also tells the sender the number of bytes it can receive from the last received TCP segment without causing overflows in its internal buffers.

Multiplexing of different TCP packets called segments (that is TCP header+TCP payload) is achieved through the use of ports in the TCP header, where the port identifies the application running over TCP (e.g. HTTP uses well-known port 80, while the HTTPS uses port 443).

TCP initializes and maintains certain status information for each data flow. Each connection is uniquely identified by the pair of sockets used by the sending and receiving processes on each of the two ends.

Finally, TCP uses full duplex, that is, it provides simultaneous data flows in both directions between the TCP client and the TCP server.

The initial TCP standard (from 1981) did not define the congestion control mechanism that became essential later for TCP. Such mechanisms were then added to TCP from the late 1980s onward, resulting in different such mechanisms in different TCP versions. The main elements of the TCP congestion control are: Slow Start, Congestion Avoidance, Fast Retransmission, and Fast Recovery. Of these four main mechanisms, Slow Start and Congestion Avoidance must be used by the TCP sender for each connection to deal with congestion. In fact, the best-effort Internet is based on TCP's congestion control mechanisms, while IP networks between hosts initially only needed to route traffic from source host to destination host without providing any guarantees for lossless transmission (TCP actually needs to provide them).

To cycle through the four mechanisms, TCP needs to define the start value of the congestion window (cwnd) and the slow start threshold (ssthresh). The stall window (cwnd) defines how many segments (in total bytes) a TCP sender can send without receiving acknowledgment from a TCP receiver for their successful reception. The Slow Start starts with cwnd set to one segment, i.e. 1 SMSS, which is the Sender Maximum Segment Size (i.e. the maximum segment size that the sender can transmit). The maximum length of IP packet directly influences the maximum segment size. Although IPv4 packet can have length of 65535 bytes, the transport technologies (below the network protocol layer) restrict further the maximum segment sized used by TCP. Since Ethernet is the most widely used LAN on the Internet, and the maximum message transfer unit in Ethernet is limited to 1500 bytes, SMSS for TCP is kept below 1500 bytes. However, larger segments reduce redundancy due to headers (e.g. TCP header, IP header), because the same headers contribute with smaller percentage in larger packets and vice versa. Then, during the Slow Start phase the TCP increments the number of SMSS bytes for each ACK received by the sender (ACK acknowledges that new data is received at the TCP receiver side). In this way, in Slow Start the cwnd grows exponentially over time, that is, it doubles in size for each RTT. Here, RTT is the time taken for TCP segments to be transmitted to the receiving end and for the ACK to be returned to the sending end.

The value ssthresh is used as a boundary between the Slow Start (congestion window grows exponentially) and the Congestion Avoidance mechanism. In Congestion Avoidance, TCP cwnd grows linearly, which means that it may remain in such phase for a longer period of time. That way ssthresh can be increased to an arbitrarily high value, but can be decreased when congestion occurs (e.g. by half). In some TCP implementations it is equal to the receiver's advertised window size.

Congestion Avoidance is used when cwnd > ssthresh, as shown in Figure 2.5. In this phase, TCP increases the cwnd by one full-size segment for each RTT. If the RTT is constant over a given period of time, then the cwnd increases linearly during the congestion avoidance phase.

How does a TCP sender detect losses? There are actually two loss detection mechanisms:

- Retransmission timeout: After a period of time longer than the retransmission timeout value (from the time the segment was sent), TCP retransmits the segment and goes into slow start with cwnd = 1 segment.

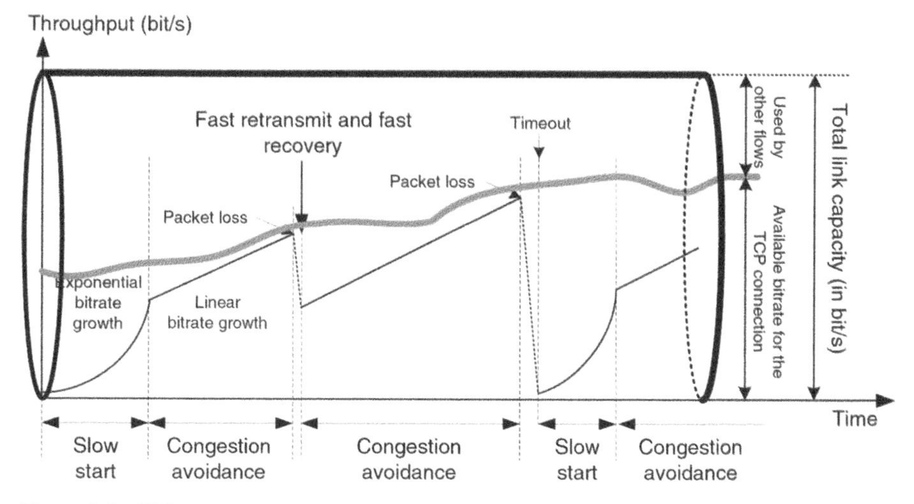

Figure 2.5 TCP congestion control mechanism.

- Duplicate Acknowledgments (ACKs): For each segment received, the TCP receiver sends an ACK that carries the sequence number of the next octet (i.e. byte) expected to be received. However, if a segment is lost and subsequent segments (after the lost one) continue to arrive at the receiving end, this indicates that congestion has occurred (and not interruption of the connection). For each segment received out of sequence, the receiver sends an ACK for the last successfully received segment in the sequence, so that such ACK will be a duplicate of the previous ACK sent back to the sender. Duplicate ACKs received by the TCP sender are treated as an indication of congestion.

Duplicate ACKs (three duplicate ACKs) are used by the fast retransmission and fast recovery mechanisms. Fast retransmission and fast recovery mechanisms ensure recovery from single losses in the cwnd. The available bitrate of a given TCP connection is directly related to the cwnd and RTT. Larger windows and smaller RTTs give higher bits and vice versa. In short, the maximum bitrate B that can be achieved by a TCP connection with the current value of cwnd = CW and RTT between the client and server (for that connection) can be calculated by using the following:

$$B = \frac{CW * SMSS}{RTT} \tag{2.1}$$

For example, for CW = 100, SMSS = 1000 bytes (where 1 byte = 8 bits) and RTT = 10 ms, the bitrate will be B = 80 Mbit/s. For CW = 10 it will be 8 Mbit/s, while for CW = 10000 it will be 8 Gbit/s using the same RTT and SMSS values. On the other hand, if RTT increases the bitrate decreases for the same size of the cwnd and vice versa.

2.2.5 QUIC: UDP-based Multiplexed and Secure Transport

One of the most important developments of the IP stack toward more secure and faster than before is the development of the QUIC, and its standardization in 2021 with IETF standard RFC 9000 [8].

What does QUIC do? First, it provides possible implementation over the existing IP stack, consisting of TCP/IP or UDP/IP, by using the lighter protocol of the two main legacy transport protocols in the Internet, and that protocol is UDP. Why? Well, because that way QUIC can be "freed" from the TCP implementations found in a given device's OS, because unlike TCP or UDP, QUIC can be installed on an existing device much like applications (e.g. through a web browser or through an application).

Comparable security in TCP/IP is offered by using the Transport Layer Security (TLS), which encrypts all content (e.g. from a Web browser that typically uses HTTP over TCP/IP). It provides strong encryption, but it is bound to use "heavy" TCP protocol. Regarding the security, QUIC also provides good encryption, but unlike TCP+TLS, QUIC operates over the simple UDP. Is that good?

Well, when using UDP the QUIC cannot rely on TCP for flow control and congestion control, so it has to implement its own data transmission algorithm. It does this by integrating connection establishment with the exchange of security keys. But, that change also brings QUIC to its most important feature, and that is zero delay (if we agree that the delay due to signals traveling at the speed of light will exist in any case proportional to the physical length of all end-to-end links). Overall, this feature allows QUIC to achieve single RTT delay (i.e. 1-RTT delay) for the full connections and zero RTT (i.e. 0-RTT) for resumption of the connections. Comparison of TCP, TCP+TLS, and QUIC is shown in Figure 2.6. In that manner, QUIC makes the IP (either IPv4 or IPv6) future-proof by reducing the delay for applications that require reliable transmission with the smallest possible transmission delay between the end points of the connection (e.g. for video streaming and VR).

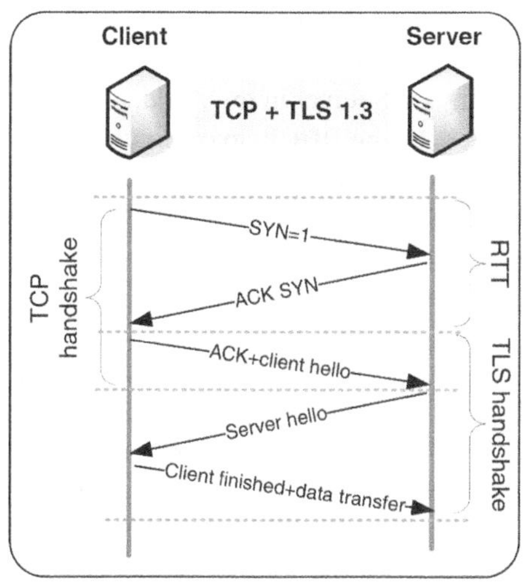

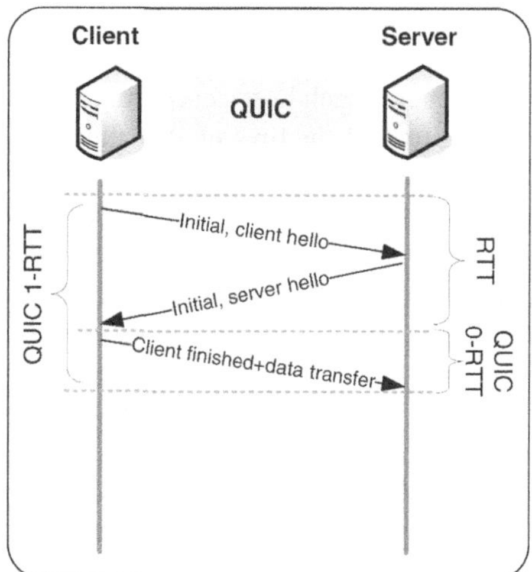

Figure 2.6 Comparison of TCP and QUIC.

2.2.6 Domain Name System (DNS)

There are exactly two so-called namespaces on the Internet, which are managed globally by ICANN and its division IANA. The two Internet namespaces are as follows:

- IP addresses, including IPv4 addresses, typically given in dot-decimal notation (e.g. 93.184.216.34), and IPv6 addresses typically given in a hexadecimal notation (e.g. 2001:0000:0000:0000:0008:0800:200C:417A).
- Domain names, e.g. "ietf.org" (the domain name of IETF, which is an organization). The domain namespace is defined as a tree structure with the root on the top. Each domain name consists of a sequence of so-called labels, which are separated with dots, and each label corresponds to a separate node in the domain tree. The label on the right side is always higher in the name hierarchy. So, in the given example, the top-level domain is the domain "org."

People use names more easily than long numbers, so the domain name space was created. However, machines on the Internet communicate only through IP addresses, so a domain name must be resolved to an IP address (corresponding to a given domain name according to DNS system records). To that end, DNS was created as a distributed, hierarchy-based system that translates a domain name to an IP address associated with that domain. So, DNS provides translation between the two namespaces on the Internet and other IP networks (e.g. carrier-grade DNS). For that purpose it has two roles:

- DNS defines domain names and rules for delegation of authority for such names.
- DNS is a distributed database system which supports the hierarchical mapping of domain names to IP addresses.

Each Internet host usually has two or three IP addresses of DNS servers. When DHCP is used for dynamic allocation of IP addresses, it is also used for configuration of DNS addresses in the client host. Typically, each telecom operator and each larger IP network (e.g. a corporate network) operates one or multiple DNS servers. DNS is implemented at the application protocol layer and initially uses UDP (as the transport protocol) because of its lower latency and lighter implementation than TCP. In general, each host can communicate with any DNS server in the world, however, typically it is configured (at the time of allocation of IP address by the DHCP) to communicate to the closest DNS servers, which ensures lowest RTT in the DNS client-server communication.

With the deployment of broadband networks around the world, as well as more processing power in all hosts (which doubles every 1.5–2 years according to Moore's Law), there is also a standardized approach to DNS communication using of TCP as a transport layer protocol (e.g. for larger DNS messages).

Overall, the correct functioning of DNS as well as the lowest possible DNS RTT and resolution time are the key performance parameters for the Internet and generally IP QoS, as they are directly related to the availability and accessibility of the Internet/IP services. Without DNS servers up and running, the hosts will need to use IP addresses for communication with other hosts, which is neither practical nor feasible, considering the need for plug-and-play Internet access (in order to be accessible to all end users with only basic digital skills requirements).

2.3 IPv6 Addressing and Implementation

Today's Internet cannot be imagined without the most fundamental Internet process: IP addressing, i.e. IPv4, and IPv6 addressing. This section covers IPv6 and the migration from IPv4 to IPv6 networks.

2.3.1 IPv4 Addressing

Each interface on every host on the Internet or other IP networks has its own IP address, which consists of two parts: a network ID (the network part of the IP address) and a host ID (the host part of the IP address) and has a total length of 32 bits for IP versions 4.

A network identifier (network ID) identifies the IP network. A host ID uniquely identifies a given host on the IP network or, more precisely, the host ID identifies the network interface on a given host (e.g. computer and smartphone) in a given IP network (e.g. fixed access network, Wi-Fi hotspot, and mobile access network). With IPv4, there are two types of IP addressing:

- Classful addressing: actually the first method of addressing in open Internet since the 1990s, based on fixed number of bits for network ID and host ID. So, for host ID, there were either 24 bits (class A), 16 bits (class B), or 8 bits (class C of IP addresses). There were also class D for multicast IP addresses (aimed for allocation for multicast IP communication, in which a copy of IP packet from a single sender is sent to multiple specified destinations), and class E (for experimental purposes).
- Classless addressing (used for Classless Inter-Domain Routing – CIDR): in this case, the number of bits for the network ID and host ID is flexible (of course, their sum must be 32 bits, the length of IPv4 address). So, this approach eliminated the IP addressing classes, which provided the possibility for finer granularity in IP address distribution. This approach prolonged the life of IPv4 until the present time.

How does the network obtain the network part of the IP address? It is assigned from IP address space of the Internet Service Provider (ISP) for the given network.

How does an ISP get a block of IP addresses? This is done through ICANN, www.icann.org. However, not all ISPs contact ICANN directly; smaller ISPs receive a block of IP addresses from large ISPs. Only the largest ISPs (which are often international organizations) communicate directly with ICANN (to assign IP addresses) through its five Regional Internet Registries (RIRs).

2.3.2 IPv6 Addressing

Each IPv6 address is 128 bits (i.e. 16 bytes) long, which provides the possibility to support more levels of addressing hierarchy, more addressable nodes, and simpler autoconfiguration of addresses.

Because of the larger address, IPv6 addresses are written in colon hexadecimal notation in which the 128 bits are divided into 8 sections, each section of 16 bits (equal to 4 hexadecimal digits). The preferred form is x:x:x:x:x:x:x:x, where an "x" can be 1–4 hexadecimal digits. It is less than 4 in cases when there are consecutive series of zeros in the address, as shown in IPv6 address example.

IPv6 address example:
2001:0000:0000:0000:0008:0800:200C:417A, also written as:
2001:0:0:0:8:800:200C:417A, and in compressed mode it is:
2001::8:800:200C:417A (the use of "::" replaces one or more groups of consecutive zeros in the IPv6 address and can be used only once).

There are several types of IPv6 addresses, from which the most important are the following:

- Unicast: This IPv6 address is assigned to a single interface of the host.
- Anycast: This type of IPv6 address is new (does not exist in IPv4), aimed to be used to send a packet to any one of a group of nodes.
- Multicast: This is an identifier to a set of network interfaces. Packet addressed to a multicast address will be delivered to all addresses in the set. IPv6 improves the scalability of multicast routing by adding a "scope" field to multicast addresses.

IPv6 address types are given in Table 2.1. Link-local IPv6 addresses are used for so-called stateless address autoconfiguration, where 64 bits of the interface ID are obtained from the interface's link address (e.g. using 16 zeroes concatenated with 48-bit Ethernet address of the given interface) [9]. IPv6 stateful address autoconfiguration is provided with DHCPv6 in a similar manner as for IPv4 (of course, DHCPv6 is a different protocol than DHCP for IPv4, considering the differences in IPv4 and IPv6 address formats and types).

Table 2.1 IPv6 address types.

Address type	Binary prefix	IPv6 notation
Unspecified	00. . .00 (128 zeros)	::/128
Loopback	0.0. . .01 (128 bits)	::1/128
Multicast	11111111 (8 ones)	FF00::/8
Link-local unicast	1111111010 (10 bits)	FE80::/10
Global unicast (includes all anycast)	All other IPv6 addresses	

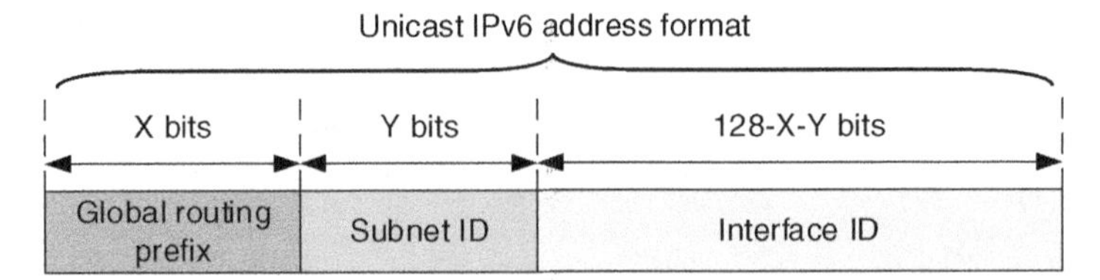

Figure 2.7 Global unicast address type of IPv6.

The general format for global unicast IPv6 addresses has three parts, which include global routing prefix, subnet ID, and interface ID (Figure 2.7). This is not really very different from classless IPv4 addressing for unicast (subnetting is also possible with IPv4), however, IPv6 has more possibilities due to the significantly longer IPv6 address (128 bits) than IPv4 and optimized header fields (when compared to IPv4).

We can conclude that the IPv6 is a well-defined protocol to support the current and future (e.g. in the 2030s and 2040s) Internet and IP network functionalities.

IPv6 provides enhanced service capabilities. For example, the Flow Label field of IPv6 header enables IPv6 flow identification independently of transport layer protocols. This means that new enhanced service capabilities can be introduced more easily. Also, IPv6 supports better mobility by removing triangle routing problem between the visited network, home network, and remote (corresponding) host. One should note that IPv6 supports secure networking using embedded IP security solutions.

Any-to-any IP connectivity is one of the vital IPv6 features in order to cope with the constantly increasing number of end devices. Such potential of IPv6 can be utilized in objects-to-objects communications.

Regarding the IPv6 address allocations, an important feature is self-organization and service discovery using autoconfiguration. IPv6 can provide autoconfiguration capability by using neighbor discovery protocol through combining addressing on the IP layer (protocol layer 3) and lower layer (layer 2, with the MAC – Medium Access Control address). Autoconfiguration enables ease of self-organization and service discovery of network management and reduces management requirements.

IPv6 addressing also provides the possibility for multi-homing based on the Next Header feature which provides the possibility of multiple IPv6 headers concatenation in a single IPv6 packet. In practice multi-homing means that IPv6 can handle multiple heterogeneous access interfaces and/or multiple IPv6 addresses over single or multiple access networks. That may provide better mobility management on the network layer (where IPv6 operates) in heterogeneous wireless/mobile environments.

2.3.3 IPv4-to-IPv6 Migration and IPv6 Implementation

The IPv4 address space is exhausted, although it has survived longer than originally thought due to classless IPv4 addressing, dynamic IP address allocation with DHCP, and the use of private IPv4 addresses (e.g. 10.0.0.0/8 or 192.168.0.0/16) which can be reused (unlike public IP addresses – all others) with a mandatory mapping between private IP addresses and one or more public IP addresses (using Network Address Translation – NAT) [2]. In the 2010s and 2020s, there is coexistence between IPv4 and IPv6, but all new blocks of IP addresses for allocation belong to IPv6, given that the IPv4 address space has been exhausted in all RIRs by 2020. However, the IPv4 will not disappear overnight, so there are required IPv4–IPv6 migration/convergence scenarios.

There are various IPv6 migration scenarios satisfying the user/application or provider requirements. The original transition plan from IPv4 to IPv6 was based on the dual stack principle.

The dual stack principle is used when there are still a significant number of IPv4 addresses and IPv6 addressing is penetrating the Internet/IP world. In this case, the IP network nodes or hosts attached to IP networks implement and enable both IPv4 and IPv6. However, this means that two separate protocol stacks exist in the OSs of the hosts considering that IPv6 and IPv4 are not compatible protocols. For example, there are differences in sockets and with that in API for IPv4 and IPv6, because IP addresses are different as well as many fields in headers between the two IP versions. The idea of a dual stack is to provide possibility to use IPv4 at a given time and then to be able to switch to IPv6 without difficulties, assuming that all modern OSs support the feature.

However, the dual stack in fact results in two separate networks, IPv4 and IPv6. But, IPv4-only systems can communicate only with IPv4-only systems, while IPv6-only systems can only communicate with their counterparts [10]. So, whether IPv4 or IPv6 will be used in a given network is determined by the chosen addressing type in that access IP network (e.g. by the operator or network administrator).

Another approach of coexistence between IPv4 and IPv6 is the tunneling principle. It typically involves encapsulation of IPv4 packet as payload in IPv6 packets, or vice versa (Figure 2.8). In the beginning, there were IPv6 "islands" in the dominantly IPv4 world, while later during the transition from IPv4 to IPv6 there will be IPv4 remaining islands in the IPv6-based Internet/IP world.

When there are interconnected IPv4 and IPv6 networks, there is the possibility of using translation mechanisms [10]. In this case, IPv4 packets are translated into IPv6 packets, and vice versa. With such approach, IPv4 networks can communicate with IPv6 networks, although there are

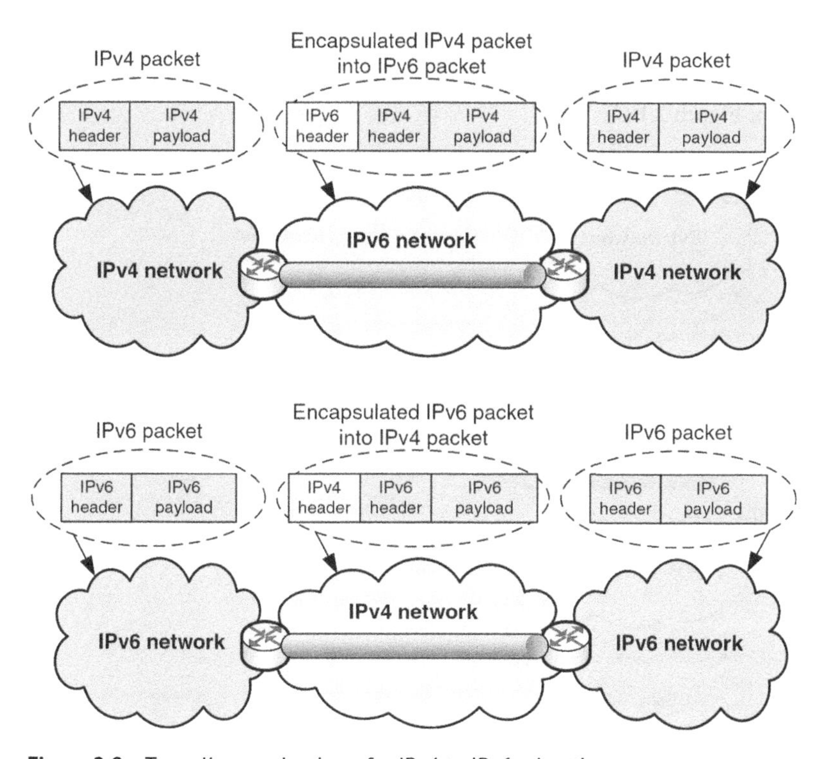

Figure 2.8 Tunneling mechanisms for IPv4 to IPv6 migration.

certain specifications that need to be followed by network nodes that do the translation of IP addresses and other IP header fields considering that they are not compatible.

The other approach for connecting IPv6 and IPv4 networks via routers between them is by using NAT. It was used in IPv4-only world for using the same private IP addresses (e.g. 192.168.0.0/16 and 10.0.0.0/8) in many private homes, enterprises, and hotspot networks by mapping such private addresses to a public IPv4 address. The distinguishing between different private IP addresses mapped to public IP address(es) is done by assignment of different port numbers at the NAT gateway, which is a router that interconnects the private and public IP networks (in respect to the type of IP addresses, private or public). This is referred to here as NAT44 (NAT for IPv4–IPv4).

In a similar manner, Mapping of Address and Port (MAP) between IPv4 and IPv6 addresses is useful in scenarios where scarcity of public IPv4 addresses would require the use of so-called stateful IPv6/IPv4 translation. So, MAP allows IPv4 addresses to be translated or encapsulated in IPv6 without a need for a stateful IPv4/IPv6 translator. There are NAT44, NAT64, and NAT46 for translation of IPv4 to IPv4, IPv6 to IPv4, and IPv4 to IPv6 addresses, respectively. Overall, there are two main possible approaches for MAP:

- MAP-E (mapping of address and port with encapsulation), standardized with IETF RFC 7597 [11], is a mechanism for transporting IPv4 packets across an IPv6 network using IP encapsulation combining with a generic mechanism for mapping between IPv6 addresses and IPv4 addresses as well as ports (given in transport layer headers such as TCP and UDP headers). Example MAP-E deployment is shown in Figure 2.9.
- MAP-T (Mapping of Address and Port using Translation), standardized with RFC 7599 [12], is a solution based on stateless IPv6-IPv4 Network Address Translation (NAT64) for providing shared or non-shared IPv4 address connectivity to and across an IPv6 network. The functionality of MAP-T is provided either by using standard NAT for IPv4 (i.e. NAT44) or with stateless NAT64 feature (defined with RFC6145) extended for provision of stateless mapping of IPv4 and ports (on transport protocol layer) to the IPv6 address space. Possible MAP-T implementation for telecom operators in shown in Figure 2.10.

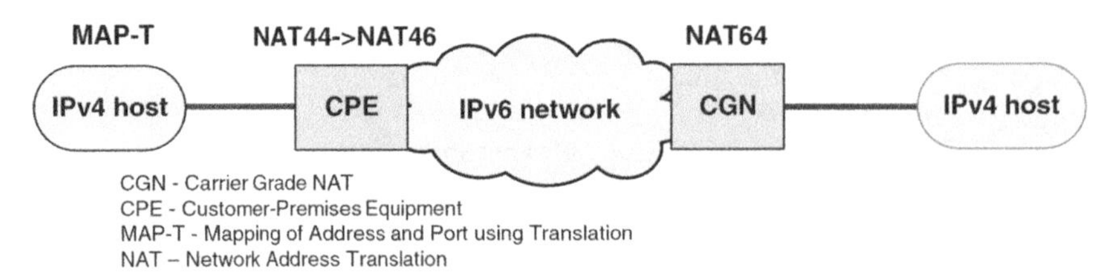

Figure 2.9 Example of a MAP-E deployment.

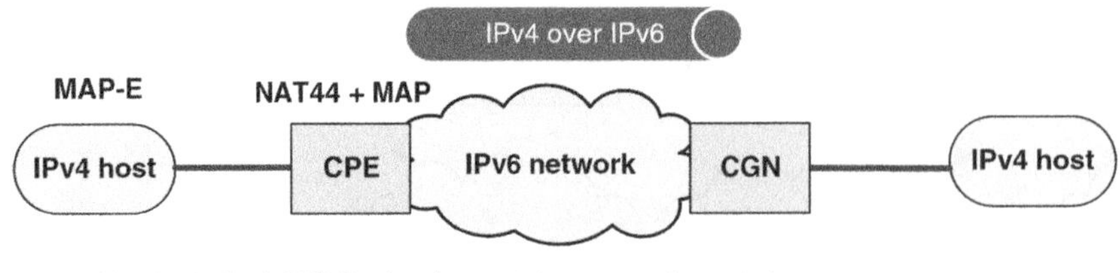

Figure 2.10 Example of a MAP-T deployment.

Where can fixed and mobile operators locate their NAT64 translators? Well, they can be located in either access or core networks. NAT64 can be deployed in various types of devices, such as routers, gateways, or firewalls [13]. There, stateful NAT64 can provide better sharing of available public IPv4 addresses for connecting IPv6 clients to IPv4 servers. However, DNS is also fundamental in the IPv4 to IPv6 migration given its role in connecting the two namespaces of the Internet and IP networks; to recall – these are IP addresses and domain names. In that way, DNS64 (which is defined with RFC6147) is recommended for use in combination with stateful NAT64. One may note that it is the essential part of an IPv6 single stack network that couples to the IPv4 network globally. On the other hand, NAT64 should normally coexist with NAT44 in dual-stack networks.

As for IPv6 deployments, new scenarios are constantly emerging from cloud and network convergence, edge computing, and 5G deployments. These scenarios impose highly demanding requirements on the IPv6 networks. However, IPv6 is expected to strongly contribute to future networks (e.g. networks toward 2030 and beyond), considering its better scalability due to very large addressing space (with countless number of IPv6 addresses; to remind – there are theoretically 2^128 possible IPv6 addresses, although they are systematically assigned and will not be used all at once), as well as better support for security and QoS (based on Flow Label and Next Header capabilities in IPv6 headers).

2.4 IP Interconnections and IP eXchange (IPX)

Global Internet and IP network (speaking as a whole) consists of network segments which are called Autonomous Systems (ASs). From a technical point of view, AS is a set of IP prefixes that are routable through the global interconnected IP networks. From a business or organization point of view, a given AS is managed by a single organization or enterprise. However, one organization or enterprise can have multiple ASs. The ASs are the main building block of the Internet and generally IP networks. Each AS is assigned a 16-bit or 32-bit long AS number. Initially, 16-bit AS number were used up by the end of the 2000s, so the 32-bit AS number was introduced. The AS numbers are distributed by IANA; however, since the beginning of the 21st century, the number of ASs increased linearly, so in 2000 there were less than 20000 ASs, in 2010 we had around 61000 ASs, and in 2020 there were around 105000 ASs [14]. With the same linear rise, the number of ASs in 2030 will be around 149000, while in 2050 we will have 237000 ASs, as shown in Figure 2.11.

Each AS consists of routers that are interconnected. Routers with various functionalities on different protocol layers are also called gateways (typically located at the edge of a given network where it interconnects to other networks). ASs are deployed and interconnected using transport networks, which connect different ports on the same network, as well as different networks at a national, regional, or global level. As far as terrestrial transport networks are concerned, today and tomorrow, the best medium is fiber given its enormous capacity (especially when using multiple wavelengths on the same fiber) and its very low error rate. So, in all cases where possible, the transport networks are made optical, so we carry IP traffic over optical transport networks.

2.4.1 IP Interconnection Approaches

With the transition of telecom operators' networks to all-IP networks, IP eXchanges (IPXs) were established in most countries. However, IPX connects national telecommunications operators to one another (usually by peering agreements), as well as connecting telecommunications operators to IP transit providers (e.g. by transit agreements).

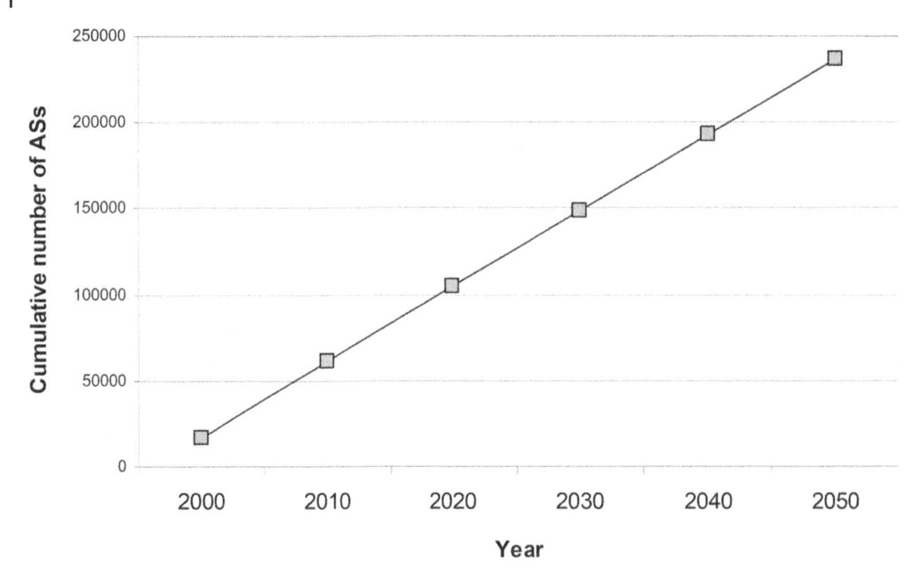

Figure 2.11 Number of autonomous systems globally in the period 2000–2050.

Peering is an agreement in which two interconnected telecom (network) operators carry each other's traffic, with no obligation to carry out traffic from a third party.

On the other hand, transit refers to a business and technical arrangement in which a given telecom operator agrees to carry Internet traffic on behalf of another operator. In most of the cases, the transit provider carries traffic to and from every destination on the public Internet, which is typically part of the transit arrangement.

While the transit is always based on some payment for the transport of IP traffic, peering can be either paid or free peering. Typically, business agreements for peering and transit are not publicly available.

In practice, large network operators typically exchange Internet traffic with comparably large network operators, based on nondisclosure arrangements and direct connections between such network operators.

In terms of Internet traffic interconnections, the impact of trends that have emerged with the advent of broadband access has increased expectations for the quality experienced by users, including bitrates and latency. So, the trend is that global service providers (e.g. Google, Apple, Amazon, and Facebook) have started deploying and managing their own globally distributed networks and/or service data centers in order to increase Quality of Experience (QoE), which refers to the quality of their services experienced by their customers, also using Content Delivery Networks (CDNs).

Additionally, major content and/or application providers (e.g. Google and Apple) operate some of the largest data transport networks in the world. On the other hand, they usually do not have local Internet access networks (fixed or mobile) which are provided by national telecommunications operators based on national regulations in each country. However, large global service providers that build their own networks also act as network operators and can therefore negotiate their own peering and transit arrangements using their role as network operators to benefit the delivery of their global services with higher QoE. High QoE is a prerequisite for global service providers to retain and expand their customer base and develop new innovative services provided over global Internet/IP networks.

2.4.2 End-to-End IP Communication

End-to-end communication over the Internet/IP is communication UNI-to-UNI (where UNI stands for User-to-Network Interface), where user may refer to a device used by a human user or a machine (e.g. IoT device). One may define the end-to-end connection as connection between the socket opened on one end (e.g. client) and socket opened on the other end (e.g. server). In end-to-end IP communication, there can be one or more IP networks on the path between the client and server, or between the two end peers. However, open Internet is a global network in which clients located in one country (connected through a fixed or mobile access network) and servers may be hosted in another country or region. In that way, IP connections typically span national or regional boundaries.

In general, each network part of the end-to-end path may have different policies and QoS approaches. End-to-end performance is affected by QoS solutions in different network parts of the path including fixed and/or mobile access networks. Most of the Internet traffic traverses through multiple autonomous networks. In such a case, the interconnection points must be designed to carry all Internet/IP traffic (in both directions) with the desired QoS and to avoid becoming bottlenecks (due to lack of capacity) in the end-to-end path.

In the past, the QoS regulation of IPX was not present, so it could appear that there is enough capacity in the access network, but there is bottleneck at the IPX, considering that unlike traditional telephone networks where a large part of the traffic remains within the same network (e.g. called and calling party reside in the same fixed or mobile network), in Internet communications, in most cases, the traffic exits the given network considering that most of the Internet users largely use well known global services (e.g. from Google, Amazon, and Facebook, i.e. Meta, Amazon), which are hosted in different networks and often in different countries in a given region. With the development of critical IoT services with 5G deployments in the 2020s, the IP interconnection and IPX further develop. There are two main future developments of IP interconnection and IPX platform:

- Decentralization of IP interconnection: In the traditional Internet world, all Internet and generally IP traffic is exchanged through a centralized IPX platform and through defined IP interconnection which is based on peering or transit agreements. With the development of services that require ultrahigh reliability and very low delays (e.g. Industry 4.0 and smart transportation), future IP interconnection points need to move toward the edge of the network (close to the access network) in parallel with the centralized IPX platform.
- High QoS at interconnection: The interconnection point and IPX platform should be able to guarantee the QoS for mission critical IP traffic via traffic engineering techniques, which are aimed to provide the needed performance (e.g. guaranteed bandwidth at interconnection, very low delay budget of IP interconnection, and in range of ms) and reliability (e.g., two or more paths between the end points that require ultrahigh reliability for the given critical service).

In general, IPX QoS parameters can be regulated and monitored in a country by the National Regulatory Agency (NRA). Then, end-to-end QoS can only be achieved through regional and global policies and agreements, which need to include IPX platforms as well as the networks attached to them from different sides. The IXPs (Internet/IP eXchange Points) can have a capacity of hundreds of Gbit/s up to many Tbit/s (1 Tbit/s = 1000 Gbit/s), with continuously increasing capacity over time due to the increasing capacity in access and core networks and increasing demands of services (including carrier grade services provided by telecom operators as well as OTT services provided by global service providers through the open Internet).

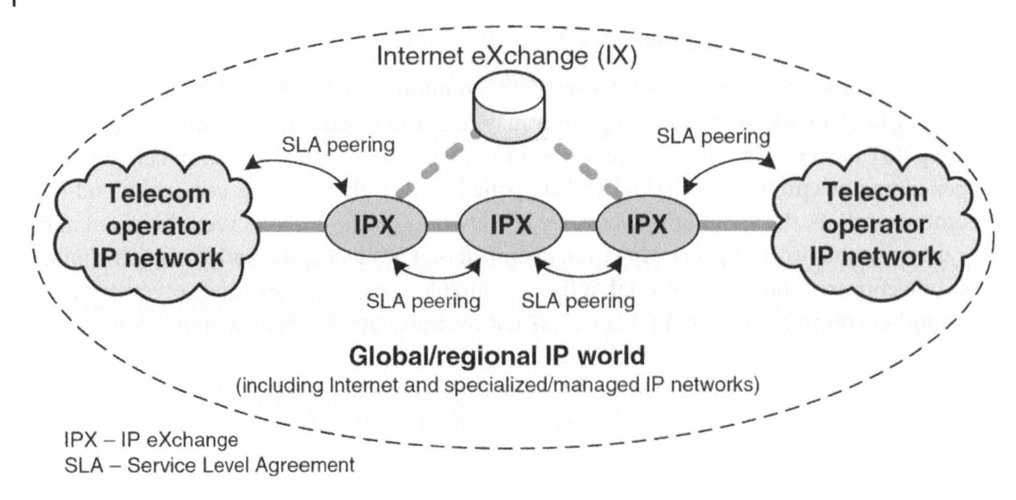

Figure 2.12 IPX in global IP world.

Figure 2.12 shows IPXs as a valuable business solution for different IP services on a global scale, including services provided by national telecom operators (e.g. carrier grade telephony, IPTV, business services, and IoT services) and services provided by the global OTT service providers (again, through the fixed and mobile access networks of telecom operators in each country). However, future services, such as critical IoT services, may be provided by new providers which will demand new locations of data centers toward the network edge which requires new locations of interconnection points. Also, future networks connected to IPX are expected to be more dynamic and more heterogeneous (e.g. private mobile networks for mission critical services), which results in a need for coherent scaling of IPX with new approaches to routing, traffic engineering, QoS, and security. In short, the future of IPX is tailored toward mastering the unexpected, with smaller locations on multiple places at the network edges, which will directly influence the future IP interconnection.

2.5 HTTP 2.0, HTTP 3.0, and Web Technology

World Wide Web (WWW, or Web for short) technologies are used to design and deliver the most popular and worldwide services over the Internet. In the basic definitions of the WWW, it can be said that it is a global information system of interconnected hypertext documents hosted on millions of servers over the Internet. Individual document pages on the WWW are called web pages and are accessed by a software application running on the user's computer, commonly called a web browser (e.g. Chrome, Firefox, and Opera). Websites may contain text, images, videos, and other multimedia components, as well as web navigation features consisting of hyperlinks.

The inventor of the WWW was Tim Berners-Lee, a computer scientist and former employee of CERN. On March 12, 1989, Berners-Lee wrote a proposal for what would eventually become the World Wide Web. The first test of the Web client-server communication was completed in December 1990 and it continued in the couple of years afterwards. The main transport (communication) protocol of Web is HTTP. In the spring of 1993, CERN decided not to patent the HTTP, so it set the path toward further standardization of HTTP by the IETF, the main standardization body for the Internet technologies. And the rest is the history of the success of the web and thus the success of the entire Internet/IP as a global networking "platform".

The WWW is essentially a huge client-server system with millions of servers distributed around the world. Each server maintains a collection of documents; each document is stored as a file (although documents can also be generated on demand). The server accepts requests to download a document and transfers it to the client. In addition, it can also accept requests to store new documents. The simplest way to refer to a document is by using a reference called a Uniform Resource Locator (URL). For example, an URL is: https://example.com/.

Most web documents are expressed using a special language called HyperText Markup Language (HTML). Being a markup language means that HTML provides keywords to structure the document into different sections. For example, every HTML document is divided into a title section and a main body. HTML also distinguishes headers, lists, tables, and forms. It is also possible to insert images, videos, animations, and other files at certain positions in the document.

2.5.1 HTTP Fundamentals

All communication on the Internet between clients and servers is based on the HTTP. The initial HTTP was a relatively simple client-server protocol; the client sends a request message to the server and waits for a response message. An important property of HTTP is that it is stateless. In other words, it has no concept of an open connection and does not require a server to maintain information about its clients.

HTTP is based on the use of TCP/IP protocol stack. Whenever the client sends a request to the server, it establishes a TCP connection with the server and sends its request message over that connection. The same connection is used to receive the response. By using TCP as the underlying protocol, HTTP does not need to worry about lost requests and responses. The client and server can simply assume that their messages are reaching the other side. If things go wrong, for example, the connection is broken or a timeout occurs, an error code is reported. However, in general, no attempt is made to recover from failure.

One of the problems with the first versions of HTTP was its inefficient use of TCP connections. Each web document is constructed from a collection of different files stored in a directory on the web server. For the correct display of a Web document including all objects (e.g. text is an object or each picture, video, or other file type is an object in Web terminology), it is necessary to transfer these files to the client as well. Each of these files, in principle, is just another document for which the client can issue a separate request to the server where they are stored.

HTTP is used in two different modes:

- Nonpersistent mode: In this case, a new HTTP connection is created for each object transfer (HTTP 1.0) [15]. This was the first version of HTTP which is noted here due to historical reasons.
- Persistent mode: In this mode, a single HTTP connection is used to transfer multiple objects between client and server (HTTP 1.1 [16] and HTTP 2.0 [17]).

In HTTP 1.0 and earlier versions, each request to a server requires setting up a separate connection. When the server responds, the connection is terminated and each next object transfer requires a new TCP connection. Such TCP behavior is called nonpersistent. The main disadvantage of nonpersistent TCP connections is an increase in the delay with the number of objects in the Web page, considering that each object requires a separate TCP connection with HTTP 1.0. As a consequence, the time to transfer an entire document with all its elements to the client can be very high, considering that each Web page usually has many objects on it.

Solving the main problem of HTTP 1.0 (separate TCP connection for each object) was done in HTTP version 1.1, which has been the standard since the end of the 1990s. HTTP 1.1 uses a

Table 2.2 HTTP methods.

Request method	Action
GET	Used to request a document from the HTTP server. It is the primary mechanism of information retrieval. This method uses URI for the request, and it enables its reuse by other applications, which on the other side creates a network effect that pushes further the expansion of the Web
HEAD	This method is identical to GET request, but the server does not need to send the whole body of the response message. So, it is used for obtaining the metadata (from server to the client) without actually transferring the representation data
POST	Used to submit, i.e. post, information from the client upstream to the server. Examples include provision of a block of data (e.g. fields entered into a Web page form) and posting a message (e.g. on a bulletin board).
PUT	Used to upload in the upstream from client to the server by using the supplied Request-URI (Uniform Resource Identifier) on the server side
DELETE	Requests the origin server to delete the resource identified by the Request-URI
CONNECT	This method requests that the client establishes a tunnel to the destination origin server identified by the request target. In this case, the client restricts its behavior to forwarding data in both directions until the tunnel is closed. Tunnels are used to create a virtual end-to-end connection through the use of one or more proxies using TLS
OPTION	This method requests information about the communication options available on the request/response chain identified by the Request-URI
TRACE	This method requests a remote, application-level loop-back of the request message, so the client can see what is being received at the other end of the request chain. This typically can be used for diagnostic purposes

persistent connection, in which case the client can issue several requests (and receive corresponding responses from the Web server) over the same TCP connection. Also, a client can issue several requests in a row without waiting for a response to the first request (referred to as pipelining).

It can be said that HTTP is designed as a general purpose client-server protocol oriented to the transfer of documents in both directions. HTTP works on the request-response principle. There are several defined HTTP methods [18], given in Table 2.2. The client sends a request (e.g. GET request, Table 2.2) and receives a response from the HTTP server. Each HTTP response consists of a three-digit response code followed by a blank space and a human-readable description of the response code. HTTP response codes (i.e. status codes) are primarily divided into five groups: Informational (1xx), Success (2xx), Redirect (3xx), Client Error (4xx), and Server Error (5xx). For example, the status code "100 Continue" means that the initial part of the request has been received (the process can continue), the status code "200 OK" means success, the status code "302 Moved Temporarily" means that the requested URL has been temporarily moved, "404 Not Found" indicates that the requested document was not found, and "500 Internal Server Error" informs the HTTP client of an error at the server location.

2.5.2 HTTP 2.0

Primary improvement of HTTP 2.0 over HTTP 1.1 is the improved performance. One should note that HTTP 2.0 extends the previous HTTP standard, HTTP 1.1, by keeping the same HTTP semantics. Also, there are no changes regarding the functionality and existing HTTP methods, status codes, URIs, and header fields. However, there are several improvements of HTTP 2.0 over its predecessor HTTP1.1:

- Streams and multiplexing: HTTP 2.0 defines as a novelty "stream" which does not exist in previous HTTP versions in that form. By definition [17], a stream refers to an independent, bidirectional sequence of frames exchanged between the client and server within an HTTP/2 connection. HTTP 2.0 provides the possibility of multiplexing multiple such streams over the same TCP connection that is established between the client and the server for the HTTP communication. There is also defined stream state model in which all streams start in "idle" state, then go to "open" state (in such state, the stream can be used by both ends to send or receive frames of any type), and the "closed" in the terminal state of the given stream.
- Header compression: While HTTP 1.1 transmits requests and responses as plain text, HTTP/2 introduces header compression that uses a binary framing layer for creating streams. During the interaction, the TCP connection remains open, and the binary format (for request and response headers) improves overall performance.
- Stream prioritization: With this functionality, HTTP 2.0 allows more important resources to be loaded first. For that purpose, web developers need to associate priority levels, because a poor prioritization scheme can result in HTTP 2.0 providing poor performance, even worse than HTTP 1.1.
- Server push: It provides the ability for the server to send data to the client before requesting it (by predicting them, and thus removing the RTT for the request-response cycle). This way, the server gets the ability to send data to the client without waiting for a request and cache such data on the client side (in HTTP 1.x, the client has to first request any data the server sends). However, the client can request that server push be disabled. Server push is semantically equivalent to a server responding to a given request from the client.

The main performance improvement of HTTP 2.0 over HTTP 1.1 is shown in Figure 2.13. Each HTTP 2.0 connection can be used to transport multiple streams that are multiplexed into it, unlike the HTTP 1.x standards which do not provide such a possibility; that is, HTTP 1.1 (without pipelining) needs to receive a response for a sent request before sending the next request. On the other hand, HTTP 2.0 provides the ability for multiple requests to be sent without waiting for a response for already sent ones (within a given TCP connection), thus eliminating the head of the line blocking due to lost TCP segments that need to be retransmitted. This is accompanied by HTTP 2's new binary framing layer, which defines how HTTP messages are encapsulated and then transmitted between the client and the server in both directions. The term "layer" refers to a new coding mechanism between the plug-in interface and the higher HTTP Application Programming Interface (API), which is exposed to Internet applications on top of the protocol stack. So, HTTP 2.0 communication is realized via smaller messages and frames, which are in binary format (unlike HTTP 1.x which uses a plain text format). Therefore, both sides, HTTP client and server, must use the binary encoding mechanism. In general, an HTTP 1.1 client cannot "talk" with an HTTP 2.0 server and vice versa. However, applications on the top are unaware of such changes, because the HTTP 2.0 client and server perform all the necessary framing work at both ends, independent of the media being transferred over the given HTTP connection.

2.5.3 HTTP 3.0

Web is based on a set of technologies used to provide web services. Regarding the underlying protocols, HTTP 2.0 also uses TCP/IP protocol stack (as HTTP 1.x versions), where the use of

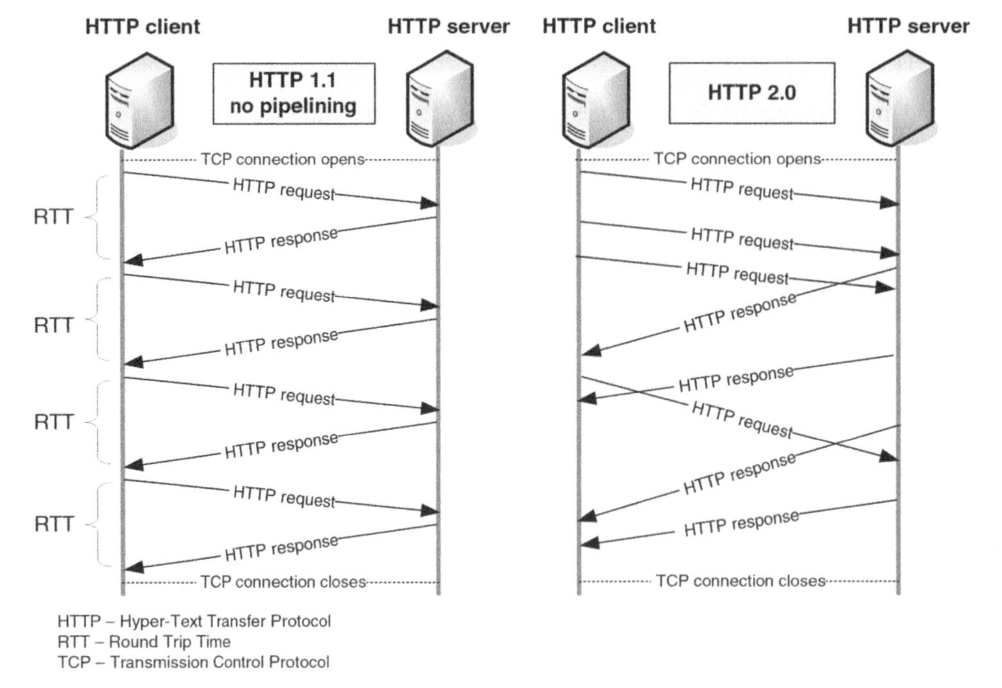

Figure 2.13 Comparison of HTTP 1.1 and HTTP 2.0.

TLS/Secure Socket Layer (SSL) is only optional. When using TLS/SSL with HTTP, one gets in fact secure HTTP (HTTPS), which is targeted to secure the HTTP end-to-end communication between the web client and the web server. With HTTPS, the data transported via HTTP is encrypted by TLS on the socket between HTTP and TCP below it. So, for secure Web communication, one uses HTTPS which is, in fact, HTTP over TLS/SSL over TCP. The main use of HTTPS is website authentication as well as protecting the privacy and integrity of web data while in transit between client and server by using end-to-end encryption (i.e. from socket on one end to the socket on the other end).

Over time, HTTPS is used more often than the HTTP (which is nonsecure), and it is primarily targeted to protect the user privacy in Web browsing and use of Web services on one side, as well as protection of authenticity of the websites (and their respective contents) on the other side. While TLS was optionally implemented in case of HTTP 1.x and HTTP 2.0, it finally became mandatory in HTTP in version 3 (HTTP 3.0), standardized in June 2022 [19]. HTTP 3.0 is designed to work over QUIC protocol on transport layer which provides stream-based multiplexing (each request-response pair is a stream) and flow control (Figure 2.14). Similar to HTTP 2.0, HTTP 3.0 uses a framing layer and within each stream its basic unit is frame. Frame multiplexing in HTTP 3.0 is performed by QUIC, which introduces mandatory end-to-end encryption in the case of HTTP 3.0 using TLS 1.3. Each request-response pair in HTTP 3.0 uses a single stream of QUIC, so most of the novelties of HTTP 3.0 actually come from using QUIC (including built encryption, zero RTT, and no head of the line blocking) instead of the standard TCP protocol as in previous versions of HTTP.

2.5.4 Web 3.0 and Metaverse

Initial development of Web technologies non-officially is noted as Web 1.0 (1990s–2000s), then Web 2.0 (2010s–2020s) and Web 3.0 (2020s and beyond). However, the years (dates) for each Web

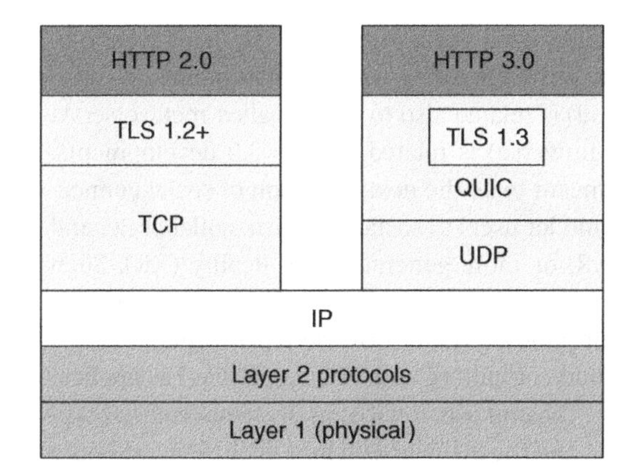

Figure 2.14 HTTP 3.0 protocol stack

generation are just approximate, because the use of Web generation is a jargon used by the industry and general ICT world, but it is not a standardized approach.

Comparison of three generations of Web (1.0–3.0) is shown in Figure 2.15. While Web 1.0 was based on "passive" Internet users which only consumed the content via various websites, Web 2.0 was characterized by increased interactivity of end users (as content creators, creating and uploading/sharing photos, and videos), easy creation and sharing of information via touchscreen of smartphones (connected to broadband Internet via mobile network, such as 5G) or drag-and-drop functions (e.g. for uploads to a cloud).

The main features of Web 3.0 include the Semantic Web (improves web technologies in order to generate, share, and link content), AI (e.g. smart and autonomous web assistants), 3D graphics (e.g. museum guides, 3D online games, electronic commerce, and geospatial contexts), connectivity (information is more connected thanks to semantic metadata), and ubiquity (the same content is available from multiple applications).

Also, blockchains and digital currencies are considered important parts of Web 3.0 services. Thus, Web 3.0 is focused on the decentralization of web services, thereby empowering individual

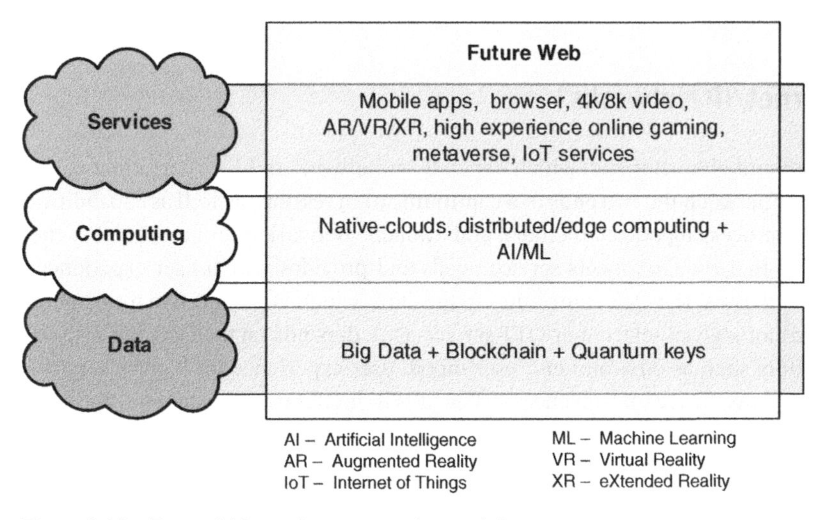

Figure 2.15 Future Web services, computing, and data.

Internet users. Thus, Web 3.0 can be seen as a decentralized Web built on decentralized technologies like blockchain, rather than centralized on servers in corporate-owned data centers.

Future Web (which includes Web 3.0 and beyond) is related also to the so-called metaverse. At the present time, the term metaverse (i.e. meta-universe) is related to Web 3.0 developments. According to the recent definitions, metaverse is meant to be the next evolution of social connection, where 3D cyber spaces in the metaverse should let users to socialize, learn, collaborate, and play in virtual ways online, e.g. by using VR, AR, or more generally Any Reality (XR). Such metaverse ecosystem includes devices and components, telecom operators as connectivity providers (who use equipment from telecom vendors for building their networks), platforms based on combination of emerging technologies such as cloud computing and Web 3.0, as well as applications, content, and monetization in the cyberspace. The end user devices in the metaverse concept appear as "portals" between the real physical world and the metaverse, where such devices should transfer the information from physical to 3D virtual world, and vice versa.

In the centralized Web (such as Web 2.0) corporations which run the Web services also provide complex cybersecurity solutions, but Web 3.0 may face new types of risks. Why? Well, because any developer who writes code can make some mistakes (simply, there is no perfect code). Such unintentional mistakes are commonly known as bugs in the program. But, in such cases, a smart attacker may exploit such bugs to carry out a cyberattack. In fact, some of the largest problems in the Web 3.0 cyberspace are caused by hackers who exploited bugs in smart contracts. As in the real world, when somewhere has invested large amounts of money, as in Web 3.0, such developments attract many developers and users, but also many cybercriminals. Considering that many users that are (or will be) engaged in Web 3.0 can make bad decisions (e.g. in the metaverse), the vast majority of security concerns regarding Web 3.0 are coming from social engineering attacks.

On the other hand, end user equipment such as VR/AR/XR headsets/devices require further development to be wearable in various everyday situations in the real physical world in order to be successful in the long term, and not be limited to only a limited number of scenarios (e.g. VR online gaming, working with AR in hazardous environments, and virtual meetings).

Considering that 5G and beyond development is aimed at wider inclusion of VR/AR/XR services as well as AI/ML use cases, many people in the telecommunications/ICT world are also linking Web 3.0 and further development of the meta-universe (i.e. metaverse) [20] with the expansion and further development of mobile technologies.

2.6 QoS in Internet/IP Networks

QoS is always an end-to-end characteristic, which depends on network and link performance on all segments on the path (between the two ends of a communication session) as well as capabilities of the end terminal (e.g. processing power, memory, and available network interfaces) [1]. For the end user, it is important that the QoS meets service needs and provides a high user experience. Regarding the human end users, the QoS contributes to the QoE, which is a subjective measure of the degree of enjoyment for a given telecom or OTT service. QoE depends on the QoS, but also on other nontechnical factors such as environment, user mood, user experience for a given service, price (e.g. in many cases, lower price for a given service results in higher user satisfaction, and vice versa), and so on. Of course, QoE refers to humans as end users. For Machine-to-Machine (M2M) communication, QoE has no meaning. However, QoS again has an end-to-end meaning and is equally important in human-to-human (e.g. telephony) or human-to-machine (e.g. web browsing) or machine-to-machine (e.g. vehicle-to-vehicle communication).

2.6.1 Legacy QoS Approaches in IP Networks

There are QoS frameworks set by various standardization organizations, of which the IETF and ITU are the leaders in terms of QoS in the ICT world. The initial QoS framework in the Internet was based on the so-called Best Effort, the main approach in initial the Internet, without any QoS guarantees. However, in the 1990s, QoS standards were defined for IP networks with so-called Integrated Services (IntServ) for per-flow QoS using resource reservation in the routers on the path, and Differentiated Services (DiffServ) where IP traffic is classified into limited number of classes (based on the ToS field in the IPv4 header and the DSCP field in the IPv6 headers), so routers only need to store information per class (not per link, i.e. per flow). While IntServ is an end-to-end QoS mechanism, DiffServ is a hop-by-hop (i.e. link-by-link path) QoS mechanism.

To avoid the dependency of QoS provisioning on the network layer protocol (which is IP in the Internet and all managed IP networks), Multi-Protocol Label Switching (MPLS) at the beginning of the 21st century, which was not only intended for use on IP networks. In that way, MPLS beside IP networks also was open for use by other packet-switching technologies (e.g. European ATM at that time), therefore, it is named "multi-protocol." Considering the protocol layering model, MPLS can be considered to be between Layer 2 and Layer 3 (where the IP is), so MPLS labels (i.e. headers) are added between Layer 2 and Layer 3 headers. The labels are then used for routing/switching in so-called MPLS network domain (that is, packet network where MPLS is implemented) as well as traffic engineering in such a network (e.g. allocation of resources/capacity to different types of traffic in the network). Also, MPLS was and still is one of the "mainstream" approaches for provision of QoS in transport IP networks, which can be combined with DiffServ principles as well as other protocols such as Border Gateway Protocol (BGP), used for routing between autonomous systems in the Internet and IP networks in general, and Virtual Private Networks (VPNs), used for logical separation of a given IP network into multiple logically separated networks over the same physical infrastructure.

Regarding the access networks, in the open Internet there is in force in most countries worldwide the network neutrality principle from the policy point of view. In short, network neutrality means that on the Internet all packets should be treated the same. Of course, that does not mean that it is forbidden to stop hazardous traffic (e.g. viruses, worms, and hostile websites); they should be filtered and stopped (by using mechanisms such as Deep Packet Inspection (DPI), which opens all protocol headers that are not encrypted; of course, all that without jeopardizing the privacy of the end users). Also, network neutrality does not prohibit traffic engineering that aims to provide needed performance metrics to different traffic types (e.g. video, audio, Web, and messages) in core and transport networks.

Table 2.3 gives the comparison for service complexity and scalability of legacy QoS approaches in open Internet and generally in IP networks, including best effort approach (no QoS differentiation), IntServ (per flow QoS), DiffServ (per class QoS), and MPLS (per class QoS).

Table 2.3 Comparison of legacy QoS approaches in IP networks.

	Best-effort	DiffServ	IntServ	MPLS
Service complexity	No isolation No guarantees	Per aggregate isolation Per class guarantee	Per flow isolation Per flow guarantee	Per aggregate isolation and per class guarantee
Scalability	Highly scalable (network nodes maintain only routing state)	Scalable (routers maintains per class state)	Not scalable (routers maintains per flow state)	Scalable (routers maintains per class state)

2.6.2 End-to-End IP QoS Framework

The transition of telecommunications networks to all-IP networks and services based on Internet technologies requires the definition of end-to-end network performance parameters. Such parameters are defined by different standardization organizations (e.g. ITU, 3GPP, and IETF) and regulatory bodies on national and regional bodies.

We refer here to the end-to-end communication as UNI-to-UNI, where UNI is the interface through which the User Equipment (UE) is attached to the fixed access network (e.g. desktops, laptops, smartphones, and IoT devices, connected through home or office wired or wireless LAN) or mobile network (e.g. smartphones, mobile/cellular IoT devices, and vehicles connected to a mobile network) connect to the telecom operator network (e.g. fixed access network and mobile access network). End-to-end interconnected IP networks can support user-to-user connections (where user is a human user who accesses services with a device connected to the network), user-to-machine connections, and machine-to-machine connections.

QoS is defined through the performances of all network parts along the track as well as the end network parts (Figure 2.16). The end parts include a local network for homes or enterprises (e.g. wired or wireless LAN devices in a home or office) for fixed access (including fixed wireless access), mobile terminals and devices in the case of mobile access, and servers in service data centers that can be located at centralized locations or network edges.

There can be defined different QoS parameters. However, if one needs to distinguish the most important standardized performance parameters for IP-based services they will be the following ones:

- Throughput in downlink and uplink: It is important for both bandwidth hungry applications, such as video with high resolutions (e.g. 4K and 8K), VR/AR/XR services, as well as interactive services such as Web services, considering that higher bitrates reduce the time to content for end users (time from a click until appearance of the content, e.g. on a Web page), which influences the QoE (lower time to content results in higher QoE, and vice versa).

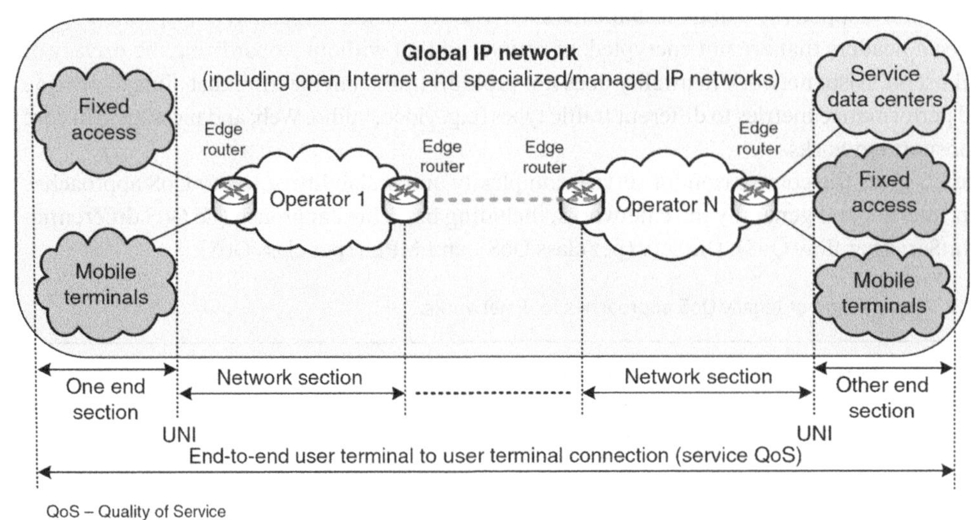

Figure 2.16 End-to-end IP network QoS.

- IP Packet Transfer Delay (IPTD): It is the time difference between the reference events of sending the packet through a network interface on one end and its receipt at the network interface of the destination host. One should note within a given sender or receiver host, all protocols up to the application layer (including also the application on the top of the protocol stack) add delays to the overall delay budget of the IP connection when it is considered application to application (or socket to socket).
- IP Packet Delay Variation (IPDV), also known as jitter: It refers to the difference between the one-way delay of IP packet and average IP packet transfer delay. It is important for real time services (e.g. VoIP and live video) and for critical services (e.g. critical IoT services).
- IP Packet Loss Ratio (IPLR): It is the ratio of the total number of lost IP packets to the total number of transmitted IP packets in a given measurement.
- IP Packet Error Ratio (IPER): It refers to the ratio of the total number of IP errors to the total number of transferred IP packages. Some applications/services are prone to errors (e.g. telephony, because the human ear can forgive up to several percent of errors or lost packages), while other services require lossless communication, which is realized in practice with TCP or QUIC on the transport protocol layer. However, the higher IPER values result on average at lower throughputs as a result of TCP retransmissions of lost (or damaged) packets.

Throughput (in bit/s) is the performance parameter that has the highest impact on the performance perceived by the end user, so higher bitrates (i.e. broadband access and transport) are normally better for all services, including real-time and non-real-time. However, the theoretical capacity of a given system (e.g. optical network, mobile network, Wi-Fi, and satellite network) is always larger than the bitrates perceived by end user applications. In practice, the error ratio is not constant; also, congestion may occur at network nodes (routers, gateways) along the path end to end, and terminal equipment may have different processing and storage facilities, as well as a different OS (where network and transport layers protocols are carried out at the end hosts).

2.6.3 Framework for Monitoring End-to-End QoS of IP Network Services

The rapid increase in the use of the Internet has changed the way we live and work, and has become an important factor in people's daily lives. Therefore, IP network service needs to be monitored based on selected KPIs that are important for end users and services [21]. In that way, Table 2.4 specifies the minimum set of parameters for monitoring the quality of an IP network service.

The QoS parameters for IP network service given in Table 2.4 van be evaluated by using measurements at national and international levels, as shown in Figure 2.17. The national level measurements evaluate the IP network QoS at the level of the telecom operator/s network (which includes access, core and transport segments), while at international level measurements the IXP is located somewhere in the global IP network (e.g. on the open Internet) and such measurements can capture the end-to-end QoS, including the interconnection points and transit and/or peering networks with which the telecom operator is interconnected. Typically, the Internet access service that is provided by national telecom operators provides access to the entire global Internet (if no filtering is applied on the national level). In that way, the international QoS measurement scenario provides the possibility for national telecom regulators to test international data transmission KPIs, such as download/upload bitrate, and delay.

So, QoS is not only about its definition, but it should be measured and monitored by different parties, which includes regulators, operators, and possibly end users (e.g. with crowdsourcing tools). There are different tools for measurement of the QoS, which may use active (generate

Table 2.4 Minimum set of KPIs for IP network service.

Parameter	Definition
IP network service activation time	This parameter refers to the allocation of IP address and configuration settings by the DHCP server in the network (also includes getting and setting up the IP addresses of DNS servers)
DNS response time	It is average Round-Trip-Time (RTT) delay metric for DNS request-response service in IP networks, which is directly dependent upon the distance between the DNS client and the DNS server, and the time needed for domain name resolution by the DNS server
Number of IP network interconnection points	This refers to the number of interconnection points to other Autonomous Systems (ASs)
RTT to IP network interconnection points	This metric measures the RTT between subscribers' service demarcation points and the interconnection points to other ASs. Typically, such interconnection points are located at the Internet eXchange Points (IXPs)
One-way IP delay variation	This is a one-way delay variation to IXP
One-way IP packet loss	This is a one-way packet loss to IXP
Mean data rate achieved (for downlink and for uplink)	This is the average of the data transfer rate achieved for transfer of a given number of samples (e.g. download or upload of file samples with a given size – the higher the bitrate, the larger sample file should be with an aim to provide realistic measurement of the data rate, because TCP with its Slow Start needs many RTTs until it reaches the maximum possible bitrate)
Percentage deviation of the mean bitrate	This refers to the deviation between the data rate contracted (advertised) to the achieved data rate
Internet IP network service availability	Represents the fraction of time probability that the end user is able to access services over the IP network
Radio coverage availability	This parameter refers to received signal strength in mobile/wireless networks (e.g. better than −95 dBm) which is however a parameter dependent upon the mobile/radio technology

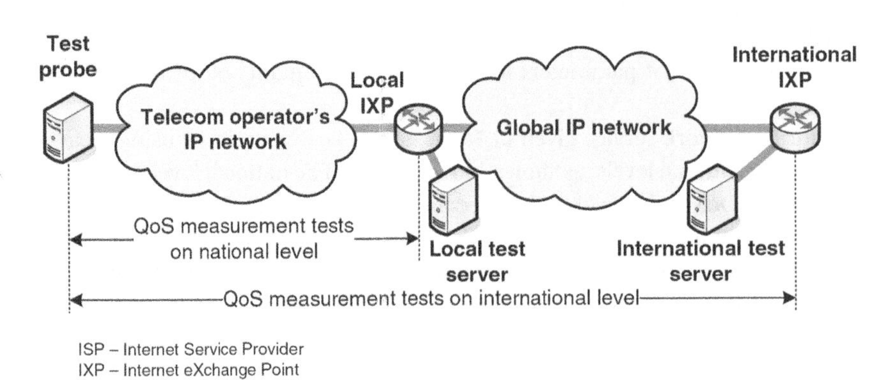

Figure 2.17 IP network QoS measurements.

traffic) and passive (monitor the traffic) measurements. However, the measurement should be typically targeted to QoS parameters that are important to the end users and services because QoS is considered always end to end.

What is most important regarding the QoS? Well, in the first place it is availability (without an available network there are no services), then comes the speed (i.e. bitrates in downlink and uplink), RTT, delay variation (also referred to as jitter), and so on (Table 2.4). For example, we have already discussed the importance of the number of IP network interconnection points (i.e. the number of ASs to the track) and also pointed to the importance of DNS as "connecting technology" between the two internet name rooms (i.e. IP addresses and domain names).

Then, what about the national regulatory frameworks for QoS? Well, they can vary between different countries. That depends on several factors, but the most important is the maturity of the telecom market regarding the competition as well as the level of cooperation between different parties (operators, regulators, and customers). Hence, the QoS is left to market forces in countries with developed and competitive telecom market, while in other countries national QoS framework (and/or legislation) requires QoS supervision and enforcement by the telecom regulators.

What about QoS in future networks? Well, future networks are expected to provide user-to-network continuum, where the data can be stored, processed, analyzed, and retrieved anywhere on the path (in the network, at the edge, at user premises). So, the number of different end points for a communication session will increase and the QoS provisioning will be more complex.

Also, there are differences in QoS in fixed and mobile networks. In general, fixed broadband can be dimensioned and planned more efficiently due to the fixed location for end user premises and higher media quality (e.g. fiber to the premises) resulting in sustainable bitrates with very low error rates. On the other hand, we have a different situation in mobile broadband access networks due to time-varying and location-dependent radio conditions that change with the user's mobility, so there is no possibility of guaranteeing certain individual bitrates at a given location in advance. However, justified predictions of user activity and mobility result in a proper design of the mobile network, and vice versa. For example, the actual bitrates for mobile Internet access depends on the distance between the base station and the user equipment (longer distance results in higher signal attenuation that results in lower level modulation and coding scheme, and that results in lower bitrates over the same physical resource – frequency and time, and vice versa), number of users simultaneously connected to a given cell in the radio access network, cell size (smaller cell gives larger overall capacity with the same resources, but may result in high handover intensity at higher user mobility); the type of mobile device may vary from one user to another (older or cheaper devices may not support certain interfaces or frequency bands, or certain needed protocols such as HTTP 3.0), and so on. In mobile broadband networks, the end users are mobile and the number of connected users connected to a base stations as well as the distance between the user device (e.g. smartphone) and base station (e.g. 5G base station) and their velocity changes over time, So, one may say that QoS provisioning is more complex in mobile environments, but it can be maintained with proper network design – e.g. using small cells in dense urban areas, larger cells in rural areas and highways, provision of newer mobile devices at affordable prices, and so on [22].

Finally, in all present and future use cases that require ultralow latency and ultrahigh reliability (as 5G uses for critical IoT cases), we can provide the desired QoS with private mobile networks deployments (e.g. managed IP networks or non-IP networks). So, there is always a workaround solution for QoS in both fixed and mobile broadband environments, which is changing and evolving over time. For example, in complex network and service environments with many heterogeneous QoS requirements (by different services over the same network).

2.7 Cybersecurity and Privacy

In the existing digital era, cybersecurity is crucial to provide universal, trusted, and fair access to the Internet. We discussed that we have an all-IP telecom world at the present time, but when talking about cybersecurity we talk about the security in the cyber world.

What belongs to the cyber world? Well, cyber world refers only to the open Internet, not to specialized networks and services. So, cybersecurity mainly refers to information security on the Internet for protection of information against unauthorized modification, disclosure, transfer, or destruction. Simply put, cybersecurity is the security of the cyberspace, and cyberspace includes all networks (e.g. routers and interconnection links) and all information systems (e.g. servers and databases) that are part of the public (a.k.a. open) Internet network.

While the use of ICT provides better management and increased productivity, the use of digital systems also generates risks. In fact, cyber threats and computer attacks increase security challenges for both public and private sector in all countries. Improving cybersecurity and protecting critical information infrastructure are essential for the social and economic development of each nation. Cybersecurity incidents can jeopardize the availability, integrity, and confidentiality of transferred information and disrupt the operation and functioning of critical infrastructure, including digital and physical ones. Overall, such threats can also threaten people's safety and entire countries.

What is cybersecurity by definition? According to the ITU's definition [23], cybersecurity is the collection of tools, policies, security concepts, security safeguards, guidelines, risk management approaches, actions, training, best practices, assurance, and technologies that can be used to protect the cyber environment and organization and users' assets.

Each element of the cyber environment can be viewed as a security risk. Threat analysis involves the task of describing the type of possible attacks, potential attackers and their attack methods, and the consequences of successful attacks. On the other hand, vulnerability refers to a weakness that can be exploited by an attacker. Risk assessment combined with threat analysis allows an organization to assess the potential risk to their network.

All critical infrastructure sectors rely on physical infrastructure such as buildings, roads, plants, and pipes. Increasingly, critical sectors also rely on cyberspace and ICT to enable this. In that way, ITU classifies cyberspace and its associated ICT as Critical Information Infrastructure (CII) [24]. CII operates and controls critical sectors and their physical assets. Consequently, ensuring the reliable functioning of cyberspace is a strategic national objective because a lack of trust and confidence in the use of ICTs can disrupt daily life, commerce, and national security.

2.7.1 Cybersecurity Fundamentals

The fundamentals of cybersecurity are specified in ITU recommendations X.800 and X.805. The cybersecurity architecture for end-to-end network security is defined in terms of three major concepts, which include security layers, planes, and dimensions (Figure 2.18) [24]. A security dimension refers to a set of security measures that are designed to address a particular aspect of network security [25].

Besides their application to the network, the security dimensions also extend to applications and end user information. The security dimensions include the following: (1) Access control, (2) Authentication, (3) Nonrepudiation, (4) Data confidentiality, (5) Communication security, (6) Data integrity, (7) Availability, and (8) Privacy.

The security primarily refers to risk management, such as defense strategy and countermeasures to possible attacks, so it includes attack detection and response to either stop it or reduce its impact, and finally formulating a recovery strategy that is tailored network to resume operation from an already known state. Of course, prevention of attacks is the main target of cybersecurity.

Similar to the crime that can happen in real life in cyberspace we have the so-called cybercrime. What is cybercrime? Cybercrime includes any crime involving computers and networks (for example, hacking) as well as traditional crimes committed over the Internet.

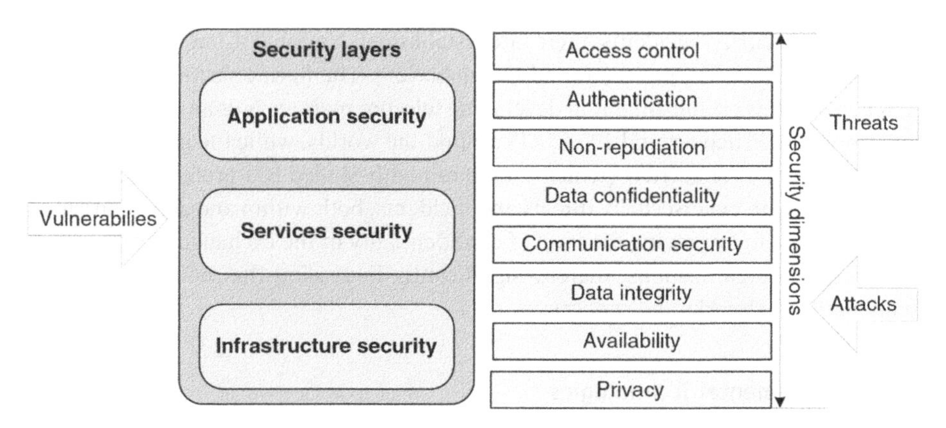

Figure 2.18 Cybersecurity dimensions.

What is the major attack vector of cybercrime? It is to exploit broken software. Of course, the cybercrime can be based also on human intrusion in the system (e.g. an employee at a service provider opens a "backdoor" for an intruder).

Then, what is one of the major parts of cybersecurity? It is to fix broken software or to decrease its presence. So, one may say (in general) that the common factor for cybersecurity and cybercrime is broken software. Software security vulnerabilities are caused by defective specifications, design, and/or implementations.

What are the cybersecurity threats and attacks? Well, cyberthreats and cyberattacks can have a wide range, going from attacks on individual targets to attacks on specific platforms; there are also possible massive scale attacks that can come from various other countries via so-called botnets. Cyberspace (i.e. the open Internet) does not have real borders, hence, there is a need for international cooperation for strengthening the cybersecurity, which goes in parallel with building cybersecurity legislation, policies, and strategies according to the national needs (which may be specific to each country).

Then, what are cybersecurity technologies for the "defense" from attacks and threats? Well, they can be of different types, such as access control (e.g. perimeter protection with firewalls, authentication, and authorization), cryptography (e.g. certificate, public key architecture, and digital signatures), system integrity (e.g. antivirus and intrusion detection systems), management (e.g. network configuration management, installing software updates, and enforcement of security policies), audit and monitoring (e.g. detection, prevention, and logging to prevent or find attackers).

So, there are cybersecurity threats, but there is also an "arsenal" for cybersecurity defense. And, there is a simple "rule" regarding cybersecurity – for every online threat there is a solution. However, when the threat is a new one and innovative, it may take some initial time to combat it (to create a solution) and then to include that solution in each subsequent version of the security technologies that are deployed, and so on.

Cybersecurity does not refer only to individual users, enterprises, governments, or countries. The same threats in the cyberspace may appear on different locations worldwide considering the open character of the global Internet network and its network neutrality principle. With the aim to strengthen the defense against cybersecurity threats, there is needed an exchange of cybersecurity information between countries via defined entities for such purpose. In that way, Cybersecurity Information Exchange (CYBEX) techniques (addressed in the ITU-T X.1500 series of recommendations) aim to improve the exchange of cybersecurity information in a way that takes into account the fact that the techniques themselves and the environment in which all that are used are constantly evolving.

To combat the cybersecurity incidents there are established National Computer Incident Response Teams (CIRTs) in each country around the globe [26]. The harmonization of national CIRTs as well as help in their establishment in developing telecom markets is done by the ITU. By year 2023, there were established around 120 CIRTs around the worlds, while the goal is for each country to have its own CIRT. The CIRTs (which consist of highly skilled ICT professionals in the cybersecurity field) deal with cybersecurity threats and incidents, both within and across jurisdictions. They also contribute to improving the level of confidentiality in the exchange of cybersecurity information between organizations, thereby significantly increasing the security of global telecommunications/ICT networks and services.

2.7.2 IP Security Fundamental Technologies

Both IP versions, IPv4 and IPv6, have similar security properties. They are both vulnerable to eavesdropping, replay, packet insertion, packet deletion, packet modification, man-in-the-middle attack (attacker somewhere on the path poses as the sender to receiver and the receiver to the sender), and denial of service (DoS) attacks in which the attacker sends large amounts of legitimate traffic to a destination to overload it, so the target cannot process other legitimate traffic [5].

The main standards for IP security are specified on network layer (IPsec) and transport layer (SSL/TLS) [2]. Both can be used in IPv4 and IPv6 networks.

The IPsec provides security on the network layer by using either Authentication Header (AH) or Encapsulating Security Payload (ESP):

- Authentication Header (AH) provides an additional AH header on the network layer besides the original IP header and the IP payload to authenticate the IP packet. It can be used in transport mode (in such a case, the original IP header remains) or in tunnel mode (a new IP header is inserted for tunneling purposes, followed by the AH header and the original IP packet).
- ESP is used to encrypt the packet payload or the entire original packet. Similar to AH, ESP can be used in transport and tunnel mode. In transport mode, the original packet header remains, while the IP payload is encrypted. In ESP tunnel mode, the entire IP packet is encrypted and a new IP header is inserted for tunneling purposes.

IPsec is applied at the network level allowing all types of Internet traffic to be protected, regardless of the specific application that initiates the connection. Applications on the top (on both client and server side) are not "aware" of security when using IPsec, and no additional changes need to be made to them to support security.

The granularity of IPsec protection is very large, so a Security Association (SA) can protect all communication between two network nodes, or it can protect only a specific type of traffic, or only an application session (on a specific port), or various other variants of traffic selection for protection (Figure 2.19).

The other solution for end-to-end security (e.g. client-server) is SSL/TLS or, in short, TLS. It provides encryption of the application data which are transferred by the application through the socket which is opened between the application layer on the top and the transport protocol layer below it. In that manner, only the payloads of transport layer packets/segments are encrypted, while the transport headers (e.g. TCP and UDP) and IP headers remain visible, so network nodes can read port numbers (from transport protocols) and IP addresses (from IP). TLS is a solution against eavesdroppers, so sensitive information, such as personal information or credit card numbers, cannot be seen by eavesdroppers and hackers on the path between the client and the server. Due to the importance of TLS, the latest version of HTTP, the HTTP 3.0, has TLS embedded in it,

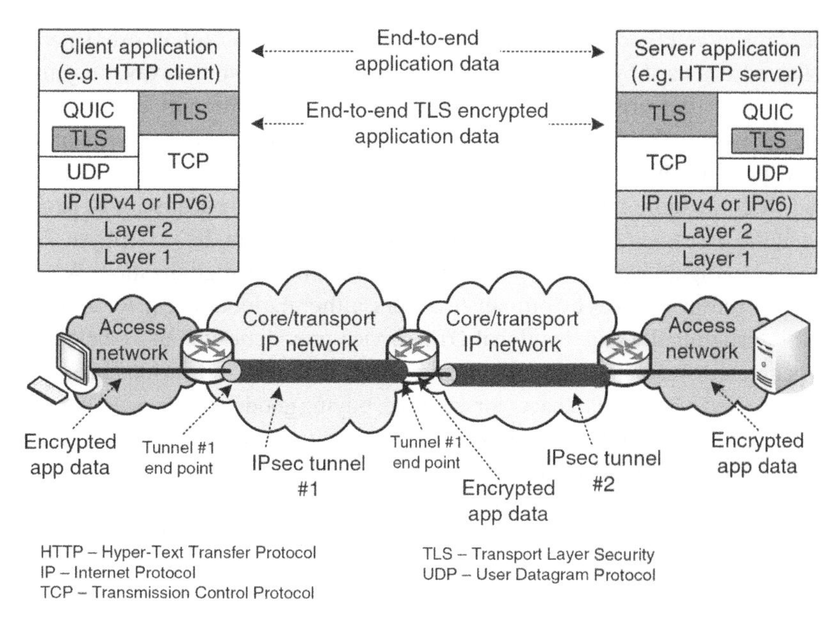

Figure 2.19 Use case of IPsec and TLS for end-to-end security.

so it is no more optional. Using TLS is mandatory for all trusted Web services, and it is one of the main technologies that builds trust in the process of digitalization (which again is based mainly on using Web services). Both TLS and IPsec can be used independently from one another, e.g. TLS encrypted application data can be transferred in telecom network by using IPsec in either transport or tunnel mode.

For enhancing security, there is also used DPI technology, which can be used by telecom operators for examination of packet headers, from Layer 2 up to the application layer (Layer 7 according to the OSI). Even if the encryption is used for the application data (e.g. with TLS), smart algorithms (e.g. based on AI/ML) can detect malicious traffic based on known traffic patterns and that way prevent the cyberthreats.

Of course, other IP security mechanisms exist, and which solution is used depends on the services provided and the type of network, as well as well-established policies and approaches from ICT equipment vendors and telecommunications operators.

2.7.3 Online Privacy Aspects

Data is the new important value for businesses and a significant subset of this is personal data. In the open Internet world, personal data has become the primary fuel for digitization on a national level, and direct marketing and sales of goods and services to individuals on both national and international levels. For example, OTT service providers (e.g. Google, Facebook, Amazon, and Microsoft) get a significant portion of their revenues from profiling and using the personal data they collect, allowing them to directly target their users with different products and services.

Privacy is one of the ICT security dimensions (Figure 2.18). So, when talking about cyberspace, cybersecurity and privacy are related to each other. But, while cybersecurity focuses mostly on protecting the data as well as networks and services from various cyberthreats and cyberattacks, privacy refers to use and governance of personal data.

What is privacy? Well, privacy refers to the right to have control over how a human user's personal information is collected and used by various service providers and telecom networks in general. However, what falls under the privacy umbrella is not the same in different cultures in different regions around the world. Thus, a simple copy-paste of privacy policies from one region to another may not always produce the expected results.

Why has online privacy become more important over the years? Well, that is because most people are online, and most of the processes and services are being transferred from the real (physical) world to the cyberspace (i.e. to digital world). Such a process is also stated as digitalization. With high consumption of various online services by human end users, either nationally (e.g. for digital government services) or internationally (e.g. by global OTT service providers such as Google), a large amount of data is generated and shared online (e.g. photos, videos, texts, and various files) as well as many activities are carried out in cyberspace (for example, buying goods and services online with a credit card or other payment options). This has created a large volume of data that is constantly increasing with the increase in broadband speeds and data caps for users in fixed and mobile telecom networks. Such a large volume of data, including personal data, is constantly collected and exchanged by various online actors. Overall, the importance of data, including personal data, is increasing in the global digital economy. Data has become an asset and a value in itself. In such rapid telecommunication/ICT technological developments, privacy has become a high priority in a short span of time.

In that way, all national cybersecurity strategies need to respect and be consistent with fundamental human rights [27]. Any such strategy will recognize the fact that the rights people have offline must also be protected online, as today's cyber (i.e. digital) world is a kind of "mimicry" of the real physical world. So, the main attention in cybersecurity strategies should be toward protection of fundamental human rights (according to UN declarations) such as freedom of expression, privacy of communication, and personal data protection. This means that arbitrary, unjustified, or otherwise unlawful tracking, monitoring of communications, or processing of personal data should be avoided.

What about privacy vs. business relations? Well, considering the point of view of businesses, it can be said that by increasing security, privacy companies can benefit from gaining the trust of their online customers by providing assurances about the safety and integrity of their personal data. Of course, this comes with certain data protection and privacy laws, such as General Data Protection Regulation (GDPR) in Europe and similar in other regions around the world.

Then, what are advantages and disadvantages of GDPR-like legislations? Well, the main advantage of GDPR is improved cybersecurity (regarding the legislation and policies), providing standardization on data protection, and increasing the trust in online services and digital businesses (which is good for online businesses). The future of digital economy is related to individual marketing, so proponents of GDPR stated that it is important to have rules on the usage of data. On the other side, there are some disadvantages for GDPR. One disadvantage is the cost of achieving compliance, which may be high burden to companies. Finally, a potential disadvantage of GDPR is possible overregulation of the privacy (e.g. a lot of bureaucracy in the cyber world, i.e. digital world), which may result in never-ending messages of consent (e.g. "Our website uses cookies – do you accept – manage" – e.g. cookies are part of how HTTP works since the beginning of the open Internet – since the 1990s, and that is technically standardized and needed because HTTP is stateless by default). Such continuous opting-in may discourage some customers and also may impact the speedy innovation that is happening on the open Internet for several decades. But, besides all cons and pros for privacy legislations (such as GDPR), there is no such thing as one size fits all, so each region and each country should proceed in the future with own pace by judging all the pros and cons.

With regard to digital services on the open internet (i.e. OTT services) they are not stand-alone but rely on OSs and Platforms (e.g. IOS and Android), smartphones and devices, telecom operators (through which they access the Internet) as well as cloud services, which facilitate the use of the online applications themselves. All these systems and (sub)-services also collect various personal data, and also use the applications to collect data as well. For example, OSs and cloud services can track the users (e.g. localization by Google services) and interact with the applications in use (which sometimes may come into conflict with transparency and consent principles). Further, many wearable devices, such as smart watches and gadgets, also connect to smartphone OSs and cloud services on Internet, potentially forwarding sensitive personal data. Then, the question is where the personal data goes, who receives it, and is it used for sale to other private companies. Therefore, it is important to ensure privacy and security across the entire digital economy value chain [28].

Regarding privacy, there is also a question about whether human Internet users have the right to be anonymous online? On one side, personal data is the primary fuel for the digital economy and most online services that are offered free to Internet users (of course, the users should bear the costs of their Internet access, which however is not related or shared between telecom operators and global service providers due to the network neutrality principle). Many online businesses certainly need personal data in order to deliver their services to the users. This is also valid for all e-government services. For example, when buying on the Internet, the user should specify credit card details and also the correct address (where she/he lives), e.g. for delivery of bought goods. Also, various national legislations (e.g. taxation rules) may require proof of the existence of physical people and companies as well as their address and country of residence.

In such a complex environment, with stricter requirements on cybersecurity and privacy on one side, and the needs of digital economy on the other, new technologies like blockchain can be used to make anonymity more possible in the future, however, that requires appropriate national legislation. However, complete anonymity online will have to be balanced with national security strategies.

Overall, cybersecurity and privacy are highly important in today's digital world, with the aim to have an enjoyable "digital life" in the cyberspace.

2.8 Future Internet Development Toward 2030 and Beyond

The existing internet has proven to be sustainable through various networks and services that have emerged over time. As with any technology, it will not last forever in its current form. But it changes over time, adapting to new technologies at different protocol layers and to new requirements from the technology, regulatory, and business side. Of course, many people work in the Internet/IP field, and with the digitalization the number of such people is increasing exponentially over time. In such a case, cyberspace becomes the "twin" of our real world (i.e. digital twin). Open Internet is very useful for our daily work and life, making them more convenient and efficient and generally more enjoyable. However, any fundamental changes to the Internet (as we know it) are unlikely to happen in a short period of time, as the Internet is already globally distributed, and most of the human population is already connected to the Internet, while the United Nation's (UN) Sustainable Development Goals are aimed at connecting everyone to the Internet, as this will be necessary to exercise their own human rights given the digitization of public administration and many non-ICT business sectors (i.e. verticals).

Usually, we use the terms "broadband" and "Internet" together. Why? In general, broadband does not only refer to Internet access service. Broadband is used both for Internet access and for specialized telecom services (which are also mainly IP-based), such as IPTV, carrier-grade telephony, and business VPN services (as a replacement for leased lines in circuit-switched networks from the 20th century). But, if we look at the amount of transferred traffic at the moment, more than three-fourths comes from IP video traffic (including online video services as well as IPTV over managed IP networks), the rest is web, file transfer, and messaging. So, one may note that in the 2020s and beyond, we are building broadband access and transport mainly for higher speeds for Internet access service, which is mainly used (today) for OTT video and IPTV services, which are accompanied with VR/AR/XR content (although video services have an advantage due to no need of additional user equipment, besides smartphone, laptop, desktop, or IoT device). And "broadband" on the open Internet can be experienced even by ordinary Internet users, because higher speeds (and lower latencies) directly result in a better user experience for the given service, that is, a higher QoE.

So, what are the expectations for the Internet in 2030 and beyond? Well, different groups (in Standards Development Organizations - SDOs, vendors, and countries) have slightly different views, as usual. There are two main ways to further develop the current Internet into something we call (here, in this course) the future broadband IP network [29]. One approach is with radical changes of the IP stack, such as development of a new IP (e.g. IP version X). However, that is not feasible, and also the transition from IPv4 to IPv6 still goes on in the 2020s and will continue in the 2030s. The other most likely path for future broadband Internet toward 2030 and beyond is further evolution of the existing Internet.

2.8.1 Future Broadband Internet Through Continuous Evolution

The evolution of the existing Internet and IP networks in general has been driven by IETF standardization and supported by other SDOs such as IEEE (e.g. Ethernet and Wi-Fi are designed for IP protocol stacks from Layer 3 above), 3GPP (for mobile networks), and ITU (for various telecom standards, harmonization of radiocommunication, as well as global development of ICT and digitization). For example, ITU's NGNs are primarily developed in the past two decades for transition of legacy (i.e. circuit-switching) telephony to carrier-grade VoIP as replacement for voice services in PSTN/PLMN, and then also for transition of TV to carrier grade IPTV (as a second major goal) [3]. By carrier grade service, we refer to VoIP and IPTV as specialized/managed services over IP networks, not over the open Internet. Such carrier-grade VoIP and IPTV are provisioned with guaranteed QoS by using the same broadband access networks (fixed and mobile) that are also used for Internet access service, but logically separated from them. In the ITU standardization roadmap, the process of convergence of telecom networks over IP-based infrastructure continues in Future Networks standardization, which includes SDN/NFV aspects, IMT-2020/5G systems, IoT, and Smart Sustainable Cities, and emerging technology areas such as AI (and its main form – ML) use in telecom/ICT networks and services.

Regarding the future of IP (e.g. new IP or evolution), ICANN as the main body for technical standardization and governance of Internet technologies stated that IP has evolved over the years and still needs to evolve [30].

In terms of continuous evolution of Internet technologies (not only via transition to IPv6), the future Internet toward 2030 and beyond will continue to use the successful cases of Internet technologies and continue to improve the technologies in the IP ecosystem. The main targets are higher speeds, lower latencies, as well as ultrahigh reliability and synchronization for critical services (e.g. motion control in Industry 4.0). The evolution of Internet is not through redesigning the IP in "one protocol fits all" terms, but through virtualization of networks when needed. For example, virtualization has already been present in IP networks of telecom operators for about two decades, as it is

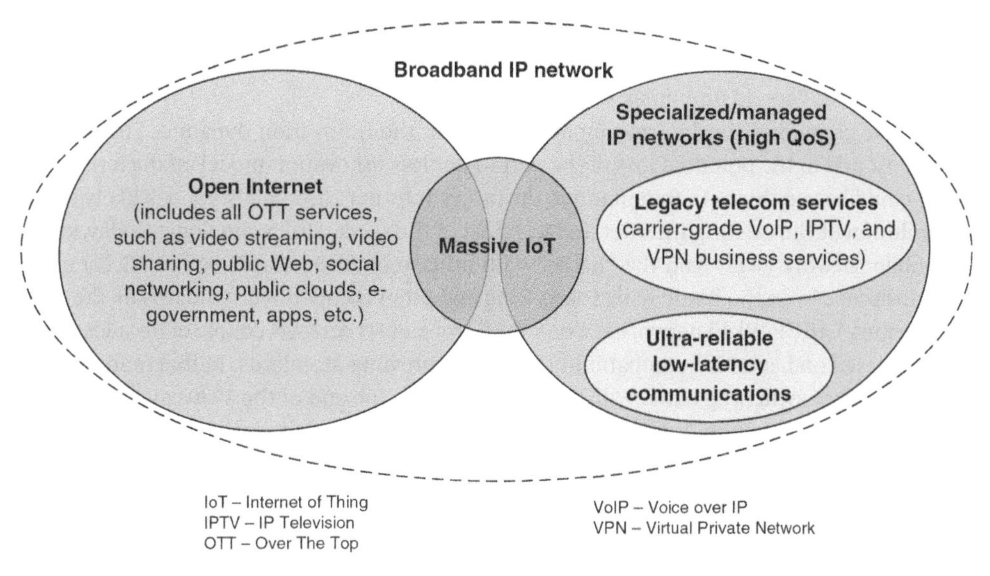

Figure 2.20 Future IP network 2030 and beyond.

required to deliver also services with QoS support. So, it can be expected that the future IP network toward 2030 and beyond will continue to be based on (Figure 2.20):

- Broadband Internet access – that is, open Internet based on the network neutrality principle.
- Massive IoT – this can be provided either over the open Internet or via specialized networks which can be IP or non-IP. However, in non-IP access networks, the IDs of IoT devices need to be translated to IP addresses at a local gateway node. Considering the Moore's law for the development of electronic components (that is, the processing power of devices doubles approximately every two years, while power consumption decreases at almost the same pace), there will no real limitations to even small and cheap IoT devices being able to use IPv6 addresses toward 2030 and beyond.
- Ultrareliable low-latency communications (the term is taken from 5G service types) – this is the most complex type of service that requires ultrahigh reliability and availability and ultralow latency (in the range of 1 ms). Of course, one should note the limitations imposed by the maximum signal speed, which is the speed of light, so for a delay of 1 ms (without considering any other path delay which normally exists, such as buffering delays at routers and switches) the maximum distance between the two endpoints of a communication session is less than 300 km (given that the speed of light is 300000 km/s, i.e. 300 km/ms). These services cannot normally be provided over the open internet (based on the principles of net neutrality), but over specialized networks, which can be operated by telecommunications operators or by private entities (e.g. an Industry 4.0 utility factory, a manufacturer of cars for vehicles for any communication, city for critical smart city services, and so on).
- Legacy telecom services (voice, TV, and VPN business services) – they are not provided over the open Internet, but over managed IP networks, with guaranteed QoS through logical isolation of services. QoS support is typically provided using QoS solutions in underlying technologies over IP, such as Ethernet, Wi-Fi, and cellular technologies (e.g. 4G, 5G, and beyond).

It is worth remembering that the success of the TCP/IP protocol suite is tied to the idea of a simple, global network connecting smart edges (which are computers, with client or server applications). As CDN caches have shown, the exact definition of what is the core and what is the edge has evolved over time, but the overall model remains the same. The TCP/IP (or lately QUIC/UDP/IP) model has led to a flourishing of new applications, accelerating innovation at an unprecedented rate in the telecom/ICT world and many verticals.

Any disruptive change of the Internet model may break the innovation dynamic. The overall opportunity cost tied to the potential loss of the permissionless innovation model characteristic of the Internet could be very high. In that manner, the future Internet 2030 and beyond is likely to be based on evolution of the existing Internet. In fact, most of the successful technologies today, such as 3GPP mobile network (who won the "battle" with other mobile standards before 5G completion), IEEE Ethernet (won the battle with Token Ring and other competitive standards by the end of the 20th century), IEEE Wi-Fi networks (won over European HyperLAN couple of decades ago), further evolve based on backward compatibility with their previous standards. In that manner, the IP has won the packet-switching battle with European ATM by the end of the 1990s due to its simplicity and bottom-up approach, which provided previously unseen innovations in the telecom world in this 21st century. In that manner, non-IP networks exist and will continue to exist (as private specialized networks), however the IP networking will continue to be the main networking principle globally toward 2030 and beyond, and to be improved as needed over time. For example, that approach continues for the TCP or QUIC (as the "smartest" protocol in the IP protocol model) with new standards based on successful proprietary implementations. For example, QUIC protocol was standardized by IETF in 2021 after previously being confirmed as a successful solution as proprietary transport layer protocol developed by Google for its own online services.

In summary, it can be said that the reality is that IP will remain, but it will certainly face more stringent service requirements in terms of QoS, especially for critical IoT services that require ultrahigh reliability and very low latency. New innovations in protocols over IP, such as QUIC, ensure lower latencies even for bandwidth-hungry services such as video with very high resolutions and provision of VR/AR services over the open Internet. Of course, delays will also depend on distance, but there is scope for further expansion of edge computing use cases (moving clouds closer to end users and machines/things) as well as the right combination of open internet and managed/dedicated IP networks, which is expected to be IPv6 dominant toward 2030 and beyond. Each new version of IP (e.g. IPv10 and IPv20) needs to be made backward compatible with previous versions of IP. Incompatibility with IPv4 has also been one of the main obstacles to the faster penetration of IPv6 since its introduction.

2.9 Governance of Broadband Internet

The main technical governance body for the public Internet and generally IP environments was and still is ICANN [31], which history traces back to the appearance and the evolution of the Internet itself, started in 1969 as ARPANET (the predecessor of the present-day Internet).

ICANN was officially founded in 1998 and it appeared as a result of the US government's commitment to transfer the policy and technical management of the DNS to a non-profit corporation with a global participation. Until 1998, the DNS and IP address allocation was administered at the University of South California on behalf of the US government, so ICANN was created to continue such governance role. Its activities for technical governance of Internet and generally IP networks were provided via its department IANA, which is one of the oldest Internet institutions [32].

ICANN through IANA primarily governed top level domain names, number resources (allocation of blocks of IP addresses and AS numbers through five RIRs) and protocol assignments (e.g. port numbers for transport protocols TCP and UDP) in conjunction with standardization bodies.

The IANA stewardship transition was completed in 2016, so nowadays its services are provided by Public Technical Identifiers (PTIs), which is the purpose-built organization for providing the IANA functions to the Internet community.

The global spread of the Internet and its technologies in the late 1990s and early 2000s prompted the activities of UN agencies to become more involved in Internet governance, given its increasing impact of Internet on daily life and work of people around the world. In that direction, after the first phase of World Summit on the Information Society (WSIS), the UN Secretary-General established the Working Group on Internet Governance (WGIG). The WGIG developed the following definition of Internet governance [33]: "Internet governance is the development and application by Governments, the private sector and civil society, in their respective roles, of shared principles, norms, rules, decision-making procedures, and programmes that shape the evolution and use of the Internet."

Further, the UN created the Internet Governance Forum (IGF) in 2006 [33] as a multi-stakeholder community which has a mission to discuss all important Internet issues at a given time, including (but not limited to) data protection digital inclusion, environment protection, trust, digital content, human rights, and interconnection.

Considering the legal and regulatory aspects of managing the Internet, most countries have established telecom/ICT sector regulators (of course, they are not intended only for the open Internet, but for the management of all telecom/ICT networks and services) and most of them allowed competitive digital markets.

2.9.1 Convergence Between Broadband IP Infrastructure and Other Sectors

The open Internet provides the possibility for many businesses to use Internet services for operations of their business, selling goods, buying materials for manufacturing, marketing, obtaining documents from the government and local administration, signing contracts, working on distance, and so on. The open Internet is provided through the IP infrastructure of telecom operators, which are in fact the main ISPs in each country globally. The infrastructure for open Internet access, as well as data centers (for hosting the applications and data on Internet) and applications (for provision of services over the open Internet) is referred in the public also as digital infrastructure. In that way, the OTT services provided over the open Internet are also referred to as digital services. The digitalization moves services from the physical world onto the digital world. However, increasing reliance on digital infrastructure and services/applications in almost every sector influences the economies and societies. That results many times in overlaps in policy issues between sectors and related government departments and ministries. Also, the digitalization process resulted in many countries specifying national digital transformation strategies in the late 2010s and 2020s.

Telecoms regulators are moving from "integrated regulation, led by socio-economic policy" to "joint regulation with metrics-supported decision-making" [34], which includes collaboration between telecom/ICT regulator and non-ICT regulators in a given country. Examples of possible collaboration points between telecom regulators and non-ICT regulators are given in Table 2.5.

Digital economy is related to doing business online. However, that is related to building trust and confidence in engaging online by different actors, including citizens and businesses as well as public administration. In that manner, the higher number of cybersecurity incidents contribute to a negative trend which requires action on national and international levels. Also, for engaging online there are required digital skills, which may span from basic skills (e.g. connecting online and using Web and messaging services) to very advanced (e.g. developing digital services). For improving the digital skills, governments need continuous investment in capacity building for ensuring people

Table 2.5 Examples for possible collaboration between telecom regulators and other regulatory agencies.

Non-ICT regulator	Topics of potential collaboration with the telecom regulator
Commerce/trade	Digital taxation, online digital services
Cybersecurity	Data use, end user devices, IoT
Education	Child online protection, digital divide
Energy	AI, blockchain, IoT
Finance	Blockchain, cybersecurity, financial inclusion, mobile financial services, privacy
Transportation	Cybersecurity, IoT, privacy

have the possibility of engagement online, with the aim to practice their rights, to have access to public services, education, health services, and buying and selling online. So, one may say that cybersecurity, digital capacity building, and digital transformation are processes that go in parallel. For example, the European Union (EU) has proposed a Digital Services Act that introduces regulatory oversight for many large platforms for online services.

2.9.2 Discussion About the Future of the Broadband Internet Governance

Future broadband is a further evolution of the existing internet with ever-increasing capacity and quality (higher bitrates, lower delays, and higher reliability and security) that is ubiquitous and accessible to all people globally, like the air we breathe. Normally, as noted before, the term "broadband" is relative to the given moment in time. For example, in this decade (2020s) broadband means individual bitrates from 10s to 100s Mbit/s up to 100s of Mbit/s and Gbit/s per individual user/home/office going toward 2030 and beyond. However, what is broadband today will be seen as narrowband several decades in the future.

Cyberspace (i.e. the open Internet) is evolving toward 2030 and beyond to become the digital twin (e.g. metaverse, supported by very high-resolution video, and VR/AR/XR services) of our real physical world. Therefore, the future governance of the Internet will certainly involve cooperation with regulatory bodies from different verticals, in addition to telecommunications regulators in each country, bearing in mind the leading role of broadband Internet in the digitalization. At the same time, the application of new technologies, for example, AI/ML, will require their gradual regulation in parallel with their evolution and use cases, in order to protect people and their safety on the one hand and on the other hand not to deter the development of new technologies.

Finally, let us recall that IP networks are used to provide open access to the Internet (used to provide all OTT services), but also specialized/managed IP services provided through managed or private IP networks (e.g. carrier-grade VoIP, IPTV, VPN business services, critical IoT services, and mission-critical services). Open Internet and specialized/managed networks and services are typically under separate service-specific governance approaches (e.g. for carrier-type VoIP, for IPTV content, for V2X services, and for Industry 4.0 services).

References

1 Janevski, T. (April 2019). *QoS for Fixed and Mobile Ultra-Broadband*. USA: John Wiley & Sons.
2 Janevski, T. (November 2015). *Internet Technologies for Fixed and Mobile Networks*. USA: Artech House.
3 Janevski, T. (April 2014). *NGN Architectures, Protocols and Services*. UK: John Wiley & Sons.
4 RFC 791, "Internet Protocol", IETF, September 1981.
5 IETF RFC 8200, "Internet Protocol, Version 6 (IPv6) Specification", July 2017.
6 IETF RFC 768, "User Datagram Protocol", August 1980.
7 IETF RFC 793, "Transmission Control Protocol", September 1981.
8 IETF RFC 9000, "QUIC: A UDP-Based Multiplexed and Secure Transport", May 2021.
9 IETF RFC 4291, "IP Version 6 Addressing Architecture", February 2006.
10 ETSI GR IP6 006 V1.1.1, "Generic Migration Steps From IPv4 to IPv6", November 2017.
11 IETF RFC 7597, "Mapping of Address and Port with Encapsulation (MAP-E)", July 2015.
12 IETF RFC 7599, "Mapping of Address and Port using Translation (MAP-T)", July 2015.
13 IETF RFC 7269, "NAT64 Deployment Options and Experience", June 2014.
14 https://www.potaroo.net/tools/asn32/, "The 32-Bit AS Number Report", Last accessed in March 2023.
15 IETF RFC 1945, "Hypertext Transfer Protocol – HTTP/1.0", May 1996.
16 IETF RFC 9112, "HTTP/1.1", June 2022.
17 IETF RFC 9113, "HTTP/2", June 2022.
18 IETF RFC 9110, "HTTP Semantics", June 2022.
19 IETF RFC 9114, "HTTP/3", June 2022.
20 Ericsson.com, "What is the Metaverse and Why Does it Need 5G to Succeed? – The Metaverse 5G Relationship Explained", April 2022.
21 ITU-T Rec. Y.1545.1, "Framework for Monitoring the Quality of Service of IP Network Services", March 2017.
22 ITU-D, "QoS Regulation Manual", 2017.
23 ITU-T Rec. X.1205, "Overview of Cybersecurity", April 2008.
24 ITU, "Security in Telecommunications and Information Technology", 2015.
25 ITU-T X.805, "Security Architecture for Systems Providing End-To-End Communications", October 2003.
26 ITU, "ITU Cybersecurity Programme: CIRT Framework", 2021.
27 ITU, "Guide to Developing a National Cybersecurity Strategy", 2018.
28 ITU, "Powering the Digital Economy: Regulatory Approaches to Securing Consumer Privacy, Trust, and Security", 2018.
29 ITU-T Technical Specification, "Network 2030 Architecture Framework", June 2020.
30 ICANN Office of the Chief Technology Officer, "New IP", October 2020.
31 ICANN, "ICANN history", https://www.icann.org/history, Last accessed in March 2023.
32 Internet Assigned Number Authority (IANA), https://www.iana.org/, Last accessed in March 2023.
33 Internet Governance Forum (IGF), https://www.intgovforum.org, Last accessed in March 2023.
34 ITU, Broadband Commission, "21st Century Financing Models for Bridging Broadband Connectivity Gaps", October 2021.

3

Future Terrestrial and Satellite Broadband

The development of broadband networks and services is directly related to fixed infrastructure for all types of access networks. However, access networks can be fixed, mobile, or satellite. The telecom world started with fixed access networks deployment for telephony, first analogs (from the end of the 19th to the end of the 20th century) and then digital (last couple of decades of the 20th century) and IP-based (since the late 1990s and the whole 21st century so far) [1–3].

Broadband Internet access began with metallic access technologies such as twisted pair (originally used to access telephone networks in the 20th century) and coaxial cable (originally used to provide TV service in the 20th century). Such a large amount of deployed metal access technologies made it affordable to easily deploy broadband in the 2000s, 2010s, and even now in the 2020s in places where fixed broadband already existed. Why? Because fixed access network costs constitute the major part of the capital investment for building a given telecom network. Of course, if one were to build a new fixed access network today or tomorrow, it should be based on fiber-optics because of the much better properties of fiber compared to copper (which is used for metal access networks).

So, fixed broadband access is provided through copper lines (e.g. twisted pair and coaxial cables) and fiber-optic cables. Then, the main technologies for fixed broadband access over copper, also called metal access, include digital subscriber line (DSL) technologies as well as cable access networks. Broadband access is offered to individual (i.e. residential) users and to business (i.e. enterprise) users.

Future metallic access technologies continue to be developed through ITU multi-gigabit fast (MG.fast) and DOCSIS 4.0 over cable. Fiber proved to be the best transmission medium, with the Fiber To The Home (FTTH) access networks bringing future optical access through XG-PON, NG-PON2, and the future 50G-PON. To reach all regions around the world, Internet/IP traffic around the world is carried over underwater fiber-optic cables. But terrestrial networks cannot reach everywhere in seas, land and air, so future access technologies include satellite broadband integrated with terrestrial networks.

3.1 Future Metallic Broadband

3.1.1 Legacy DSL Technologies for Fixed Broadband Access

There are multiple technologies developed through the metallic approach. It was the first true broadband technology for fixed access anywhere landline telephone lines (based on twisted pair

Future Fixed and Mobile Broadband Internet, Clouds, and IoT/AI, First Edition. Toni Janevski.

copper) existed. Given that the initial network was asymmetric, Asymmetric Digital Subscriber Line (ADSL) which supports asymmetric transmission [4] was the first choice in the 2000s. Typically, the downstream bitrate is up to eight times the upstream bitrate, although the exact ratio depends on the length and noise of the local loop. ADSL generally operates in a frequency band over Plain Old Telephone System (POTS) or, in some regions of the world, over Integrated Services Digital Network (ISDN). Thus, ADSL enables simultaneous use of the telephone and broadband connection. Originally conceived as a means of providing video on demand, ADSL has become the world's DSL of choice for copper broadband access. Historically ADSL was first specified in the 1990s in USA, followed by European standardization by ETSI, and later it continued under ITU-T standardization from 1999. The ADSL was originally defined to support downstream bitrates up to 8 Mbit/s and upstream bitrates up to 1.3 Mbit/s. The maximum range of ADSL was approximately 5–6 km, which is standard length of local loop in PSTN in the 20th century.

The ADSL2 specification, ITU-T Recommendation G.992.3, specified a number of additional modes that extend the service diversity and reach of ADSL. The various ADSL2 annexes allow downstream bitrates of up to approximately 12 Mbit/s and upstream bitrates as high as 3.5 Mbit/s. ADSL2+ [5] nearly doubles the downstream bandwidth, increasing the maximum bit rate to 24 Mbit/s. Further, ADSL2+ also supports all the same upstream options as ADSL2.

ADSL and ADSL2 use the same frequency bands (from 0.14 to 1.1 MHz) for downlink and below 0.14 MHz for the uplink, however, ADSL2 uses more efficient modulation techniques than ADSL to achieve higher data rates. In ADSL2+, the frequency range for the downlink is doubled (from 1.1 to 2.2 MHz), so proportionally the data rates achieved are twice as high.

Table 3.1 gives an overview of DSL standard development from initial ADSL standard until the multi-gigabit MG.fast standard for the metallic access networks. By 2023, the DSL technologies (all ITU standards) have been used to connect over 600 million homes and businesses to the Internet [6]. Very high speed digital subscriber line (VSDL) was the successor of ADSL for provision of higher bitrates than ADSL, going from several tens of Mbit/s of initial VDSL standard.

3.1.2 Future Multi-gigabit Fast Access to Subscriber Terminals

G.fast is standardized by the ITU to fill an access network technology gap, regarding the metallic access with gigabit speeds. The standard refers to provision of gigabit speeds over the metallic access, which is used in cases where fiber may not always be possible for deployment into the home. G.fast supports Fiber To The distribution point (FTTdp) and Fiber To the Building (FTTB)

Table 3.1 DSL standards and theoretical maximum bitrates.

DSL type	Downlink (theoretical max)	Uplink (theoretical max)	Distance	Standard
ADSL	8 Mbit/s	1 Mbit/s	6000 m	ITU-T G.992.1
ADSL2+	24 Mbit/s	3.3 Mbit/s	6000 m	ITU-T G.992.5
VDSL2 (35 MHz)	300 Mbit/s	100 Mbit/s	2500 m	ITU-T G 993.5
G.fast (212 MHz)	Combined downlink and uplink 2 Gbit/s		100–200 m	ITU-T G.9701
MG.fast (424 MHz)	Combined downlink and uplink 4 Gbit/s (TDD over twisted-pairs) and 8 Gbit/s (FDX – full duplex mode over coaxial cable or CAT5 local network wiring)		50–100 m	ITU-T G.9711

ADSL – Asymmetric DSL; G.fast – Gigabit fast; VDSL – Very high bit rate DSL; MGfast – Multi Gigabit fast.

architectures, where the metallic access is used in the last few hundred meters, while fiber architecture is deployed in the rest of the access network part. Further development of metallic access technology, with the aim to prolong the life of twisted pairs (where they exist and cannot be replaced with fiber within the building or homes/offices), is the MG.fast, where "MG" stands for multi-gigabit speeds. In fact, according to the ITU-T standard for MG.fast [7], MG.fast is a technology capable of transmission with an aggregate bitrate of up to 8 Gbit/s if in Full Duplex (FDX) mode (best suited to coaxial cable and CAT5-type wiring) and up to 4 Gbit/s with Time Division Duplexing (TDD) over twisted pair legacy telephone wires. It provides Far-End Crosstalk (FEXT) and Near-End Crosstalk (NEXT) cancellation between multiple wire-pairs with the aim to provide such high bitrates.

If one compares MG.fast with previous DSL standards, it appears that ADSL2+ uses approximately 2 MHz of the spectrum, VDSL2 uses up to 35 MHz of the spectrum, G.fast uses up to 212 MHz of the spectrum, while MG.fast is using spectrum up to 424 and 848 MHz. The 424 MHz profiles are fully specified and with them is derived the theoretically maximum bitrate of 8 Gbit/s. So, multi-gigabit speeds are achieved in DSL by increasing the spectrum more than 400 hundred times (up to 424 MHz and beyond) from the initial 1 MHz in initial ADSL standard in 1999. MG.fast can operate over twisted pairs in range between 30 and 100 m, depending on the quality of wires and connectors (bad connectors may introduce more noise, which negatively influences the bitrate, i.e. reduces it). Besides higher bitrates, MG.fast also provides ultra-low latency for applications that are very sensitive to delay, as well as possibility for QoS optimization in line with the requirements by different applications, based on the possibility of use of four simultaneous QoS classes. MG.fast (ITU-T G.9711) appeared in 2021 as successor of G.fast from 2014 (ITU-T G.9701). Both G.fast and MG.fast provide a combination of the best aspects of fiber and DSL technologies, which provides an alternative for fixed broadband in all cases where the metallic infrastructure in the last tens or hundreds of meters already exists, as shown in Figure 3.1.

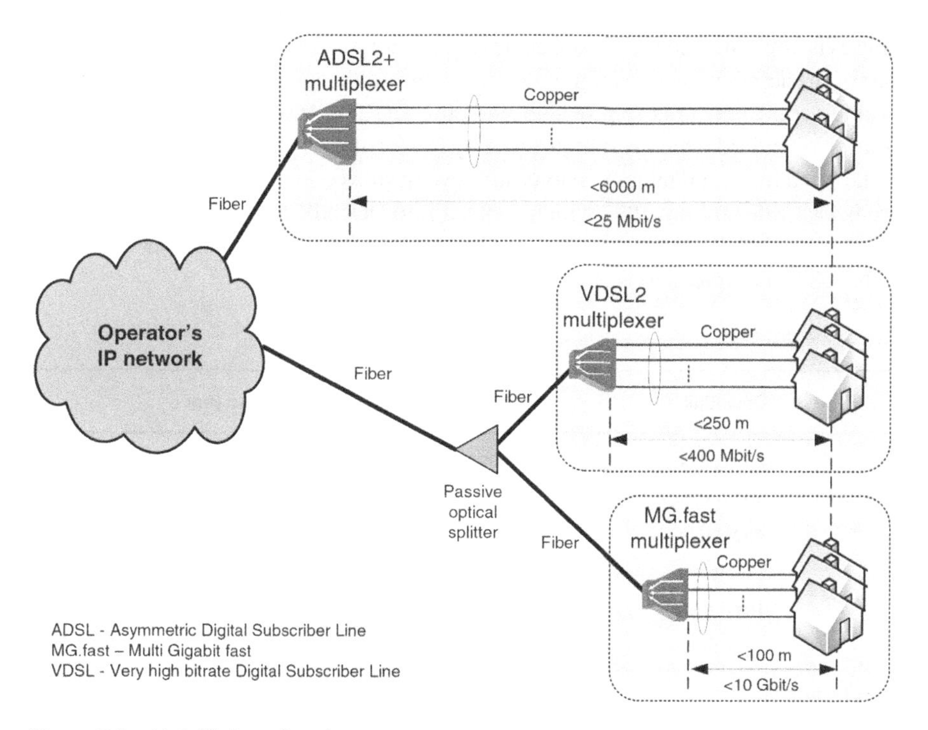

Figure 3.1 Metallic broadband access.

Another improvement of MG.fast over xDSL technologies (e.g. ADSL, VDSL) and its predecessor G.fast is in support of point-to-multipoint communication, hence, multiple devices can be connected over copper wires, e.g. can be used in homes and offices that need multiple WiFi access points to cover the whole area. Further improvement of MG.fast is power saving by using discontinuous transmission at the times and frequency, so there is transmission only when there is a traffic demand by the customer equipment. That approach contributes to reduction of crosstalk and complexity of vectoring. There is similar possibility in MG.fast (as in G.fast) for reverse power feeding, that is powering the distribution point (where copper and fiber meet) from customer premises, which is convenient in all cases for implementation (no power supply required at the distribution points, which are typically outside the home or office which uses MG.fast for fixed broadband access).

With the two ITU standards for metal (i.e. copper) access, G.fast and MG.fast, bitrates go up to 1–2 Gbit/s (with G.fast) and up to multi-gigabit speeds up to a maximum of 10 Gbit/s in the downstream direction (with MG.fast), which is comparable to the aggregate bitrates of optical access networks in the 2020s. Of course, this comes at a price, which is that Gbit/s speeds can be achieved on shorter local loop lengths, such as a few hundred meters (for G.fast) and no more than a hundred meters (for MG.fast). So both technologies must use a hybrid copper approach, with fiber going to the cabinet and a copper line for the last hundred meters. What is possible use? Well, a possible use is the reuse of copper when the fiber cannot reach the end user (either residential or business user) due to some obstacle (e.g. if it is not possible or affordable to deploy new fiber infrastructure at the given time).

3.2 Future Cable Broadband

In addition to DSL, another copper-based broadband access technology that has been widely used since the late 20th century is coaxial cable access. However, where cable access is used, it is usually implemented as Hybrid Fiber Coaxial (HFC). This means that primarily the last segment toward the end user premises is implemented via coaxial cable (e.g. last tens or hundreds of meters), while the remaining part of the access networks is an optical one (toward the telecom operator). Internet access through cable access networks is provided with the DOCSIS (Data Over Cable Service Interface Specification) standard which is standardized by CableLabs.

DOCSIS is standardized in several major versions and several subversions, all listed in Table 3.2. The first standard is DOCSIS 1.0 from 1997 which evolved into DOCSIS 2.0 which has improved the upstream direction by providing higher bits up to 30–40 Mbit/s and enabling VoIP services over DOCSIS to emulate PSTN-like voice services.

Table 3.2 DOCSIS standardization.

DOCSIS version	Downlink	Uplink	Release year
DOCSIS 1.0	40 Mbit/s	10 Mbit/s	1997
DOCSIS 1.1	40 Mbit/s	10 Mbit/s	2001
DOCSIS 2.0	40 Mbit/s	30 Mbit/s	2002
DOCSIS 3.0	1 Gbit/s	100 Mbit/s	2006
DOCSIS 3.1	10 Gbit/s	1–2 Gbit/s	2013
DOCSIS 4.0	10 Gbit/s	6 Gbit/s	2020

Source: From DOCSIS 1.0 to 4.0.

A further improvement in bitrates over cable access was made possible by DOCSIS 3.0 with aggregate speeds of up to 1 Gbit/s in the downlink, which also included support for IPv6 for the first time. Further advances in capacity and efficiency were made with DOCSIS 3.1 in 2013, which further evolved into the FDX version DOCSIS 3.1 in 2017 [8]. The latest version of DOCSIS is 4.0 as of 2020 [9]. DOCSIS 3.1 could theoretically provide maximum speeds of 10 Gbit/s in the downlink and 1 Gbit/s in the uplink, but such speeds are not available in practice. In reality, aggregate bits of 10 Gbit/s became possible with DOCSIS 4.0, which is available to telecom operators from 2021. An overview of the DOCSIS standardization is given in Table 3.2, including their downlink and uplink theoretical maximums and release years.

High bitrates with DOCSIS are obtained by so-called channel bonding which uses multiple physical TV channels to provide broadband access (either in the downlink or uplink direction). Also, the bitrates given in Table 3.2 are the aggregate bitrates of a single coaxial cable, so they are shared between all users using the same cable in the DOCSIS access network. However, for example, DOCSIS 3.1 FDX can provide speeds of over 100 Mbit/s per cable modem in both directions.

3.2.1 DOCSIS 4.0

Similar to MG.fast, DOCSIS 4.0 provides ultra-low delays, such as less than 1 ms delay for traffic traveling across networks and up to 5 ms delay on fully utilized cable access networks (all cable bandwidth used). Such low-latency DOCSIS uses pushing less demanding and intermittent traffic (e.g. games, chats, etc.) ahead of queued traffic (e.g. video streaming) to optimize traffic flows according to their performance requirement. DOCSIS 4.0 with low latency can be used also for fronthaul, midhaul, and backhaul (i.e. Xhaul) in cellular networks, including future 5G deployments involving critical services.

What are the changes in DOCSIS 4.0 compared to previous versions of the standard? One of the main changes is enhancing the upstream direction as well as downstream, and leaving the concept of channel bonding as previous DOCSIS versions. The bonding is in fact using 6 MHz (in US) or 8 MHz (in Europe) wide channels on coaxial cable bandwidth, which were initially (in the 20th century) used for delivery of TV, thus DOCSIS versions 1 to 3.1 were created to coexist with traditional bandwidth allocation per TV channel. However, with moving the TV to IPTV and higher resolutions (e.g. 4K, 8K, and so on), the 20th century cable channels are not sustainable for the future. This causes high asymmetry (downstream : upstream = 10 : 1) of the cable access, which is suitable for high-resolution video content that is transmitted in the downstream direction, but on the other hand is not suitable for symmetric services, such as video conferencing with high-resolution video (for example, Zoom, Microsoft Teams, etc.). In that direction, one of the most important novelties of DOCSIS 4.0 is better balance of downstream and upstream capacity, which is achieved by using two methods: Extended Spectrum DOCSIS (ESD) and FDX.

DOCSIS 4.0 uses the upstream frequency range 5–84 MHz for upstream and 684–1218 MHz for downstream, while the cable spectrum in the range 108–684 MHz can be used for downstream (as 3×192 MHz) or upstream (as 6×96 MHz) by using FDX approach, providing different downstream-upstream split ratios depending upon the cable operator's strategy and planned offer to the customers (Figure 3.2). The novelty in DOCSIS 4.0 is using FDX in the range 108–684 MHz which provides possibility for more symmetrical bitrates in downstream and upstream.

The other option for increasing the bandwidth on physical layer is using ESD. It is initially defined as an extension from 1.2 to 1.8 GHz, however, the future cable networks may use extended spectrum up to 3 GHz for shorter lengths of the coaxial cable. However, coaxial cable has high signal attenuation at higher frequencies, therefore they can be utilized only for inside building

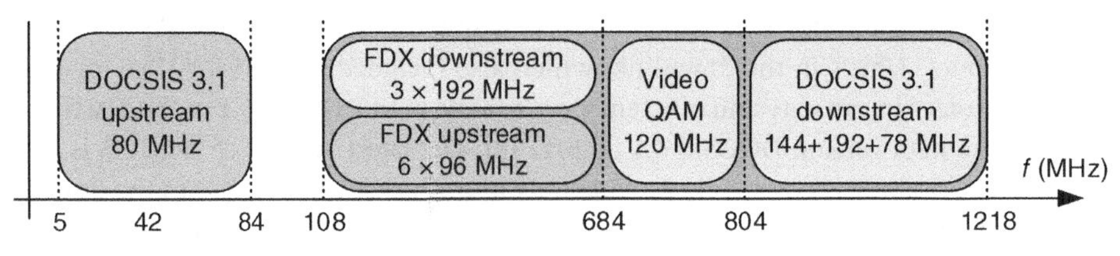

Figure 3.2 DOCSIS 4.0 Frequency Duplex (FDX) spectrum.

cable use, where fiber is deployed to the premises. One may also note that DOCSIS 4.0 moves to Orthogonal Frequency Division Multiplexing (OFDM) carriers only in all new bands that were not used in previous DOCSIS versions, considering the advantages of OFDM over other multiple access techniques, proven in various other systems (e.g. in 4G and 5G mobile systems).

Regarding the network architecture future cable networks that will use DOCSIS 4.0 will be implemented in Distributed Access Architecture (DAA) network, as shown in Figure 3.3. The change from HFC architecture to DAA does not mean topology change in existing cable networks, except when changes are applied regarding the bandwidth provided or node splitting.

From a business perspective, deploying new versions of DOCSIS, such as DOCSIS 4.0, has greater benefits in brownfield deployments (as upgrades to existing cable networks) than in greenfield deployments (first-time cable deployments). It is highly unlikely that there will be significant greenfield deployments of cable in the future due to the better characteristics of fiber for such purposes. So, the biggest use of DOCSIS 4.0 is the reuse of existing cable inside a building (e.g. high-rise buildings) where it already exists.

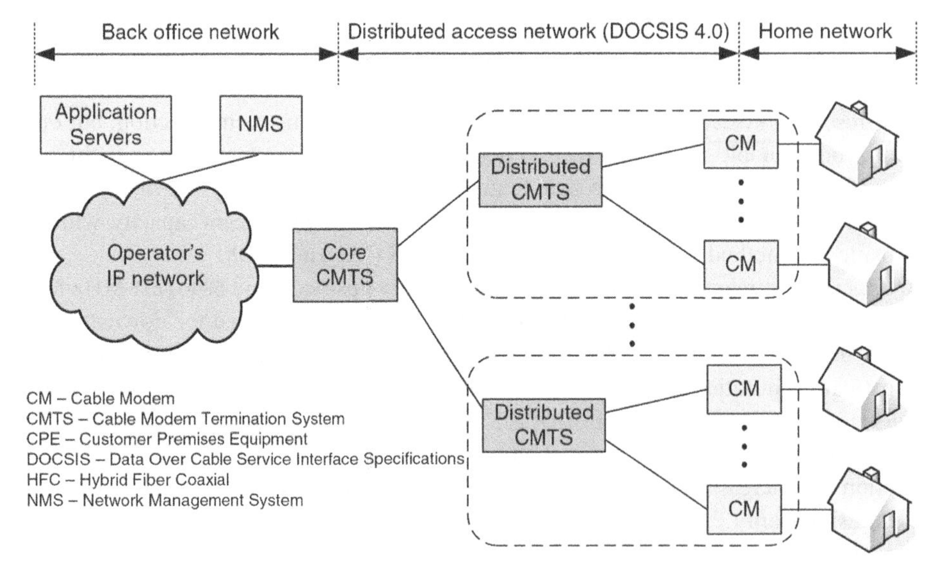

Figure 3.3 Future cable networks with Distributed Access Architecture (DAA).

What are the competing technologies for DOCSIS 4.0? Well, the last-mile technologies considered to compete with DOCSIS in the 2020s and 2030s are Passive Optical Networks (PONs), Mobile 5G or Fixed Wireless Access (FWA), and satellite broadband access.

3.2.2 Discussion on Future of Copper Access Technologies

New DSL standards (VDSL2, G.fast, and MG.fast from the ITU) and new cable standards (such as DOCSIS 4.0 from CableLabs) show that metallic access is still in the broadband "game", supporting gigabit and multi-gigabit speeds to end users. The DPoE technology specified in 2020s by CableLabs aims for a convergence between two different technologies, DOCSIS from cable networks and EPON for PON access. For example, the DOCSIS Provisioning of EPON (DPoE) is a solution standardized initially in 2012 (with its first release) to suit telecom operators that had implemented DOCSIS and use also EPON solutions, such as 1G-EPON and 10G-EPON [10]. DPoE provides IP networking by using so-called DOCSIS Internet service toward the user equipment and Metro Ethernet services as defined by MEF (Metro Ethernet Forum) [11].

Why are we still talking about metal broadband? Well, that is because although fiber technology has better features than the metal approach, it is not feasible (either from technical or business aspects) to deploy fiber everywhere in a short period of time (in every home and every office in every country where metallic access was deployed in the 20th century). Therefore, existing metal infrastructures (based on PSTN-era copper phone lines or coaxial cables deployed primarily for TV and Video-on-Demand services) provide broadband access in the short term to those who have metal access and cannot get fiber access. For example, in Europe and North America there is a large amount of deployed metallic access that proves it still has value in the near future broadband telecom/ICT world, because it provides value of scale due to the high cost of deployment of fixed access infrastructure, which is the major part of capital investments in telecom networks for fixed broadband. However, it should be noted that all future fixed broadband deployments will be fiber-based, i.e. optical access networks, due to much better characteristics of the fiber in all performance characteristics of the transmission media, including capacity and bit error probability.

3.3 Future FTTH/FTTx Optical Access

Optical access network is a long-term way to achieve very high broadband speeds per fixed broadband subscriber. Although copper-based access networks are still in the field and with VDSL2, G.fast, and MG.fast which extend their life in the telecom world, future physical layer access is aimed at all optical networks due to better transmission characteristic of optical access compared to copper (i.e. metallic) access.

As an evolution, fiber (i.e. optical) transmission technology moves from transport networks (in the 1980s and 1990s) to core networks and access networks in the last mile. The most popular technology for FTTH solutions for fixed broadband access is PON.

What is PON? A PON is a point-to-multipoint optical communication system consisting of fiber cables interconnected by passive network elements (called optical splitters) between an Optical Termination Line (OTL) on the telecom operator's side and an Optical Network Unit (ONU) on the end user side (as shown in Figure 3.4). The term "passive" refers to splitters (1 : N, that is 1 input fiber from OLT and N output fiber links toward ONTs at the customers), which work in optical signal domain by using prisms, lenses, and mirrors (packed in a small volume of the splitter) without a need for power supply. That simplifies the deployment of PONs because there is a need only

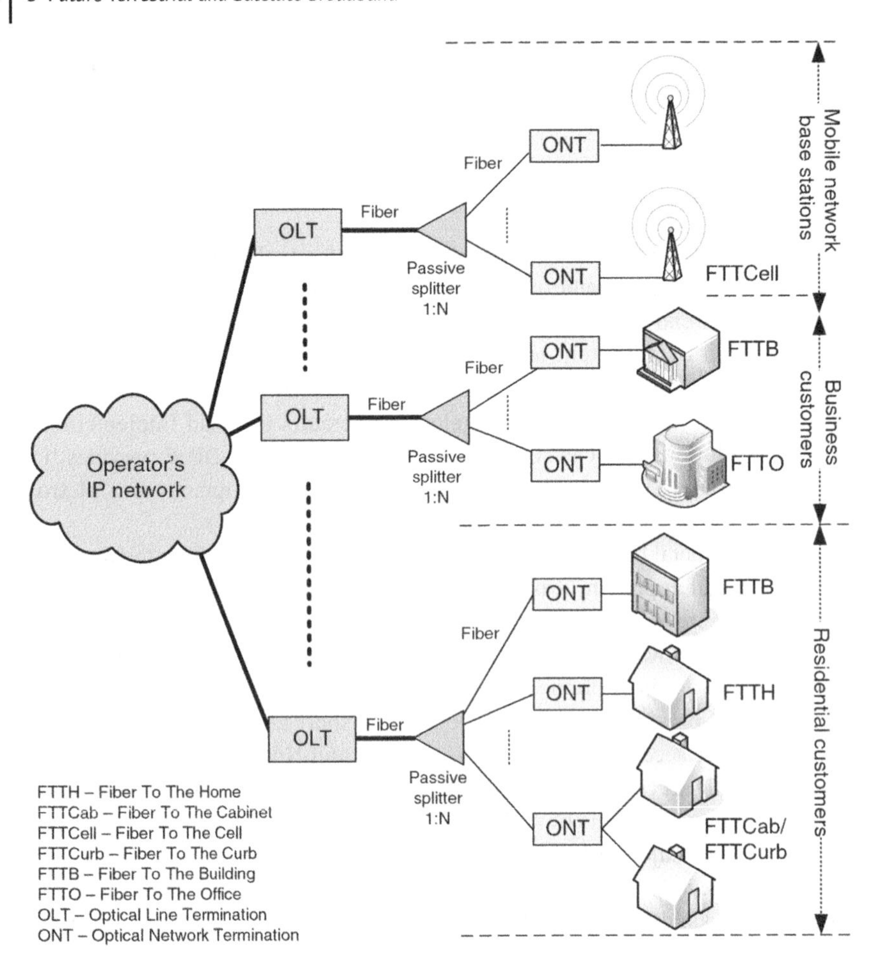

Figure 3.4 Passive Optical Network (PON) architecture.

for optical infrastructure in the access network part (toward the homes) without a need to have power supply (i.e. electricity at the splitter locations). These PONs are deployed since the beginning of the 21st century, starting in dense urban areas and progressing (with the deployments) in suburban and rural areas. Typically, PONs use single-mode fiber which has a radius of a few millimeters and hence can provide transmission over 10s of kilometers between adjacent optical regenerators.

The main work on the standardization of optical access networks is primarily done by the ITU and IEEE. There are three legacy ITU-T standards for PON:

- Asynchronous Transfer Mode - Passive Optical Network (ATM-PON), based on ITU G.983.2: It provides maximum downstream bitrates of 622 Mbit/s and upstream bitrates of 155 Mbit/s by using ATM technology between the telecom operator network and the user equipment. Normally, the Optical Distribution Network (ODN) is passive (i.e. PON). However, ATM as a technology lost the packet-switching battle to IP and Internet technologies by the end of the 1990s, so all PONs are used to carry IP traffic, including Internet and managed IP traffic.
- Broadband Passive Optical Networks (BPONs), ITU G.983.1 [12]: In this case the ODN is in fact PON, with nominal downstream bitrates of 155.52, 622.08, and 1244.16 Mbit/s, and nominal upstream bitrates of 155.52 and 622.08 Mbit/s.

- Gigabit-Capable Passive Optical Networks (GPONs), ITU G.984 [13]: In this PON standard, the nominal bitrate is 2.4 Gbit/s in the downstream direction and 1.2 and 2.4 Gbit/s in the upstream direction. GPON provides the possibility for symmetric bitrates in the upstream and downstream, and was the main approach for new deployments of fixed broadband access via fiber in the 2010s.

ITU-T defined GPON which provides 1 Gbit/s upstream and 2 Gbit/s downstream. Many telecom operators with GPON have upgraded their networks to a higher speed version called NG-PON1 with speeds of 10 Gbit/s downstream and 2.5 Gbit/s upstream. In XG-PON, the letter 'X' denotes the maximum speed of 10 Gbit/s. ITU-T later defined a symmetric version of NG-PON1 of 10 Gbit/s downstream and a 10 Gbit/s upstream version of GPON called XGS-PON. The typical path of telecom operators in the second half of the 2010s was to upgrade GPON with a symmetrical NG-PON1 called XGS-PON (XGS = 10 Gbit/s symmetrical in upstream and downstream). Thus, XGS has become the dominant solution in the early 2020s for fixed broadband access in countries with a developed fixed broadband telecommunications market, and will be one of the preferred solutions for residential fixed broadband access with PON.

3.3.1 Architectures of Optical Networks for Fixed Broadband Access

Regarding the PON deployment, fiber access networks' topology can be either Point-to-Point (P2P) or Point-to-Multipoint (P2MP). Based on the topology, there are four basic architectures for FTTx (Fiber-To-The-x) access network, i.e. ODN architectures:

- Point-to-Point (P2P) architecture uses a dedicated direct fiber connection between the telecom operator's Central Office (CO) and the end user's home. In this architecture, the number of fiber links is equal to the number of homes.
- Typical PON architecture is in the form of passive star Point-to-Multipoint (P2MP) architecture which uses a shared point-to-point link between the OLT and passive splitters (splitting is done using optics only, no power unit) between the shared fiber and the Optical Network Terminals (ONTs).
- Wavelength Division Multiplexing (WDM) based PON architecture uses several wavelengths per fiber in order to further increase bandwidth to the end user. This is also P2MP architecture. For example, NG-PON2 is based on achieving 40 Gbit/s by multiplexing traffic from four wavelengths, each carrying up to 10 Gbit/s.
- Active Optical Network (AON) architecture (e.g. active star, with a switch in a central position) is not a passive optical network (the switch requires a power supply). However, it can be used for provision of active fiber-based broadband access, typically for business users, and typically is based on using 1 or 10 Gbit/s Ethernet in existing AON deployments. It uses a shared point-to-point link between an OLT and an active remote switch (i.e. a curb switch) and then Point-to-Point (P2P) connections between a remote terminal and an ONT in homes.

Deployment of a PON can be either a migration of an existing PON network to a newer PON standard or deployment of a PON network for the first time on a given location by a given telecom operator (also referred to as greenfield deployment). The migration can be done mainly in two possible ways:

- Deployment of new fiber and splitters for each technology (higher capital investment costs for the telecom operator);
- Reusing the existing fiber by wavelength coexistence via a single feeder fiber, so the changes are done only at both ends of the fiber (OLT and ONT).

Table 3.3 Main ITU and IEEE PON standards.

PON type	Max. downstream	Max. upstream	Standardization
GPON	2.4 Gbit/s	2.4 Gbit/s	ITU-T G.984
XG(S)-PON (NG-PON1)	10 Gbit/s	2.5 Gbit/s (XG) 10 Gbit/s (XGS)	ITU-T G.987 ITU-T G.9807
XLG-PON (NG-PON2)	40 Gbit/s	40 Gbit/s	ITU-T G.989
25G/50G-PON	50 Gbit/s	12.5 Gbit/s/25 Gbit/s/50 Gbit/s	ITU-T G.9804
EPON	1.25 Gbit/s	1.25 Gbit/s	IEEE 802.3ah
10G-EPON	10 Gbit/s	10 Gbit/s	IEEE 802.3av
25G/50G-EPON	25 Gbit/s/50 Gbit/s	25 Gbit/s/50 Gbit/s	IEEE 802.3ca

3.3.2 Next Generation High Speed PONs

PON standards have been created by ITU-T and IEEE, as given in Table 3.3. The most used standard of the above in the 2010s was GPON. These ITU standards have been further developed into the Next Generation PON (NG-PON) and High Speed PON (HS-PON) series of standards:

- NG-PON1 – Next Generation PON 1, i.e. XG-PON, standardized with ITU-T G.987 series [14–16];
- NG-PON2 – The next generation of PON 2 (i.e. passive optical networks capable of 40 Gbit/s XLG-PON series, G.989) [17].
- HS-PON (High-speed PON), i.e. 50G-PON, which aims to provide 12.5, 25, and 50 Gbit/s per single wavelength, standardized in G.9800 series [18], is the future use ITU PON standard which is recently developed. For example, the 50 Gbit/s downstream carrier wavelength is in the range of 1340–1344 nm downstream, while the upstream has 12.5 or 25 Gbit/s per wavelength in the range of 1260–1280 nm (option 1) or 1290–1310 nm (option 2).

In addition to ITU, IEEE also has several standards for PONs with Gigabit Ethernet-based PONs including the following:

- Ethernet Passive Optical Network (EPON) is defined with standard IEEE 802.3ah [19]. This standard is based on the use of the Ethernet family of protocols between the Central Office (CO) and end users. It is also known as Ethernet in the First Mile (EFM).
- 10G-EPON, standard IEEE 802.3av [20]: This standard provides bitrates up to 10 Gbit/s, a similar ITU standard to NG-PON1.
- 50G-EPON, standard IEEE 802.3ca-2020 [21]: this Ethernet PON standard in the 2020s is 10G/25G/50G-EPON (i.e. standard IEEE 802.3ca), which provides both symmetric and asymmetric operation for the following data rates downstream): 25/10, 25/25, 50/10, 50/25, and 50/50 Gbit/s.

Future PON and Ethernet standards are developing to provide speeds of 100, 200, and 400 Gbit/s based on 100 Gbit/s signaling. In this direction, bitrates of up to 100 Gbit/s per wavelength are possible in Ethernet standards (as of 2021). Thus, the future of EPON is toward achieving 100G-EPON – Ethernet PON with bitrates gradually up to 100 Gbit/s. After that, future developments of both ITU and IEEE PONs are targeted to provide multiple hundreds of Gbit/s, such as 200 and 400 Gbit/s in the access part, by using multiple wavelengths over single-mode fiber.

3.4 Carrier-grade Ethernet for Telecoms

In the past, the core and transport networks of the telecommunications operator were primarily built on SDH/SONET technologies, which were based on Time Division Multiplexing (TDM) and were well standardized in all aspects.

On the other side, Ethernet was started as LAN technology at the time when SDH was in full power in the telecom operator's world in the 1980s and 1990s. At that time in the LAN segment there were several different standards (e.g. Token Ring from IBM, and others), which lost the market share to the emerging Ethernet at the end of the 1990s and beginning of 2000s. In the 1990s, the Ethernet as a whole was not as good as it is nowadays in the 21st century (speaking about the LAN segment), because the topology was a line topology (one erroneous computer could cause all computers connected on that line to experience traffic problems, due to congestion avoidance mechanisms used in the Ethernet before accessing the media). And, today, we have LAN with a typical star topology with an Ethernet switch in the middle (so, today's Ethernet is more trouble-prone from individual hosts).

3.4.1 The Rise of Ethernet from Local to Carrier Technology

What has happened to Ethernet technology since then? Well, Ethernet as a unified local area network technology since the 1990s has expanded from local to metro and finally into telecom carriers' transport network environments. So the Ethernet, standardized by the IEEE 802.3 family of standards, has done tremendous work from the 1980s and 1990s (when it first appeared) to the present time, and can expect this to continue at least for the near future (e.g. 2030s).

Why did the Ethernet become what it is now? Because the telecom world has changed from legacy telecommunications (digital circuit-switching at the end of the 20th century) to all-IP world (in the 21st century), so IP networks are used for Internet traffic and for all managed services (including telecom legacy services, such as voice, TV, and business services, as well as many new emerging services).

From its beginning, Ethernet is standardized by the IEEE with the idea to be used under the IP stack above it (transport protocols such as TCP or UDP, and applications on the top). Hence, IEEE standardizes only physical layer (protocol layer 1) and MAC layer (that is protocol layer 2) for Ethernet and assumes that IP will be on layer 3 (the network layer) and then transport protocols on layer 4 (e.g. TCP, UDP) and applications on the top (e.g. Web applications). So, in short, the Ethernet was built particularly for IP traffic to be carried over it. Initially, it was used for access to the Internet and generally IP networks in enterprises, which typically have larger number of computers that need to be connected (depending on the size of the company, small, medium, or large). With the spread of broadband to residential environments (to homes) since the 2000s, it started to be a major choice for implementation of home networks, together with Wi-Fi (IEEE 802.11 family of standards) which initially was used as wireless extension of Ethernet in the local network. Then, with the success of IP and Internet technologies, becoming the main packet-switching technology in telecommunication networks, contributed the Ethernet also to make a "breakthrough" into the telecom carriers' world.

On the other side, the organization which supports the expansion of the Ethernet from LAN to MAN (Metropolitan Area Network) and further to WAN (Wide Area Network) and international networks is Metro Ethernet Forum (MEF). In fact, IEEE standardizes OSI protocol layers 1 and 2 for the Ethernet, while MEF defines network architectures and approaches for the use of so-called Ethernet services by telecom operators.

In general, the Ethernet has become unified LAN due to lower cost of the equipment, lower operation and maintenance costs, as well as dedication to transport IP packets on the network layer. Initially, the Ethernet had no QoS built-in and no TDM support, which were normal in legacy telecommunication networks (e.g. for voice services).

3.4.2 Carrier Ethernet Characteristics

The evolution of the Ethernet toward the requirements by the service providers and telecom operator's networks has led to the definition of "Metro Ethernet Network" or "Carrier Ethernet" (both names are interchangeable) by the MEF. With moving to Carrier Ethernet (that is, Ethernet for telecom operators or carriers, according to terminology that we use in this book), certain end users that were using the Ethernet were demanding the same level of performance that they had in the past with TDM-based WANs. So, Carrier Ethernet (when compared to "ordinary" Ethernet) has several additional fields in the layer 2 headers which provide QoS support over such Ethernet networks.

Regarding the development of the Carrier Ethernet, the requirements were set to provide network scalability, QoS for different user types (with the same performance as in TDM-based access and transport networks), reliability (in the similar manner as found in SDH transport networks, e.g. 50 ms link recovery after a failure), possibility to guarantee the SLA, as well as to provide TDM emulation services (e.g. for VoIP as carrier service), and so on.

What about carrier Ethernet bitrates? Well, the bitrates of the physical interfaces are the same as those defined for LAN Ethernet standards which include 10, 100 Mbit/s, 1, 10 Gbit/s, and further up to 100 Gbit/s in the 2020s.

Further, Carrier Ethernet standards define layer 2 services that can be consumed directly by the subscribers (that is the name used for enterprises, i.e. companies) and by telecom operators. Such Carrier Ethernet services are intended to be used between in access, core, transport and transit networks. Of course, different network segments on the path of certain IP packets can be based on different transport technologies (e.g. OTN), but also it is possible for all segments end to end to be based on Carrier Ethernet services.

Regarding the enterprise (i.e. customer), there are several types of Carrier Ethernet (CE) services, according to the Carrier Ethernet 3.0 specifications from MEF, which include the following:

- Carrier Ethernet E-Line service, which provide point-to-point connectivity;
- Carrier Ethernet E-LAN services, which provide multipoint-to-multipoint connectivity; and
- Carrier Ethernet E-Tree services, which are based on rooted multipoint connectivity.

On the other side, Carrier Ethernet services for telecom operators also include Access E-Line and Transit E-Line:

- Access E-line Service is for connecting the subscriber sites via point-to-point connections.
- Transit E-Line service enables point-to-point connectivity, which is targeted to connect multiple provider networks by using the Carrier Ethernet network.

For provision of Carrier Ethernet services, there are tags added in regular Ethernet frames. In that manner, Virtual LAN (standard IEEE 802.1Q) is used for provisioning of Carrier Ethernet E-Line services, which is suitable for connecting offices of corporate LAN networks on different locations. In this case, VLANs are created to provide virtualization of corporate Ethernet infrastructure, so the traffic to different user groups is separated into different VLANs, where each VLAN has unique 12-bit long Q-tag in layer 2 header.

As shown in Figure 3.5, customer network that uses CE services is connected to provider bridges – PB network (based on standard IEEE 802.1ad), which adds additional 12-bit long tag in the Ethernet frame, called S-tag, and it is used as provider ID. Providers that provide CE services to customers are tier-3 CE service providers, while on the other side they may purchase transit CE services from tier 2 providers with which they are interconnected via Network-to-Network Interface (NNI) also noted as External NNI (ENNI). On the other side, customer network is connected to provider's network via a User to Network Interface (UNI), and there is a separate UNI for each customer's network. So, in CE provider's network there are two tags, one used as provider ID and the other as VLAN ID; therefore, this approach in Carrier Ethernet is called Q-in-Q network.

Provider Backbone Bridges – PBB (IEEE 802.1ah standard) are used to transfer Ethernet frames from one provider to another provider. An additional 24-bit long tag, which is referred to as I-tag, identifies services in the PBB network.

Between any adjacent pairs of UNI and NNI, or NNI and NNI is set up Operator Virtual Connection (OVC), as shown in Figure 3.5. Used by wholesale service providers, OVC can work in P2P or P2MP mode. Multiple OVCs can be merged into an EPC, e.g. to connect remote offices of an enterprise. In that way, connecting remote Ethernet LANs through Carrier Ethernet is done via Ethernet Virtual Connections (EVCs), which can be point-to-point (P2) or point-to-multipoint (P2P).

3.4.3 QoS for Carrier Ethernet

When a certain network is used as a transport technology for telecom operators it needs to have mechanisms for QoS provisioning. How is QoS provided in Carrier Ethernet?

Well, the extension of the Carrier Ethernet QoS is via use of bandwidth profiles and traffic management techniques based on quantification of delay, delay variation, then Ethernet frame loss ratio, as well as network availability. Definition of Class of Service (CoS) provides the ability to define an SLA on UNI. There are several so-called CoS phases, from which the last ones specify so-called CoS Performance Objectives (CPOs), which refer to several Performance Tiers (PTs) for

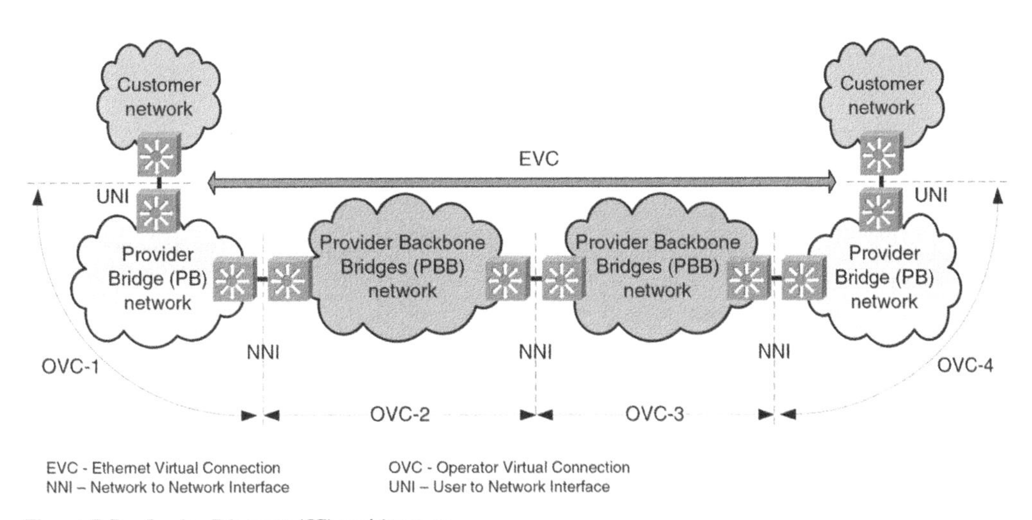

Figure 3.5 Carrier Ethernet (CE) architecture.

Table 3.4 Performance Tiers (PTs) for Carrier Ethernet.

Performance Tier	Performance Tier name	Distance	Delay
PT 0.3	City PT	<75 km	0.6 ms
PT 1	Metro PT	<250 km	2 ms
PT 2	Regional PT	<1200 km	8 ms
PT 3	Continental PT	<7000 km	44 ms
PT 4	Global PT	<27500 km	172 ms

Carrier Ethernet networks. There, as given in Table 3.4, we have PT0.3 for city areas, PT1 for Metro areas, and also PT2, PT3, and PT4 for regional, continental, and global reach, respectively.

Looking at Table 3.4, the global reach means a distance of up to 27500 km (with performance level 4). Considering that the Equator (around the Earth) has a length of approximately 40000 km, and we need to travel halfway to reach any point on the Earth (though not in a straight line, we need more than half, i.e. more than 20000 km), with PT4 and its 27500 km we can actually reach every location on our planet with Carrier Ethernet. And it's simply a great development for a technology that started out solely as LAN technology in the 1980s.

The MEF defines three particular CoS names, which are also noted as CoS labels, given as following (as given in Figure 3.6):

- H (High) label, which is targeted to applications which are very sensitive to delay, delay variation, and losses, e.g. signaling traffic, VoIP, critical IoT services, etc.
- M (Medium) label, targeted to applications which are more tolerant to delay and/or delay variation, but are sensitive to losses, e.g. near real-time services, and application with critical data.
- L (Low) label: It is targeted to non-critical applications, which are more tolerant to delay, delay variation, as well as losses.

In the 2010s and 2020s, mobile operators that operate 4G and 5G mobile networks, respectively, are increasingly using Carrier Ethernet as a technology for their core and transport networks.

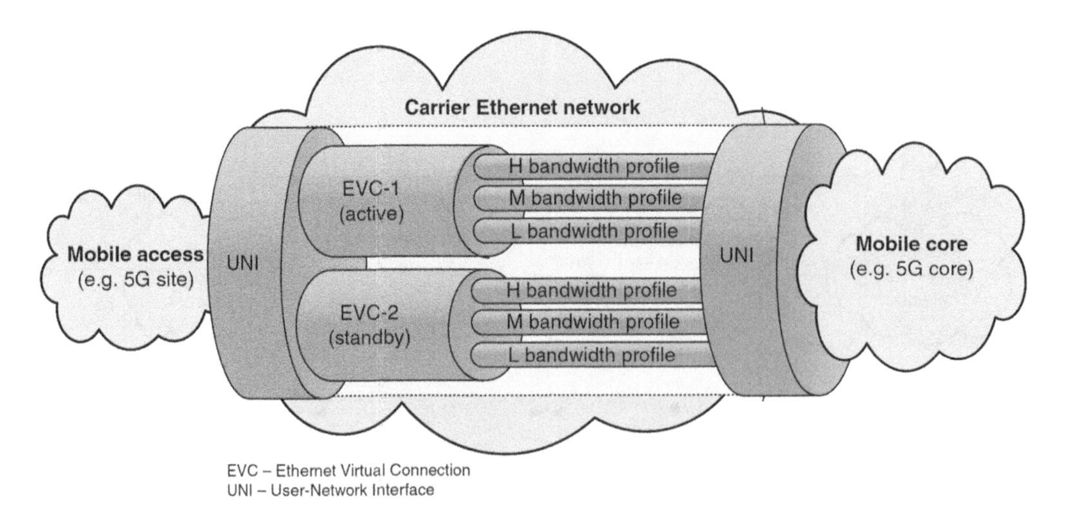

Figure 3.6 QoS provisioning in Carrier Ethernet.

Of course, the use of Carrier Ethernet in mobile networks requires mandatory synchronization over the Ethernet, which is needed for the TDM emulation and particularly for handovers in the mobile networks (which occur more often in newer mobile generations than before, due to smaller cells in urban environments).

3.5 Software Defined – Wide Area Network (SD-WAN)

When talking about the transport network, it is always good to have some solution in between the optical underlying network and IP protocol stack above. And one such well-proven solution is MPLS, which was initially deployed at the start of the 2000s and it continues to be used for several decades for traffic engineering and QoS provisioning in packet switching networks. With the introduction of software defined networking approach in transport networks, the MPLS evolves to Software Defined Wide Area Network (SD-WAN), although both technologies are not excluding each other.

3.5.1 IP/MPLS for Telecom Transport Networks

The MPLS can be implemented over optical transport networks, as well as below different networking protocols in packet switching networks, which is the origin of the term "multi-protocol." For example, MPLS can be used for IP networks, but also for ATM networks [3], which were the European candidate for packet switching technology in the 1990s. However, the spread of Internet by the end of the 1990s resulted in development of the all-IP telecom world since the beginning of the 21st century, so here we talk about MPLS application in IP-based transport networks, also noted as IP/MPLS.

In fact, MPLS is the main IETF (Internet Engineering Task Force) approach which is standardized for QoS traffic management in transport networks, defined in 2001. In IP transport networks with implemented MPLS (which is a typical scenario in transport IP-based networks since the beginning of the 21st century), label switching is used instead of routing based on IP addresses. MPLS is implemented between layer 2 (data control layer) and layer 3 (network layer), as shown in Figure 3.7. With MPLS, all QoS provisioning is made without changing IP packet (and all protocol headers above the IP), but adding labels between layer 2 and layer 3 headers, which are used only in the network that has implemented the MPLS (also called MPLS domain). Someone may say that MPLS is protocol layer 2.5 (as a jargon). Typically, a given MPLS domain is operated by a single telecom operator in a given country, so labels are added to IP packets at the entry into the MPLS domain and deleted on the way out of that network. The QoS is not implemented with DiffServ fields in the IPv4 or IPv6 headers, but by adding additional information via label headers (inserted between MAC header on layer 2 and IP header on layer 3).

In IP/MPLS networks, the Label Switching Path (LSP) technology is used to pre-provision a Virtual MPLS Transport Network (VMTN) for each service type (e.g. VoIP, IPTV, best-effort Internet traffic, corporate VPN, IoT services, and so on).

Regarding the MPLS network architecture, in short, fixed-length labels are associated with streams of data. In the ingress edge MPLS router, called Label Edge Router (LER), labels are added to IP streams (based on different constraints for traffic differentiation at the network edges). Packets belonging to those streams are forwarded along the Label Switched Paths (LSPs) based on their labels (not on IP addresses), so there is no IP header examination in an Label Switching Router (LSR). There, LSRs are used for packet switching in the MPLS network domain.

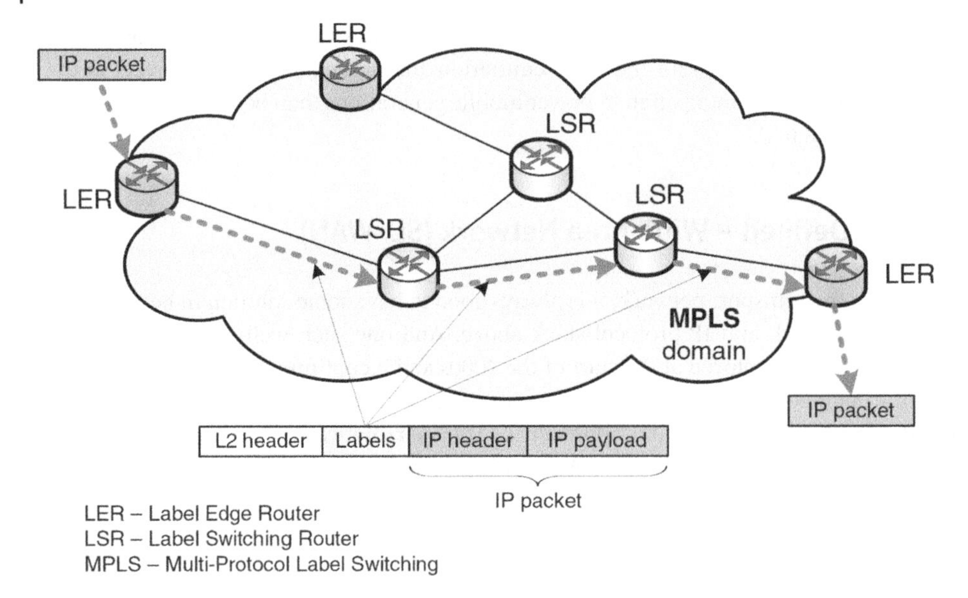

Figure 3.7 IP/MPLS architecture.

MPLS Transport Profile (MPLS-TP) has been developed jointly by the ITU-T and the IETF in 2011, with the aim to provide QoS provisioning based on a set of parameters, without a control plane [22]. Why MPLS-TP? The major innovation over standard MPLS-TP is the addition of Operations, Administration, and Maintenance (OAM) functionality to MPLS in order to monitor each packet and thus enable MPLS-TP to operate as a transport network protocol. These features are aimed at making MPLS comparable to SDH and OTNs in terms of reliability and tracking capabilities, that is, MPLS-TP is a carrier-type packet transport technology, which provides different options for deployment, regarding access, aggregation, and core networks of the telecom operators (as shown in Figure 3.8).

Thus, the MPLS-TP management network is a subset of the Telecommunications Management Network (TMN) that is responsible for managing the portions of the network element that contain

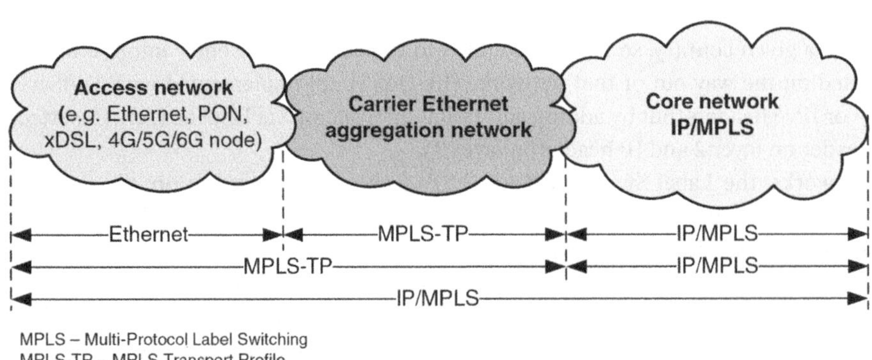

Figure 3.8 Options for MPLS-TP deployments.

network entities at the MPLS-TP layer. The ultimate goal is to transport MPLS-TP packets over a variety of different physical technologies ranging from SDH and OTN to Carrier Ethernet.

3.5.2 Software-Defined WAN for Telecom Networks

The SD-WAN appearance was driven with the development of Software Defined Networking (SDN) in the 2010s, especially by ITU (as umbrella standardization) for IMT-2020 (i.e. 5G) mobile networks and fixed mobile convergence [23]. The idea of SDN was to provide simplified and flexible management of different services by using heterogeneous hardware and software components in telecom networks. So, SDN is in the root of the SD-WAN standardization. In that way, let's define what SDN is? By definition [24], SDN is a set of techniques that enable direct programming, orchestration, control, and management of various network resources in respect of the design, delivery, and operation of network services via dynamic and scalable approach. With the significant impact of cloud services over the public Internet and continuous digitalization in all countries worldwide, many enterprises and operators are moving to so-called SD-WAN approaches. In such solution, the data may be stored on different clouds accessed via the public Internet via secure connections end to end, with centralized control regarding the management and orchestration of such distributed access.

The high-level view of SDN layered architecture is given in Figure 3.9, which includes the following layers [24]:

- Application layer that includes all SDN applications, which in fact refers to applications that control and manage network resources, not applications or services that are used by the end users (for example, application-based routing in the SDN-enabled network).
- Control layer provides abstraction of underlying resources to the application on the top, orchestration of resources, as well as application support.
- Resource layer includes control, support, and data transport and processing by network devices.

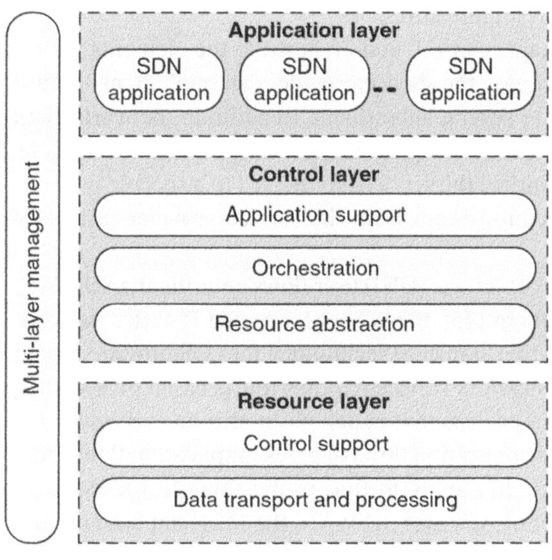

Figure 3.9 SDN architecture.

In general, SDN is based on the separation of the user plane and the control plane in the network, which provides an opportunity to improve QoS provision to different types of services. Typically, the QoS is provided via a set of Key Performance Indicators (KPIs), which may include (but are not limited to) bandwidth, delay, delay variation, and packet losses.

So, what is SD-WAN? It is a software-defined wide area network that provides intelligent service aware transfer of data over the network. As noted above, many enterprises are moving from their own corporate data centers to cloud services. Unlike in the past, when users were connected to corporate data centers, with the move to cloud computing, there are increasing requirements to adapt (i.e. "program" in some ways) the network to be more cost and bandwidth efficient and to be more flexible than before.

The first standard on SD-WAN came from MEF in 2019 [25], although ITU in parallel also completed work on SD-WAN requirements [26], particularly focusing on IMT-2020 (i.e. 5G) and its Fixed-Mobile Convergence (FMC) services [23]. Why does MEF have standardized SD-WAN?

Well, Ethernet is continuously expanding into all kinds of networks, from local to metro, regional, and global. But, like other technologies in the late 2010s and early 2020s, software-defined networking is also having an impact on Ethernet technologies. Thus, MEF standardized SD-WAN in 2019 and updated it in 2021 [25].

From MEF perspective, SD-WAN service is end-to-end connectivity that creates an overlay network over one or multiple existing networks that provide connectivity, therefore called Underlay Connectivity Services (UCSs). These UCSs may be instantiated on network slices. In term of MEF, network slice is an organized and structured subset of the network of service provider. It should be noted that different organizations have slightly different definitions of network slicing, which has gained popularity with its application to 5G mobile networks with the ability to implement services with highly heterogeneous requirements for quality, security, and business constraints. Hence, network slicing gives the possibility of scalable and flexible use of the deployed network infrastructure and its resources for existing and future services in a given period of time. Of course, the use of resources can be optimized with network slicing, but additional capacity requires investment in new resources (or upgrade of existing ones, e.g. use of DWDM on existing fiber). Network slicing can be and actually is applied in SD-WAN architectures.

The network slice includes functions to manage, control, and orchestrate the elements within that subset of service provider resources. In doing so, the service provider can implement and use a network slice dedicated to one subscriber or to several subscribers. In addition, network slices may also exist for the provider's internal purposes of managing, orchestrating, and connecting its employees. In doing so, the subscriber can instantiate the services connected to a specific network slice on the network slice (on that slice) or request them from the service provider (SD-WAN operator).

The SD-WAN service recognizes User Network Interface (UNI) user flows entering the SD-WAN, based on certain policies and rules applied to them [25]. SD-WAN as a routed IP-based network (e.g. it can use IP/MPLS, Carrier Ethernet, or other transport technology for QoS provisioning at underlying connectivity services, which are IP network services) routes traffic based on so-called Application Flows (AFs). Each AF consists of IP packets that match given criteria and associated policies (which can be applied to either inbound application flows or egress application flows) that define the requirements as well as restrictions that apply to that traffic flow. In order for an SD-WAN service to provide the desired behavior of AFs over SD-WAN, the MEF standard defines service attributes. The attributes of network services according to MEF are given in Table 3.5. However, standards from other SDOs may use different sets of network service attributes.

Table 3.5 Network service attributes.

Attribute	Description
Network service identifier	The unique identifier of the network service
Network profile descriptor	This is labeled as category description of Network Slice used for the network service, such as string, (e.g. "IoT low bandwidth," "mobile broadband," etc.)
Network profile	Describes the characteristics of the Network Slice used for this Network Service
Supported service types	List of service types supported by the network slice (physical layer, Ethernet, IP, SD-WAN, network service)
Network service topology	Describes the topology for network services
Service instantiation capability	Refers to ability of subscriber to request service instantiation on the given network service (TRUE/FALSE)
Instantiated services configuration capability	Refers to ability of subscriber to be able to request modification of instantiated services (TRUE/FALSE)

An example of SD-WAN network service is shown in Figure 3.10. In the given figure, SD-WAN orchestration functions provide a centralized configuration of SD-WAN Edge functions through which policies are implemented to manage the flows of SD-WAN subscribers. Moreover, the connections of different SD-WAN subscribers are typically realized through an IP routed network, so UCSs in SD-WAN are actually IP network services that use a given technology in the transport network. Because each telecom operator through which the traffic passes may have a different transport technology, such as IP/MPLS, Carrier Ethernet, OTN, and so on, the UCS as underlying connectivity service for SD-WAN as overlay network can be based on different transport technologies in different segments on a given path between any two SD-WAN edges (SD-WAN subscribers are connected via one or multiple SD-WAN edges). Such UCS can have instances on network slices that are created in the IP network.

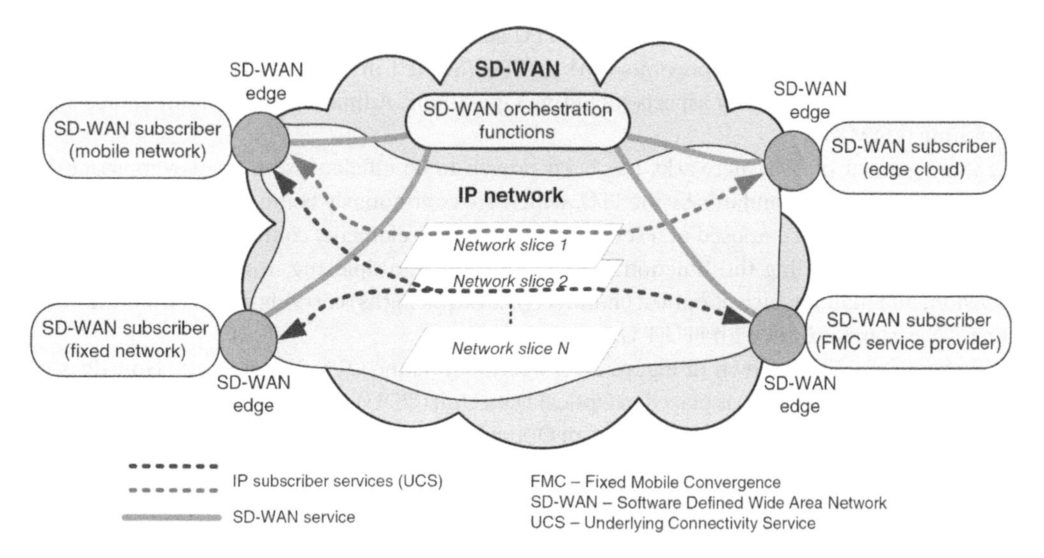

Figure 3.10 SD-WAN use case example for fixed-mobile convergence.

SD-WAN enables service providers to launch such a service for enterprises as they move from their own data centers to cloud operations with digitization. For uniform management of the entire network in SD-WAN, a centralized controller is introduced to implement orchestration functions such as rapid deployment and flexible configuration. With mobile technologies that support network slicing, such as 5G and next (6G), an SD-WAN service supports mobile network access as one of several connection types. When needed, load sharing can be applied across WAN connections by provisioning of a simple WAN management interface for fixed broadband access (e.g. PON, Carrier Ethernet, etc.) and mobile broadband access (e.g. 5G, 6G, and beyond). In this way, SD-WAN network services can be used for enhancement of the FMC [23, 26].

The security of underlying connectivity services in SD-WAN, which are IP-routed network, can be provided via the deployment of Virtual Private Network (VPN) services.

Using an SD-WAN approach can improve FMC (which has existed in mobile networks for several mobile generations, and increases as mobile broadband speeds approach fixed broadband at a given time) by enabling each operator's node to be able to transmit data traffic on dual links (that is, via mobile access network and via fixed access network) simultaneously to provide ultra-wide coverage of certain critical services.

3.6 Optical Transport Networks

Today's global communications world has many different definitions of OTNs, which are supported by different technologies. In general, the transmission of information through optical media in a systematic way is an OTN. An OTN consists of network capabilities/functionalities and the technologies required supporting them. Here, we refer to OTN transport networks standardized by ITU, while the most used technologies over OTN are MPLS and VPN.

3.6.1 Optical Transport Network

Most of the terrestrial transport networks nowadays are optical networks, so OTN as a terminology can fit all of them. However, in there is also an ITU standard named OTN (i.e. ITU G.709 series), developed as continuation of the previous work on SDH/SONET in the 1990s, which has proven to be a quality one regarding many aspects including Operations, Administration, Maintenance and Provisioning (OAM&P).

The management of SDH networks has been proven to be efficient, so a similar approach is applied to modern OTN standards by the ITU, which are continuously being upgraded.

By definition, OTN is composed of a set of optical network elements connected by optical fiber links, capable of providing the functionality of transport, multiplexing, routing, management, supervision, and maintenance of optical channels (i.e. wavelengths) carrying signals from the customer. OTN is standardized with ITU-T G.709 [27].

An overview of OTN is shown in Figure 3.11, which shows the OTN digital layers. How does it work? Well, the data (any data) is placed in Optical Data Unit (ODU) digital channels of OTN. Then, such optical data unit (the ODU) is placed in an Optical Transport Unit (OTU) by adding overhead (OH) needed for multiplexing and demultiplexing in the transport network and also Forward Error Correction (FEC) for better performances. Further, the OTU is carried via the optical channel, where the term "optical channel" refers to a single wavelength in the fiber link. Optical Multiplexing Sections (OMSs) and Optical Transmission Sections (OTSs) are constructed using the additional overheads along with the optical channel.

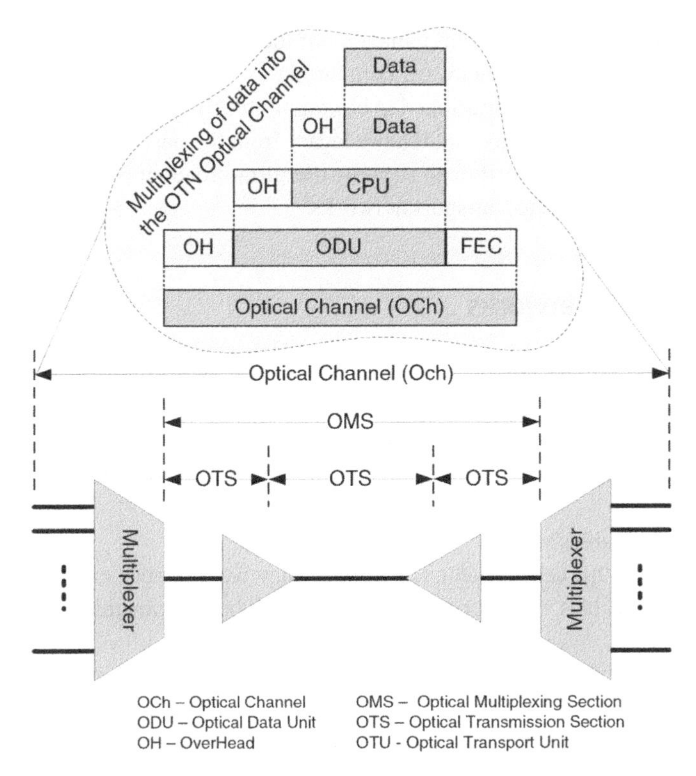

Figure 3.11 Optical transport network (OTN) structure.

OTN uses a layered structure that is very similar to SDH (as past optical transport standardized by the ITU) [28]. Figure 3.11 shows the basic OTN layers that are visible in the OTN transport structure. They consist of an Optical Channel (OCh), an Optical Multiplex Section (OMS), and an Optical Transmission Section (OTS).

With such an approach, OTN provides functionality for managing DWDM (Dense WDM) transport networks that use many wavelengths over a single-mode fiber. The management approach for traffic engineering (i.e. allocation of portions of bandwidth to different customers or services via a centralized operating and support system) is comparable to the approach of managing SDH, which on the other hand has used only a single wavelength in a fiber. Hence, it can be said that the main advantage of the OTN standard by the ITU is in its backward compatibility with SDH and its management functionalities, while on the other hand OTN is completely transparent to network protocols such as IP that will be placed within its optical data units. Also, MPLS (as one of the most popular solutions for IP transport networks) can be used together with the OTN.

What are the bitrates of OTN G.709 standard series from ITU? The OTN in the 2000s started with backward compatibility with SDH. In that manner, OTU1 refers to 2.5 Gbit/s (i.e. STM-16 in SDH, where STM stands for synchronous transport module), OTU2 refers to 10 Gbit/s (which is equal to a multiplex consisted of 4 OTU1s), and OUT3 refers to 40 Gbit/s (i.e. 4 OTU2s). Since 2017, OTN includes so-called flexible OTN (FlexO), which defines long-reach interfaces that provide the possibility of bonding (i.e. grouping) via multiple optical interfaces (i.e. optical channels), and bitrates provided are in the range of multiples of 100 Gbit/s. Such flexible approach gives available bitrates to customers (of OTN) of 100, 200, and 400 Gbit/s, compatible with the same data rates from IEEE 802.3 (i.e. Ethernet) standards family.

What is the OTN future? Well, the OTN provides needed management functionalities in optical transport networks. Looking to the past, it is made backward compatible to its predecessor, the SDH/SONET. Looking toward the future, the OTN is developed to be compatible (in data rates and framing) with Carrier Ethernet standards. One may say that it has a legacy from the past (from the SDH era in the 1980s–2000s) and it is developing further to support the ultra-broadband future, which requires hundreds of Gbit/s and Tbit/s in the transport networks.

3.7 Submarine Cable Transport Networks

Optical and satellite networks are the main approach to connect different continents and islands. While satellite networks were originally the first to provide global connectivity, optical terrestrial networks provide the best experiences in most cases due to lower delays as well as low bit and packet losses. To provide an optical connection between different countries separated by water, we use submarine cables.

What kind of cables are the submarine cables?

Well, when they first appeared in the 19th century (for telegraphy) they were copper cables. However, today, all deployed submarine cables are fiber-optic cables, due to better transmission characteristics of fiber over copper.

Where is the submarine cable placed? It is placed on the bottom of the ocean or sea, while it is buried near the shore for protection. Most of the cables deployed in the oceans are deployed between Europe and North America (on the bottom of the Atlantic Ocean), and between Asia and North America (on the bottom of the Pacific Ocean). Then, there are submarine cables deployed all around Africa, Americas, and south shore of Eurasia land, as well as in Australia and the Pacific islands [29].

Why there are so many cables from/to North America? Well, that is because most of the large global OTT service providers (e.g. Google, Facebook, Amazon, Microsoft, etc.) are coming from North America and all of them have large data centers there. Of course, there are also Content Delivery Networks (CDNs) of these large OTT service providers, with data centers close to end users everywhere (with the aim to provide lower round trip delay and with that better user experience). However, such data centers on different sides of the globe are also connected via submarine cables and they carry a very large volume of traffic from/to billions of Internet users worldwide as well as traffic exchange between data centers of service providers.

What is the capacity of submarine cable systems? With WDM technology being used in the cables the newly deployed fibers carry more bits per second than submarine cables deployed in the past (e.g. in the late 20th century). So, the newly deployed submarine cable systems can carry hundreds of Tbit/s per fiber, and there is deployed a lot of fiber at once, including dark fiber ("dark" denotes fiber state when it is deployed but not yet used to carry traffic – of course, it is deployed to be used in the future). However, one may note that the capacity is not a fixed value here because there are two main approaches to measure the capacity of the submarine cables. What are they?

Well, they are potential capacity and lit capacity. The lit capacity is in fact the currently running capacity while the potential capacity is the maximum capacity that can be achieved with the deployed submarine cables if the owner installs newer available equipment at the both ends of the given cable.

Who owns the submarine cables? In the past, they were primarily owned by telecom network providers, i.e. telecom carriers (telecom operators). But, in the 2010s and 2020s, there is a change in the type of companies that invest in submarine cables (which carry almost all international IP traffic today over seas). The "new faces" in submarine cable business are in fact well known content and application providers, i.e. large OTT providers. And the submarine cable further development is also changing in that direction, toward higher participation in the ownership of cables by large global OTT providers. Why?

Well, the majority of IP traffic is due to Internet traffic, and the main role in the Internet application scene is played by the largest OTT providers, so they are interested in gaining control of the path between their data centers. Why? To provide a better user experience that further contributes to their further business growth. This encourages further expansion of their services, resulting in more traffic to be carried. All this results in the need for more capacity on submarine cables, and the "wheel" continues to roll toward the future as a never-ending "game" because newer services will have greater bandwidth requirements.

Finally, many of you reach this book on the platforms of online sellers by passing through submarine cables, with which IP packets "dive into the seas and oceans" on their way.

3.7.1 Deployment of Submarine Cable Systems

The submarine fiber-optic cable system has specific technical characteristics [30], so they should be planned to have a long lifetime and to be very reliable because its installation should last for several decades. The main reason for this is that its construction and maintenance is long and expensive due to the difficulty of accessing it. Also, submarine cables are of strategic importance in the transmission network and the interruption of the connection usually results in a significant loss of traffic and thus a loss of revenue. For island countries, the damage to such systems would mean an interruption or a significant reduction in capacity and thus in quality of broadband services such as broadband Internet access.

The submarine cable system should have mechanical characteristics that allow:

- Precise installations with considerations of the seabed and physical specificities, as installations can reach up to 8000 m depth. Generally, submarine cable systems should be installed buried in the seabed, which are laid by specially designed cable vessels and appropriate equipment.
- To take into account the environmental conditions of the seabed that exist at the depth at which the underwater cable installation is laid, especially taking into account the hydrostatic pressure that increases with depth, the temperatures that are different at different depths, abrasion, corrosion, and marine life.
- Underwater optical cables should be adequately protected by armoring or burial from damage caused by fishing in the sea, anchors from ships, or other obstacles.
- Enabling the possibility of repair and maintenance at such depths, with appropriate safety considerations.

The material characteristics of the submarine cable system should enable the functionality of the optical fiber in order to have the desired reliability during its lifetime, as well as to tolerate the specified loss and aging mechanisms, especially in relation to bending of the optical fiber, stretching under external influences over time, earthquakes on the seabed, etc.

Figure 3.12 shows the main design concept of optical fiber submarine cable systems. Each such system consists of submarine portion and land portions on each end of the cable. However, optical signals also attenuate over the distance (although much less than copper-based cables), so there are required repeaters along line, e.g. on each 60 km in length of the fiber. The repeater regenerates the optical signal in the fiber, so they need power supply, which is provided via submarine cable stations. The repeaters may be per fiber (traditional approach) or per group of fiber.

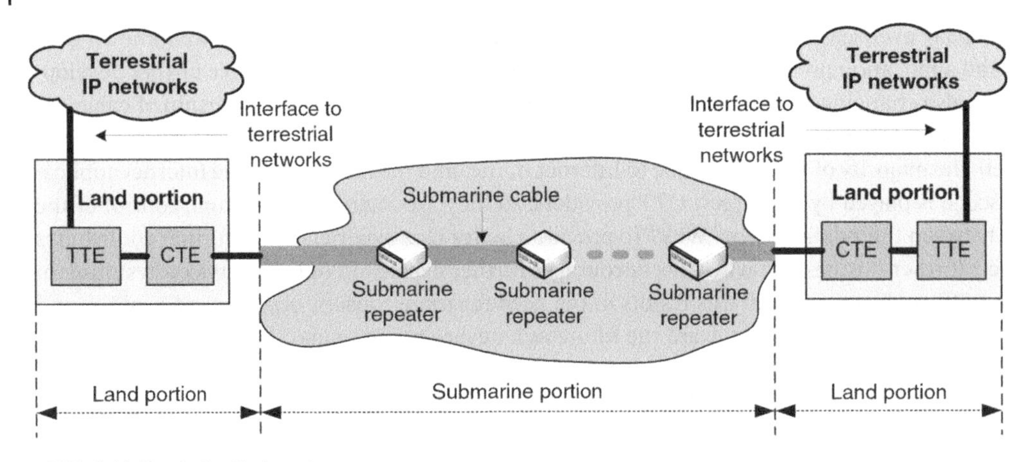

Figure 3.12 Optical fiber submarine cable systems.

The terminal station is typically the place where the submarine cable system is interfaced to terrestrial digital links or to other submarine cable systems, while on the other side are beach joints or landing points. All procedures for laying down the submarine cables and their connectors and repeaters should be completed with stringent quality procedures. One may note that such procedures are dependent upon each optical fiber submarine cable supplier, but certain general principles may apply.

One part of the development and deployment is demonstration of the performance under different possible conditions in practice, such as high-stress testing with an aim to estimate durability of components or subassembly. Besides such quality assurance procedures at submarine cable deployment, there is needed long-term life testing, with the target to evaluate the lifetime and/or the reliability of the technology as well as component or assembly.

The assembly of a submarine link consists of cascading the cable sections by use of repeaters (or branching units) between adjacent fiber sections. Such link assembly is typically performed in the cable factory. Then, the next step is ship loading which consists of installing the submarine portion onboard the cable ship, which is performed prior to laying. At such stage the submarine fiber link is not powered. All steps are accompanied with testing and quality assurance procedures on a periodic basis.

The installation of the submarine cable system starts with sea route survey (studying the sea/ocean depth, including the profile on the bottom, sea temperature and its variations, pressure, seismic activity, fishing activities along the route, applicable laws of countries that are on the route, etc.), which results in appropriate choice of fiber protection, armor, and burial approach needed for laying out the submarine cable on that route. In that way, prior to the installation of the submarine cable, there are required discussions with local authorities and fishing bodies.

The process of lying of the cable is typically performed at appropriate weather and sea conditions with an aim to avoid any damages of the cable sections or injury to the staff. The submarine cable can be buried in the seabed for higher protection.

In general, there are the following types of optical network topology for submarine cable systems [31]:

- Point-to-point topology: It refers to a direct submarine link between two Terminal Transmission Equipment (TTE) placed in two separate Terminal Stations (TSs).
- Star topology: It consists of a main TS that is connected via separate cables to several other TSs.
- Branched star topology: It refers to a star in which the splitting of traffic is done underwater, so there is minimization of the number of submarine cables to connect different TSs.
- Trunk and branch topology: It is used to connect several TSs, including TTEs, with the help of branching units that allow extracting part of the carried traffic over submarine cables in the branching TS direction.
- Festoon topology: It is a series of loops between major coastal sites and is usually without repeaters.
- Ring topology: It refers to multiple connected point-to-point cables with twice the transmission capacity.
- Branched ring topology: It expands the submarine cable ring by adding a branching unit, which retains the self-healing nature of the ring (i.e. when one link fails, the traffic is carried via working links in the opposite direction of the failed one).

After installation of the submarine cable system, it needs to be maintained as all other telecom systems. Routine maintenance consists of periodic monitoring of system parameters and, if necessary, preventive switching to redundancy. It is carried out by the terminal stations with the help of the deployed surveillance system from the submarine cable installation.

3.7.2 Business and Regulatory Aspects for Submarine Cables

There are several hundred active submarine cables (about 450 as of 2022). However, laying them requires years of route research, large investments (in the billions of dollars), and large ships capable of holding cables several thousand kilometers long. However, despite the cost and difficulty of installing these submarine cables, they are considered to be far cheaper and more efficient than satellites (in terms of available capacity, i.e. cost per bit/s, and have much better performance such as lower latency and lower losses). The main reason for laying new submarine cables is the demand for bandwidth all over the world, and often submarine cables are the only possible way to terrestrially connect countries from different continents and regions. Of course, satellites are still used in remote parts of the world to cover those areas not covered by terrestrial networks. On the other hand, submarine cables are primarily used to transfer data between terrestrial networks (which are all IP nowadays), including fixed and mobile broadband access.

Companies prefer to install newer cables because they are much more technologically advanced, so in the end the unit cost seems to be cheaper for new cables than old cables. One may say that new cables have a better economy of scale.

Various large content/OTT providers (e.g. Google, Facebook, Microsoft, Amazon) are investing heavily in submarine cable on major routes, mainly trans-Atlantic and trans-Pacific, in order to connect their data centers with sufficient capacity. As of 2018, international traffic carried by content provider networks has displaced Internet backbone providers as the largest source of international bandwidth used [30]. In the future, it is possible that several large content providers (or in other words, large OTT players) will further expand in terms of submarine cable networks globally.

Further, submarine cables are laid in international waters, but they always end in national waters, which span in the coast line according to United Nations Convention on the Law of the Sea. That directly includes the geopolitical situation across the world, which directly influences where the cables are laid (e.g. between which end points) [32]. For example, as noted before, most of the trans-Atlantic cables end on European west coast and United States east coast. On the other side trans-Pacific cables are laid between the US west coast and Japan, and smaller portion toward Australia. Most of the Pacific Islands are connected to these submarine cables.

Overall, politics always plays a role in telecommunications [33]. While in the 20th century, there was competition mainly between US and European technologies (e.g. ATM was European standard for packet-switching, IP was US standard – IP won, Europe had HyperLAN while US had Ethernet and WiFi – US standards won; but, on the other side, 2G GSM and its successor 3G UMTS, 4G LTE, and 5G NR are European standards, while mobile technologies such as IS-95 in 2G, cdma2000 as 3G as well as WiMAX as 3G and 4G were US standards – European mobile standards won, and so on), currently in the submarine cables field, especially in the Pacific, there is competition between state-owned Chinese telecom operators which increasingly deploy submarine cables and on the other side telecom/provider companies from other countries that have similar business and regulatory environments (e.g. US, Australia, Japan, Singapore, etc.).

Also, submarine cables are often a better option than land-based fiber cables (over long distances, e.g. many thousands of kilometers), because land-based cables may transit through different national jurisdictions with different political and regulatory environments that can change with the course of time.

Finally, submarine cables should be shorter in all cases where possible, with the aim to limit both the costs and end-to-end delay (the signal travels with maximum speed of light in the cables, so longer cables result in longer end-to-end delays).

3.8 Satellite Broadband

To meet ubiquitous broadband access, it is clear that broadband coverage with a terrestrial system is not sufficient. Why? This is because they cannot reach distant points on our planet where we do not have a landline or mobile access network. So, satellite broadband is entering the future broadband "game," of course, along with terrestrial systems, including submarine cables.

New innovations have enabled impressive broadband speeds and extended reach via satellite broadband. However, satellite internet services have come a long way in the last few decades.

What satellite networks do we have? Well, we have Low Earth Orbit (LEO) satellites at elevations between 400 and 2000 km, then we have Geostationary Earth Orbit (GEO) satellites which orbit at about 36000 km, and in between there are Medium Earth Orbit (MEO) satellites flying in orbits from 8000 to 20000 km above the surface of earth.

Normally, satellites closer to the Earth's surface have smaller one-way and round-trip delays, while GEO satellites have the largest delays (for GEO they are over 100 ms one-way either forward or backward, given that radio waves propagate at approximately the speed of light, which gives a 100 ms delay every 30000 km, so one cannot go below it – that is the minimum delay with the speed of light as the maximum possible speed known to mankind so far).

In the past, the satellites have been used for traditional telecom services, which include telephony and TV broadcasting (as main telecommunication services in the 20thcentury, and still are in the third decade of the 21st century), covering large geographical areas. While telephony is a two-way service, the TV broadcast (which is the first broadband service in the telecom world) requires only the downstream (i.e. downlink) direction. However, with broadband Internet/IP access through satellite networks, which is two-way traffic (e.g. client–server request-response approach as main principle in the Internet/IP world), there appeared high demand for two-way satellite broadband access over large geographical areas that are not served by land-based (i.e. terrestrial) telecom infrastructure.

Furthermore, the ubiquity and resilience of satellite networks make them critical for broadband coverage of uncovered areas around the world (not covered by terrestrial fixed or mobile networks), oceans and islands (in addition to undersea fiber-optic cables), deserts, high mountains, and many other places in countries that are not accessible by land routes and hence do not have land networks. Also, satellite broadband access can play a key role in various disasters that occur in the world (natural or man-made, such as damage to telecommunication infrastructures caused by hurricanes, fires, wars, and so on).

Satellite companies are also leading innovations in communications technology, developing powerful next-generation satellites that increase coverage areas and use cutting-edge technologies to provide secure connections to consumers, businesses, and governments. Such constellations can keep people connected anywhere in the world, whether they are at home, in the office, or on the go.

Of course, besides the advantages of omnipresent access, there are disadvantages for satellite IP communication, such as possible variable quality of the bitrates (e.g. due to obstacles in the path of radio wave, processes in the atmosphere, and so on), and of course the round trip delay, which is always higher than the delay experienced via terrestrial networks (with servers being closer to the user by using CDNs from today's large OTT service providers over the open Internet network).

So, there are two disadvantages of satellite broadband that need to be addressed. The first one, delay (i.e. latency), depends on the distance between the satellite and the Earth. Regarding the second one, satellite signals may be attenuated by rain or other atmospheric conditions. This problem especially occurs in tropical areas and primarily affects the higher frequency bands.

On the other side, the costs of equipment for provision of satellite broadband, including the cost for launching satellites, have been decreasing over the years, so the satellite broadband in 2020s and toward the future is becoming competitive with other land-based broadband access technologies. That is accompanied with increasing throughputs of satellites for the broadband (Internet/IP) access.

Traditional satellite technology in the past typically used a broad single beam to cover entire continents and regions, especially for services that need only the downstream direction, such as TV (first analog and then digital). Later, in the 2010s and 2020s, the use of multiple narrowly focused spot beams and the reuse of frequencies have made it possible to increase bandwidth of satellite broadband access by a factor of 20 or more (and obtain high throughputs), when compared to traditional satellites. By the end of the 2010s, the satellite broadband services evolved and were offered in the following basic technology categories [34]:

- C-band (4–6 GHz) Fixed-Satellite Service (FSS);
- Ku-band (11–14 GHz) FSS;
- Ka-band (20–30 GHz) FSS;
- L-band (1.5–1.6 GHz) Mobile-Satellite Service (MSS).

3.8.1 Fixed-Satellite Service (FSS)

Fixed satellite technology enables the availability of broadband services to all parts of the world, including least developed countries, inaccessible land areas (poles, deserts, high mountains), and island countries, then seas, oceans, and airspace.

FSS newer generation systems enable broadband Internet access at higher data rates than before through small user terminals which communicate with satellites orbiting in the sky, where satellite systems are connected to terrestrial networks through ground stations. FSS systems are designed to use different frequency bands.

Broadband connectivity directly to end users with small satellite user terminals is provided with High Throughput Satellite (HTS) systems which are operating in the 20/30 GHz bands in the FSS.

For the purpose of provision of high capacity and high spectrum efficiency, HTS systems use many satellite spot beams which give higher frequency reuse. Overall, there are short term and long term FSS approaches. In this manner, "short term" refers to frequency bands for which satellite technology has already been developed, which include 4/6, 11/14, and 20/30 GHz FSS allocations, and partially the 40/50 GHz FSS allocations. FSS allocations above 50 GHz are also included in the ITU Radio Regulations (RRs), but there is required more time for their significant development.

Within the 20/30 GHz FSS bands where HTS systems are mainly deployed, there is a portion of the 500 MHz spectrum (bands 19.7–20.2 GHz in space-to-Earth and 29.5–30 GHz in Earth-to-space) for which satellite services do not share with other primary services in the ITU frequency table. User terminals operating in these bands can generally be deployed without the need for satellite earth station coordination beforehand. This is so-called "exclusive Ka-band," for which the user terminals can be sold and installed in homes as well as offices (in large numbers), which on the other side requires applicable regulatory regime.

Considering the future, 20/30 GHz FSS allocations are likely the most suitable for provision of satellite broadband Internet access over user terminals, considering that at such frequencies the wavelength is consistent with very small antennas on user terminals. However, one should note that the individual Internet access is incompatible with the legacy regulatory approaches for international use of the FSS bands (in the past decades), in which the coordination of individual earth stations is set as mandatory (such coordination is not possible with many individual users, such as homes or offices with satellite broadband access).

The frequency bands which are targeted for use by high-density broadband applications/services in FSS are given in Table 3.6 [35]. Future HTS systems are expected to be deployed also in the 40/50 GHz FSS bands. These bands are initially planned to be used for gateway feed links for FSS systems where end user terminals operate in the 20/30 GHz bands. But, in the future it can be expected that HTS systems will also deploy ubiquitous user terminals in parts of the 40/50 GHz FSS bands.

3.8.2 FSS Technical Characteristics

The most interesting bands for satellite broadband Internet access are the 20/30 GHz bands, where much of the development of HTS satellites has been focused to date (to provide regional and global broadband Internet access). Moreover, with the evolution of satellite technologies over time, the number of beams per satellite has increased from about tens of beams to hundreds of beams. Thus, in the 2020s, a typical value for HTS operating at 20/30 GHz is to have 200 beams and more per satellite.

Also, the capacity per satellite increased over time, from several Gbit/s to tens of Gbit/s, and for HTS deployments is measured in hundreds of Gbit/s (and more) per satellite. In the future, Very High-Throughput Satellite (VHTS) in 20/30 GHz (considering that 4/6 and 11/14 GHz frequency bands are already heavily utilized) is expected to provide Tbit/s capacity through a single GSO FSS satellite.

The design of satellite systems affects capacity and cost per Mbit/s or Gbit/s. Frequency reuse on a given satellite is achieved through multiple beams. However, reducing the beam width increases the requirement for the accuracy of pointing the satellite antennas to the earth's surface, so the larger capacity results in higher costs for controlling the beam footprints.

Table 3.6 Spectrum for use by high-density applications in FSS.

Frequency band	Region
Space to Earth	
17.3–17.7 GHz	Region 1
18.3–19.3 GHz	Region 2
19.7–20.2 GHz	All regions
39.5–40 GHz	Region 1
40–40.5 GHz	All regions
40.5–42 GHz	Region 2
47.5–47.9 GHz	Region 1
48.2–48.54 GHz	Region 1
49.44–50.2 GHz	Region 1
Earth to space	
27.5–27.82 GHz	Region 1
28.35–28.45 GHz	Region 2
28.45–28.94 GHz	All regions
28.94–29.1 GHz	Region 2 and 3
29.25–29.46 GHz	Region 2
29.46–30 GHz	All regions
48.2–50.2 GHz	Region 2

On the other hand, higher frequencies allow reducing the size of the terminals because signals can be received from smaller antennas (for example, reducing the diameter of the antenna to 60 cm for a fixed terminal). By reducing the diameter of the satellite antenna, the implementation and thus the availability of broadband satellite systems is improved. By moving to higher frequencies for future satellite systems, antenna sizes can be further reduced (as a positive aspect), but the requirement for pointing accuracy will also increase.

3.8.2.1 Example for Global Broadband Internet Access Via FSS Systems

Let's consider the design of an FSS system that should accommodate use of low-cost user terminals for large-scale deployment. In this example, we will use legacy user terminals operating in Ka-band for both transmitting and receiving. The goal is to design a system that offers low cost per bit/s.

We assume a star network topology, in which terminals are transmitting to the satellite in Ka-band, while their signal is re-transmitted to gateways in Q-band. In the opposite direction, gateways are transmitting in V-band and their signals are being re-transmitted to user terminals by using the Ka-band. The architecture for this FSS deployment for provision of satellite broadband Internet access is shown in Figure 3.13.

Appropriate FSS frequency ranges for this design in this case is the full Ka-band with the following assumptions for both directions:

- Transmission within 27.5–30.0 GHz;
- Reception within 17.3–20.2 GHz.

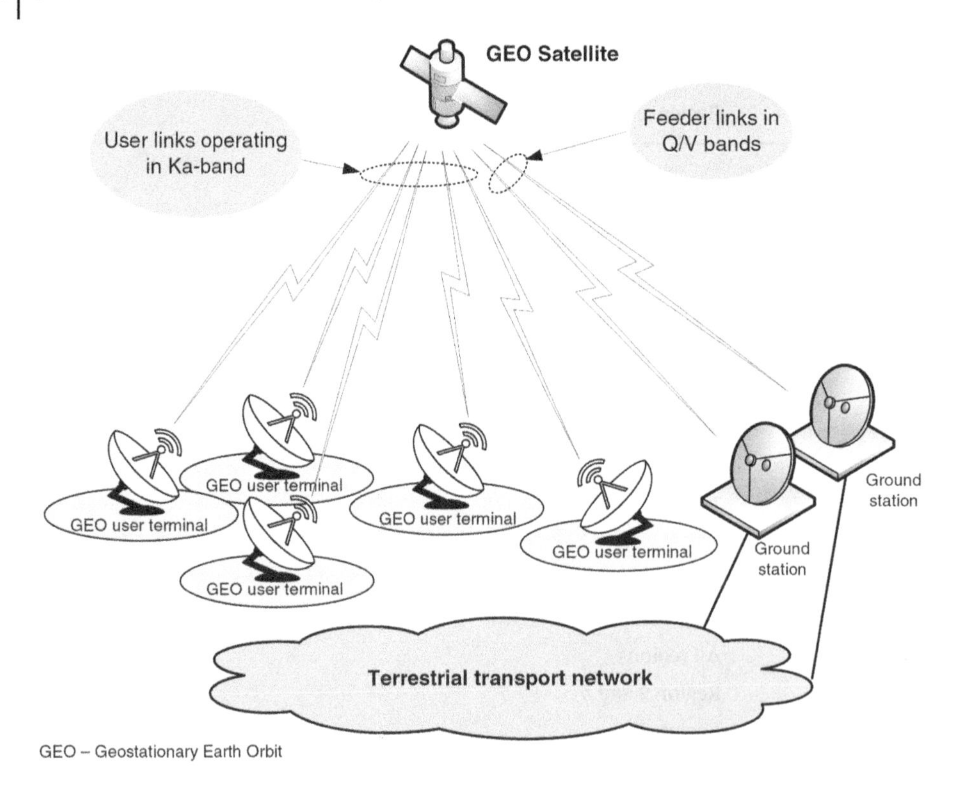

Figure 3.13 FSS architecture for broadband Internet access.

Considering the huge demand of bandwidth for this example, the gateways shall use V-band for transmission (that is feeder-to-satellite direction) and Q-band for reception (that is satellite-to-feeder direction) or more specifically by using:

- Transmission in band 47.2–50.2 GHz and 50.4–51.4 GHz (that is, V-band);
- Reception in band 37.5–40.4 GHz (that is, Q-Band).

The advances in satellite design, manufacturing, and launch service capabilities have created new possibilities for high-bandwidth connectivity around the world to provide broadband Internet access to the most remote regions.

Filings for frequency assignments to Non-Geostationary Orbit (NGSO) satellite systems composed of hundreds to thousands of satellites have been received by ITU since 2011, in particular in frequency bands allocated to the FSS or the MSS.

3.8.3 Earth Stations in Motion (ESIM)

Earth Stations in Motion (ESIM) refer to provision of reliable and high-bandwidth satellite broadband Internet services via FSS bands to moving stations. So, ESIM provide broadband Internet connectivity to platforms in motion, which include the following:

- ESIM on aircraft (aeronautical ESIM),
- ESIM on ships (maritime ESIM), and
- ESIM on land vehicles (land ESIM).

These ESIM stations are targeted to connect people on aircraft, ships, and land vehicles (e.g. trains, buses, cars). For example, when ships are at sea or ocean or an aircraft is in the air, they are far from the reach of terrestrial networks, because typical base stations in mobile networks do not cover more than approximately 30 km in radius for downlink and uplink. For moving stations, the ESIM systems can provide a continuous service with very wide or even global coverage over almost any location on Earth.

Typical bitrates in the 2020s from ESIM networks to the individual end user terminals are in the range of 100 Mbit/s, which is higher than the bitrates provided by MSS in the first two decades of the 21st century.

It is no surprise, then, that the demand for radio frequency spectrum that can be used by ESIM is increasing. For example, in the mid-2010s, there were over 20000 vessels connected via FSS, and in one decade (in the 2020s) this number increased to over 50000 [36]. Due to the increasing demand for ESIM, the ITU WRC 2019 outcome resulted in bands 17.7–19.7 GHz (space-to-Earth) and 27.5–29.5 GHz (Earth-to-space) [37], to be used by the three types of ESIM which communicate with GSO space stations using FSS. However, the ITU radio regulations (used to harmonize also the use of spectrum for satellite services) also take into account avoiding harmful interference for existing systems operating in these bands used in terrestrial and space services [38].

What about the future of ESIM? Well, the future of ESIM is establishing and maintaining a technical and regulatory framework for Ka-band, for ESIM communications with non-GSO FSS space stations on a global basis (according to the ITU WRC 2023) [39]. However, ESIM in the non-GSO mode should coexist with other spectrum users, which include GSO FSS services and others.

In cases when vessels from one country are visiting seas in another country (which is a regular case), the ESIM does not communicate with the national telecom networks in the visited country but directly with the satellite, because they are authorized for their service by their home country. However, ESIM stations should not cause any interference to any system in the visited country. Regarding the regulatory part, in cases when ESIMs are in a visited country, there can be national telecom regulations in terms of exemption from licensing for visiting ESIM terminals from another country, or (as another possibility) the national regulator/administration can provide authorization on a request ad-hoc basis.

3.8.4 Non-GSO vs. GSO Satellite Service

While the GSO satellites (which include only GEO type) are Earth-orbiting satellites placed over the Equator at an altitude of about 36000 km, non-GSO refers to all satellites that are not GSO, i.e. to all satellites that are not orbiting with the same speed as the Earth (as GSO satellites do). In that manner, non-GSO includes LEO and MEO satellites, which move across the sky while they orbit the Earth, so their position is not fixed over the Earth surface. Therefore, all non-GSO satellites require a constellation which consists of many satellites that are launched in a given constellation and have links between one another (adjacent satellites in the constellation) and some of them (or all of them) have links to ground stations on Earth for connecting to terrestrial telecom networks (e.g. to global Internet network). Multiple satellites over a given region on Earth are needed for non-GSO satellites due to their lower attitudes and hence lower footprints on the surface or Earth. The radius of the footprint is smallest for LEO and highest for GEO satellites, as shown in Figure 3.14.

GSO and non-GSO satellites differ also in velocities and Round-Trip Time (RTT), i.e. latency, to and from the Earth's surface, where the human users are located. In that manner, GSO satellites have angular velocity of 0 rad/s, which means that they do not change their position in the sky,

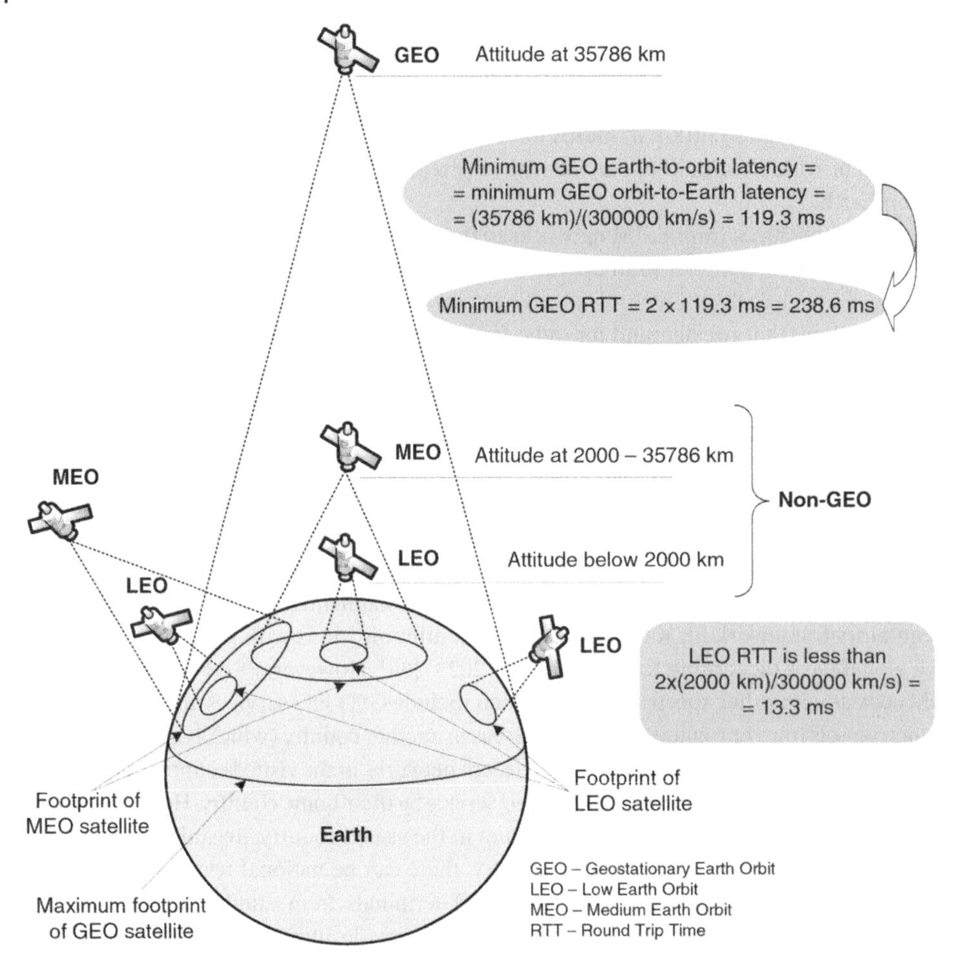

Figure 3.14 GSO vs. non-GSO satellites.

which is the main goal of selecting their orbit at an altitude of 35786 km. However, the RTT for GEO is large, so they can be used for non-real-time services (it means that the delay is not critical) such as messaging, Web, or massive IoT communication, or unidirectional video content (e.g. satellite TV broadcast); however, they are not convenient (due to large delays) for critical IoT services (e.g. demanding few ms or 10–20 ms maximum latency) or for legacy voice communication (considering that voice can accommodate maximum delays in one direction of 400 ms), and the GEO satellite introduces minimum or around 240 ms delay (depending on the position of the ground terminal, e.g. the delay is even higher closer to the edge of the footprint of the satellite), without considering the delays that certainly exist in ground networks. Non-GSO satellites are placed at much lower attitudes, especially modern constellations such as Starlink from SpaceX [40]. Considering that their altitude is several times lower than 2000 km, the LEO RTT is approximately less than 13 ms, which provides close to terrestrial experience for broadband access by end users.

Regarding the orbiting, Starlink satellites which are non-GSO (in this case LEO) positioned at an altitude of 550 km complete 70 roundtrips around the Earth during a day (24 hours), while GSO satellites complete only 1 roundtrip (of the Earth itself). The Earth stations as well as user terminals need to track the non-GSO satellites as they move across the sky. For non-GSO services, the user terminal uses Electronically Steerable Antenna (ESA) which is an array of antennas that can track the non-GSO satellites without moving the antenna itself. The ESA consists of many small identical antennas working together for transmitting and receiving the signals; they are called phased array antennas. Due to constant beam movement needed for communication with ESA on user terminal, it does not transmit signals (called blockages) unless directed via commands received from the satellite, with an aim to satisfy the Satellite Earth Stations and Systems (SESs) regulations and standards [41].

3.8.5 Regulatory and Business Aspects of Satellite Broadband

Satellite service providers and government entities are trying to meet the ever-increasing demand for broadband services everywhere, so satellites are the main and often the only option for coverage of uncovered areas of the planet, including isolated and remote areas, aircraft in the air, and ships in the seas. Satellite broadband can also be used to provide broadband coverage to underserved areas such as rural areas. In these cases, contrary to the philosophy that satellite bandwidth is more expensive than terrestrial infrastructure, satellites appear as a cost-effective upgrade to the performance of many times limited terrestrial networks (in deserts, high mountains, jungles, savannahs, steppes, etc.).

Satellite communications include orbits and spectrum that span the space over many countries in different regions. Accordingly, satellite broadband is international in nature and consequently needs harmonization at a global level. Such global harmonization of the use of satellite radio resources is carried out by the ITU Radio Regulations [42] that are updated over time to ensure interference-free operation of satellite networks through appropriate international coordination procedures.

Satellite footprint is the area of Earth illuminated by the satellite, and that does not coincide with national borders. The ITU coordination process is designed to mitigate interference between satellite networks. In that way, if a satellite service operator is licensed to use a satellite by a country that owns the satellite and has coordinated it through the ITU, no duplicate licensing requirement should be imposed for using that satellite to provide services in any other country. Such nondiscriminatory approach for domestic and nondomestic satellite service providers to have direct access to all available satellite resources and markets constitutes the so-called "open skies" policy. It aims to enable seamless access to orbital resources via globally agreed coordination and procedures, regardless of the satellite operator's country of origin.

The countries can reserve spectrum for LEO satellites (or generally, for non-GSO) in advance, which is called warehousing radio frequencies. However, to avoid reserving the spectrum and not using it, the radio regulations specify that the satellite fillings must be used within seven years with the aim that their validity does not expire. Since ITU WRC 2019, there is also a specified minimum dynamic which includes 10% of the constellation to be launched within two years and 50% within five years.

There are several big global satellite broadband providers that have launched or have plans to launch many satellites in the orbit (e.g. Starlink, OneWeb, Amazon satellites, Kepler satellite networks, and so on).

However, gaining market share is not an easy task for satellite broadband service. Why? Because satellite broadband cannot compete in terms of QoS with terrestrial broadband based on fiber transport networks (either land or undersea). Why? Because the delays are high for GEO, i.e. GSO satellites, while non-GSO satellites (which have lower and acceptable delays in today's telecommunications world) are moving in the orbit and hence require special antennas.

So, what will be the role of satellite broadband access in the near future? Will they be complementary to the competition of terrestrial fixed and mobile operators?

Well, they will be complementary, not really competition. Overall, satellites in the future can be used for the following four use cases:

- One such use case is already noted covering the unserved areas.
- Second use case is backhauling and multicasting (e.g. one-way video transmission, content delivery to local caches, and so on).
- Third use case is satellite communication on the move, which is targeted at broadband access service to moving vehicles in all cases where terrestrial mobile networks are absent (or not reliable), which includes passengers on airplanes, ships, and even trains. For example, Internet access onboard airplanes is provided to passengers via Wi-Fi onboard the plane, and the plane connects its local Wi-Fi network with the satellite network in the sky.
- Fourth use case is hybrid multiplay, which is in fact a broadband access provision to households. And this is the use case where terrestrial providers can face a competition from satellite broadband providers. However, that is not a novelty, considering that the satellite TV provisioning is working since the 20th century.

Finally, satellites have a certain lifetime (for example, until now, the lifetime of GEO satellites was about 15 years), and new satellites are constantly being launched every month. But there is a trend for new satellites to be more adaptable to change once they hit the sky, which can be achieved by using softwarization and virtualization technologies in their initial design and then in operation.

3.9 Business and Regulatory Aspects of Fixed Broadband

Broadband access is a key priority of the 21st century as the basis of digitalization of the society and digital transformation of businesses, which aims to boost economic and social growth everywhere.

The digitalization refers to moving every person and every process and everything online, that is, on the open Internet. For that purpose, we need high speeds and that is called broadband Internet. So, digitalization needs broadband Internet, and broadband Internet needs modern regulatory frameworks and strategies that are targeted to clear goals, and are feasible regarding the required funding for building high-speed networks or increasing the speeds of existing networks by updating or changing the network elements and functions.

3.9.1 Business Aspects of Future Broadband Internet

Increased adoption and use of broadband in this decade (2020s) and beyond will be driven by the extent to which bandwidth demanding services and applications will be available and affordable to all consumers on the global scale.

However, broadband development has its own issues and challenges. The first challenge is the construction of broadband infrastructure. While access networks can be fixed, mobile, or satellite, the transport networks are typically terrestrial and optical, considering the much better characteristics of fiber compared to other types of media, including metallic media or radio such as mobile, wireless, or satellite access. However, building optical transport and access networks requires capital investments in the network. The investment in broadband infrastructure has also an effect on the economy. On one side, the deployment of broadband networks initiates creation of jobs and impacts the economy by several multipliers. On the other side, there are also "spillover" externalities, which have an influence on individual end users and enterprises. For example, citizens are able to access many administrative services via open Internet, to purchase and sell goods, to socialize, to work with their coworkers on a distance (e.g. in different regions or even continents), to entertain in a personal way, to read, listen, or watch at any time any content that they like or need to consume, and so on. This is not an exhaustive list; however, the main idea is that it is becoming available to more people due to higher penetration of broadband access and its availability.

Further, the adoption of broadband within companies, especially including small and medium firms, results in a multifactor productivity gain, which on the other side contributes to Gross Domestic-Product (GDP) growth in the given country.

What is the main use of broadband? It is for open Internet access and for IPTV as a carrier grade service. In the 2020s, the majority of traffic is video traffic [43], which at the beginning of the 2020s is around 70% of the total traffic, and it is expected to be around 80% toward the end of the 2020s. However, such videos include online video content (accessed over the public Internet), including video sharing sites, (e.g. YouTube), online video on demand services (e.g. Netflix, etc.), online meeting platforms (e.g. Zoom, MS Teams, etc.), social networking platforms (e.g. Facebook, TikTok, Twitter, etc.), but also the video content of carrier grade services such as IPTV (as replacement of legacy broadcast TV service). One may easily conclude that most of the video content is accessed through the open Internet access.

Regarding the businesses, broadband is one of the main contributors to economic growth. On one side, broadband improves productivity in enterprises by making more efficient many of the business processes which transfer to use of digital channels instead of the legacy ones. For example, a telecom operator offers and sells services via digital channels (i.e. via specialized applications or through its website, in both cases accessible over the open Internet) facilitating more efficient business processes including marketing, selling, cooperation, supply chain, etc. On the other side, the deployment and continuous investments in broadband (considering that broadband means different speeds at different times, as technology evolves the meaning of broadband also evolves) accelerates the innovations for customer-oriented applications. Why? Because when all citizens have broadband access to Internet, they can choose and use new innovative applications and services based on their interests and needs. Also, there are possible new forms of commerce and financial operations and intermediation. Further, regarding the enterprises, broadband is maximizing their reach to labor including global reach when it is applicable (e.g. outsourcing the software development and support, remote customer call centers, etc.).

In the developed economies, development of future broadband access networks and infrastructures is typically commercially funded, but additional investments are required in countries with large territories and spread population. Regarding the telecom infrastructure, the broadband access network is typically the most expensive part of providing a broadband service, which was one of the main reasons for reusing existing copper-based deployments (such as twisted pairs and coaxial cables) in places where they already existed. All new fixed broadband investments are tailored toward optical access networks, either as passive (i.e. PON) or active. As with copper-based access deployments in the 20th century, the deployment of fiber access costs depends on the following:

- Distance between the customer premises and network nodes, which depends on the density of customers.
- Deployment costs, which may vary from one country to another depending on the cost of civil engineering works and availability of existing ducts for the fiber.

- Broadband access technology type, which can be based on dedicated or shared fiber (for example, shared with splitters in a PON deployment), but can also use already deployed copper access in order to reduce initial costs and speed up broadband deployment, although copper comes with poorer performance than fiber. Hence, all future greenfield deployments will be optical (for fixed broadband access), but the technology used will continue to evolve and change over the years and decades, so that end user equipment (e.g. ONT) will need to be updated in 5- to 10-year cycles.

The entire future broadband economy depends on individual bitrates that are sustainable. Thus, legacy copper solutions will in the future limit service growth and limit revenue for service providers, the speed of digitization, as well as the entire service economy that relies on ultra-broadband internet (by ultra-broadband, here, we mean bitrates higher than average at a given time).

There is a specific circle in broadband development. On one hand, with an increase in demand for a given service, capacities increase and costs decrease. On the other hand, reducing prices increase demand. As the number of users increase in a given region, the service and content providers will create platforms to or near that region, and with that will reduce the RTT between the end users and cloud servers which host the applications.

The creation of digital content on the Internet and ICT systems for their delivery to users are being innovated, developed, and produced in more advanced economies. On the other hand, the implementation of broadband networks is carried out at the local level and in the long term it is realized with local workforce, which acquires the necessary skills through training or by working on telecom networks and services. However, each country has jurisdiction over the telecommunications networks and services implemented in its territory. All countries have already established national regulators for the telecom sector, which are further responsible for providing regulation and policies for the promotion of digital services and content, where the prerequisite is the spread of broadband Internet, which in turn has the potential to contribute to economic growth based on the digitization of all spheres of people's lives.

However, for every development there is required a strategy and targets, and the number one target is making the broadband policy universal. This requires a national broadband plan in each country. Such broadband plans need to be updated over time as technologies evolve and the environment changes.

The next target is making broadband affordable to all worldwide. This is a critical step in achieving the targeted universal connectivity. Although the prices per bit/s or per GB are declining over the years, there are still affordability gaps that persist in certain countries and regions.

But what is the essential target for broadband Internet development? Well, digital skills are essential for the meaningful use of broadband and Internet-powered resources. Lack of digital literacy and digital skills, at least basic ICT skills (e.g. using a browser), is one of the main barriers to further penetration of broadband Internet access in certain developing economies.

Finally, the future target is making broadband omnipresent on Earth, either in urban areas, in a desert, high mountains, in a ship on the sea or in a plane in the air. It is not possible for fiber to be deployed everywhere on Earth including all lands and seas. So, the coverage of fixed and mobile networks has high penetration in all areas where the majority of people live (e.g. cities) and move (e.g. roads). But, for omnipresent broadband access on the planet, the terrestrial networks must be accompanied with non-terrestrial satellite networks for broadband access in order to cover uncovered areas. Such future goals, as well as the integration of terrestrial and non-terrestrial networks, which began with 5G-Advanced standardization in the mid-2020s, are expected to open many new business opportunities for various IoT services for various verticals, including agriculture, forestry, animal tracking, monitoring the transport of goods across seas and oceans, unmanned vehicles in the air (e.g. drones, aircraft), seas (e.g. unmanned ships, boats), and on land (e.g. drones, driverless trains, and driverless vehicles), and more.

3.9.2 Impact of Broadband on Economy

Fixed and mobile broadband have significant impact on the economy via digitalization, but it is not the same everywhere in all countries [44].

Where is the biggest impact of fixed broadband? Well, fixed broadband has a bigger impact in developed economies, which also have copper infrastructure (which enabled faster implementation of fixed broadband from the first decade of the 21st century onwards) and thus have higher fixed broadband penetration, so they benefit more from technology than developing economies which lacked copper infrastructure from the 20th century and hence have initially lower penetration of fixed broadband. It should be noted that the penetration of fixed broadband access is related to total number of households and companies in a given country, unlike mobile broadband access which is expressed through the number of individual users from the total number of citizens. When fixed broadband penetration is low, its economic impact is minimal. On the other hand, when fixed broadband infrastructure increases penetration it will at some point reach a critical point (typical of developed economies) when it starts to have a significant impact on the national economy (Figure 3.15).

Where is the highest impact of mobile broadband? The impact of mobile broadband is lower in countries with higher penetration of fixed broadband, because most of the broadband traffic traditionally travels through fixed access networks due to lower prices and larger data caps of fixed than mobile broadband. On the other side, in developing economies which lack fixed broadband infrastructure, mobile broadband has a higher impact on the national economy, because it is in many countries the main or only way for accessing the Internet and with that online services and applications.

The economic impact of broadband in a given country reaches its maximum when investment in telecom/ICT infrastructure approach a critical point after which the impact is at a lower level.

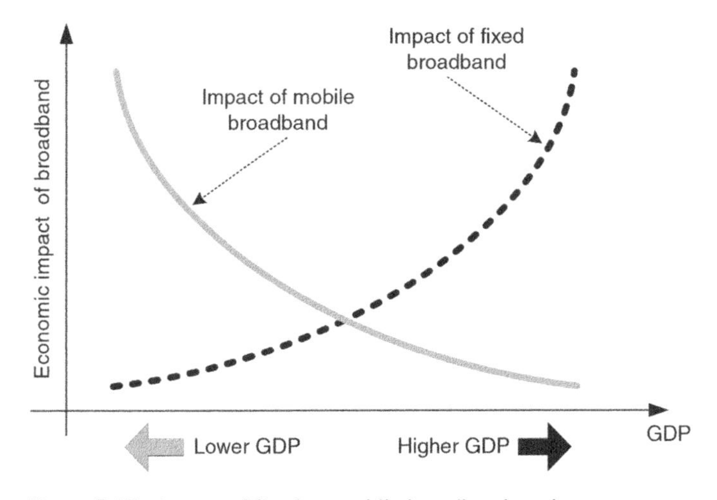

Figure 3.15 Impact of fixed vs. mobile broadband on the economy.

In general, mobile broadband has higher economic impact in countries with low fixed broadband penetration (usually those countries also have low GDP per capita), because it substitutes the absence of fixed broadband infrastructure (e.g. due to copper-based fixed infrastructure from the 20th century, which is usually used with xDSL and cable standards for speedy broadband deployments).

Global networks across different technologies and dimensions, including land, space, and undersea, combine to reach and cover every part of the world. However, the overall challenge is constantly ensuring the availability of broadband access everywhere on the globe with sufficient broadband networks capacity, continuous support to competition in digital/telecom markets, and affordability of digital/telecom services.

3.9.3 Infrastructure Sharing

The main aspect for development is terrestrial fiber infrastructure which provides high performance transport of traffic from/to different fixed and mobile access networks to/from local/national or international destinations. The needed capacity of the infrastructure depends on the average capacity used in peak hour from different access networks, in each of the two directions, upstream and downstream. Also, real end-to-end capacity of a given access network depends on the proportion of national and international traffic. If most of the traffic is established between local clients and servers of global OTT service providers (e.g. Google, Amazon, Facebook, Microsoft, Netflix, etc.), which are not hosted in the telecom operator's network or on servers in that country, then most of the traffic will be in fact international traffic. In such cases, the capacity (i.e. throughput) of interconnection points (IPXs – IP eXchange) between the broadband infrastructure and other IP networks, nationally or internationally, is needed to be proportionate to the traffic needs at the given time (considering the aggregate traffic to and from customers, including residential and business ones), with the aim of interconnection point not to become bottlenecks.

However, the volume of traffic increases over the time in parallel with the increase of number of customers and particularly with the increase in bitrates provided to customers (such increases are also motivated by national broadband strategies). This means that infrastructure is not done once for the future, but it needs upgrades of the capacities and technologies being used (e.g. for QoS, security, etc.) that may be high burden to operators in some countries. In all cases where infrastructure is not available to all operators or not affordable (e.g. due to low density of population and large territory to cover), infrastructure sharing is a possible option. However, in such cases, there are required certain regulations by the national telecom regulator. Also, regulators can use financial incentives to support infrastructure sharing; for example, by reducing universal service requirements when infrastructure is shared with market competitors [45].

The sharing of infrastructure requires efficient regulation to avoid overdesign when there is a lot of capacity that is not used, or in the opposite case, insufficient design (insufficient capacity in the infrastructure) which will result in the appearance of bottlenecks in the given parts of the infrastructure (e.g. in high-density urban areas that generate the highest traffic volumes or at interconnection points). Also, infrastructure should be compatible with sustainable business models, including Business-to-Business (B2B) and Business-to-Customer (B2C), based on competitive market dynamics.

In some cases, it is useful to enable infrastructure sharing as a result of a voluntary process and adaptation of the telecom/digital market [46]. So, infrastructure sharing should not be mandatory for all telecom market players in a given country, but market forces should encourage infrastructure sharing in order to reduce traffic transport costs and increase competition. But, in case of failure of market mechanisms, telecommunications regulators may need to enforce infrastructure sharing (for example, for broadband coverage in rural areas).

Infrastructure sharing may become more important for future technologies, given that evolution (e.g. 5G network development) is moving toward the appearance of private networks and services for fixed and mobile access, in addition to the existing networks of telecom operators that provide legacy services such as telephony, IPTV, and Video on Demand (VoD) services, broadband Internet access (i.e. data service) and business services (e.g. Ethernet or IP/MPLS services using VPNs). The emergence of a larger number of private telecommunication networks, such as for IoT services, for V2X services, Industry 4.0, etc., will certainly impact both business models and the regulation of broadband infrastructure through greater interest in its sharing. From a technical point of view, infrastructure sharing is becoming more feasible than before, considering that by the 2020s, network virtualization and softwarization have already become the main approach in designing telecommunications networks and services.

References

1. Janevski, T. (April 2019). *QoS for Fixed and Mobile Ultra-Broadband*. USA: John Wiley & Sons.
2. Janevski, T. (November 2015). *Internet Technologies for Fixed and Mobile Networks*. USA: Artech House.
3. Janevski, T. (April 2014). *NGN Architectures, Protocols and Services*. UK: John Wiley & Sons.
4. ITU-T Rec. G.992.1, "Asymmetric digital subscriber line (ADSL) transceivers", July 1999.
5. ITU-T Rec. G.992.5, "Asymmetric digital subscriber line 2 transceivers (ADSL2)– Extended bandwidth ADSL2 (ADSL2plus)", January 2009.
6. ITU hub, https://www.itu.int/hub, Last accessed in April 2023.
7. ITU-T Rec. G.9711, "Multi-gigabit fast access to subscriber terminals (MGfast) – Physical layer specification", April 2021.
8. Cablelabs, http://www.cablelabs.com/, Last accessed in April 2023.
9. Cablelabs, "DOCSIS 4.0 – Physical Layer Specification", 2020.
10. CableLabs, "DOCSIS Provisioning of EPON Specifications: DPoEv2.0 – DPoE Architecture Specification", March 2022.
11. CableLabs, "DOCSIS Provisioning of EPON Specifications: DPoEv2.0 – DPoE Metro Ethernet Forum Specification", March 2023.
12. ITU-T Rec. G.983.1, "Broadband optical access systems based on Passive Optical Networks", January 2005.
13. ITU-T Rec. G.984.1, "Gigabit-capable passive optical networks (GPON): General characteristics", March 2008.
14. ITU-T Rec. G.987, "10-Gigabit-capable passive optical network (XG-PON) systems: Definitions, abbreviations and acronyms", June 2012.
15. ITU-T Rec. G.987.1, "10-Gigabit-capable passive optical networks (XG-PON): General requirements", March 2016.
16. ITU-T Rec. G.987.2, "10-Gigabit-capable passive optical networks (XG-PON): Physical media dependent (PMD) layer specification", February 2016.
17. ITU-T Rec. G.989.3, "40-Gigabit-capable passive optical networks (NG-PON2): Transmission convergence layer specification", October 2015.

18 ITU-T Rec. G.9804.1, "Higher speed passive optical networks – Requirements", November 2019.

19 IEEE 802.3ah, "Ethernet in the First Mile", 2004.

20 IEEE 802.3av-2009, "Physical Layer Specifications and Management Parameters for 10 Gbit/s Passive Optical Networks", October 2009.

21 IEEE 802.3ca 50G-EPON Task Force, "Physical Layer Specifications and Management Parameters for 25 Gb/s, 50 Gb/s, and 100 Gb/s Passive Optical Networks", 2020.

22 IETF RFC 6370, "MPLS Transport Profile (MPLS-TP) Identifiers", September 2011.

23 ITU-T Y. 3139, "Fixed mobile convergence enhancements to support IMT-2020 based software-defined wide-area networking service", September 2022.

24 ITU-T Y.3300, "Framework of software-defined networking", June 2014.

25 MEF, Standard 70.1, "SD-WAN Service Attributes and Service Framework", November 2021.

26 ITU-T Q.3741, "Signaling requirements for SD-WAN service", July 2019.

27 ITU-T G.709/Y.1331, "Interfaces for the optical transport network", June 2020.

28 Viavi Solutions, "G.709 – The Optical Transport Network (OTN)", 2021.

29 ITU, Submarine Cable Map in the world, https://www.itu.int/, accessed in April 2023.

30 S. Mathi, "The future of undersea Internet cables", APNIC, April 2019.

31 ITU G-series Supplement 41, "Design guidelines for optical fibre submarine cable systems", February 2018.

32 European Parliament, "Security threats to undersea communications cables and infrastructure – consequences for the EU", June 2022.

33 G. Huston, "The politics of submarine cables in the Pacific", APNIC, June 2022.

34 ITU News Magazine, "Evolving satellite communications – ITU's role in a brave new world", No. 2, 2019.

35 ITU-R Report S.1782-1, "Guidelines on global broadband Internet access by fixed-satellite service systems", September 2019.

36 ITU, "Delegates reach agreement on spectrum for Earth stations in motion (ESIM)", January 2020.

37 ITU-R S.2464-0, "Operation of earth stations in motion communicating with geostationary space stations in the fixed-satellite service allocations at 17.7–19.7 GHz and 27.5–29.5 GHz", July 2019.

38 ITU, "Satellite issues: Earth stations in motion (ESIM)", March 2022.

39 ITU, "WRC-23: International regulation of satellite services", February 2023.

40 L. M Langiewicz and D. R. Novotny, "Planar Arrays Communicating With Non-GSO Satellites", SpaceX, March 2023.

41 ETSI EN 303 980 V1.3.1, "Satellite Earth Stations and Systems (SES); Fixed and in-motion Earth Stations communicating with non-geostationary satellite systems (NEST) in the 11 GHz to 14 GHz frequency bands; Harmonized Standard for access to radio spectrum", October 2022.

42 ITU-R, "Radio Regulations", edition 2020.

43 Ericsson, "Ericsson Mobility Report", 2022.

44 ITU, "How broadband, digitization and ICT regulation impact the global economy – Global econometric modeling", November 2020.

45 Davide Strusani and Georges V. Houngbonon, "Accelerating Digital Connectivity Through Infrastructure Sharing", International Finance Corporation – World Bank Group, Note 79, February 2020.

46 ITU Broadband Commission, "The State of Broadband 2022: Accelerating broadband for new realities", September 2022.

4

Mobile Broadband

Mobile broadband is available in most countries of the world, and in some developing economies it is the only way to access telecommunications services. What does mobile broadband refer to? In the 2020s, it refers to bitrates (i.e. speeds) of tens of Mbit/s or hundreds of Mbit/s (and more) per individual end user. In the 2010s, mobile broadband referred to bitrates on the order of Mbit/s up to a few 10s of Mbit/s toward the end of that decade. Mobile broadband actually started in the 2000s with the then-appearing 3rd generation mobile systems, with bitrates of the order of 100 kbit/s. Based on the development so far, it can be easily concluded that the term mobile broadband has a relative meaning, similar to fixed broadband, so that what is mobile broadband today may be considered narrowband in 1–2 decades in the future [1], e.g. in the 6G era and beyond.

Nowadays, mobile networks are largely used for access to open Internet, which is usually referred to as "mobile data." But, how did it start?

In fact, the Internet and mobile networks both started separately, each with different targets at the beginning. Internet started as a research project in the USA in the 1960s and became a globally opened network in 1995 when it stopped being a project and started to be what it is today [2]. The appearance of the Web in the early 1990s made the final step of Internet technologies to become the telecom packet-switching "choice" since the end of the 1990s. However, the Internet was not built for mobile networks, because Internet Protocol (IP) initially appeared in 1981 (as a standard) when there were no mobile networks, so the Internet was built for fixed access networks such as Ethernet Local Area Networks (LANs). What does it mean?

It means that the Internet was developed for fixed networks only, to which are attached fixed hosts (i.e. computers) which are not mobile (e.g. connected to LANs such as Ethernet). There, each IP host sends IP packets to a destination IP host by using the IP address of the destination host (of course, when URL is used it is first resolved into an IP address of the destination server via DNS, because hosts on the Internet and IP networks communicate only on the basis of IP addresses; they are either IPv4 or IPv6). Then, where is the problem here with IP and mobility of the end user's devices?

Well, to remind us, the IP address has two parts: one identifies the IP network (the network ID) and the other identifies the network interface on a given host (e.g. computer, smartphone, and so on) attached to the IP network and that part is called host ID. Then, both parts together (network ID and host ID) in fact give the whole IP address. According to the Internet routing principles, network ID is the same for all hosts connected to the same IP network, while the host ID is different for different hosts connected onto the same IP network. What is the problem regarding the mobility of hosts?

When a host moves from one IP network (identified with one network ID) to another IP network (identified normally with another network ID), it means that the host must change the IP address

Future Fixed and Mobile Broadband Internet, Clouds, and IoT/AI, First Edition. Toni Janevski.

on its network interface when moving from one IP network to another one. And, this changing of an IP address is in fact the problem here. Why?

Because IP addresses (both IPv4 and IPv6) have a dual role. What are those? Well, one role is to identify the location of the host's network interface in the IP network (and globally on the Internet if the IP address is a public one), that is locator role. Due to this role the IP address must change when the device is changing the IP network. On the other side, the IP address internally within the host identifies the network interface to the application on the top (e.g. to the Web browser, Viber application, email application, and so on). Then, when the IP address will change (due to mobility of the host from one IP network to another one), the established connection from the application (toward the remote end) will be terminated and should be established again because IP address is part of the definition of the "socket" between application on the top and TCP (Transmission Control Protocol)/IP or UDP (User Datagram Protocol)/IP protocols implemented in the operating system of the host. The result is that IP connection is broken when changing the IP network, and in fact we have no mobility. Why?

Because mobility (in mobile networks) means that a given call/connection/session is maintained all the time from its start until its end regardless of the mobility of the end user (i.e. mobile device) during the whole duration of that connection. So, the IP has a problem regarding the mobility, and there are several possible solutions for it. What are they?

They are standards, from 3GPP and IETF, which handle the mobility management on different protocol layers. Each of the two Standards Development Organizations (SDOs) has developed its own mobility management standards, which started with GSM/GPRS (by 3GPP) and with mobile IP protocols (standardized by IETF) [2]. In fact, the mobility was first implemented successfully in the digital mobile world in 2G, where GSM is the most successful technology so far (from the 2G mobile standards "family"), which appeared in the 1990s. However, mobility in GSM was targeted to circuit-switching networks, not to IP, and was implemented on lower protocol layers (i.e. layers 1 and 2).

So, the mobile technologies started to spread globally in the 1990s in parallel with the public Internet, however, initially not related to one another. Due to its obvious dominance on the mobile market (e.g. GSM) and data services (e.g. Internet network) they have started to converge – that is, mobile network included IP connectivity (starting with 2.5G, such as General Packet Radio Service GPRS in GSM networks). Then evolution to 3G UMTS/HSPA, then to 4G LTE/LTE Advanced/LTE Advanced Pro, and recently to 5G NR mobile networks and beyond. The development of 3GPP mobile networks in the 21st century is shown in Figure 4.1.

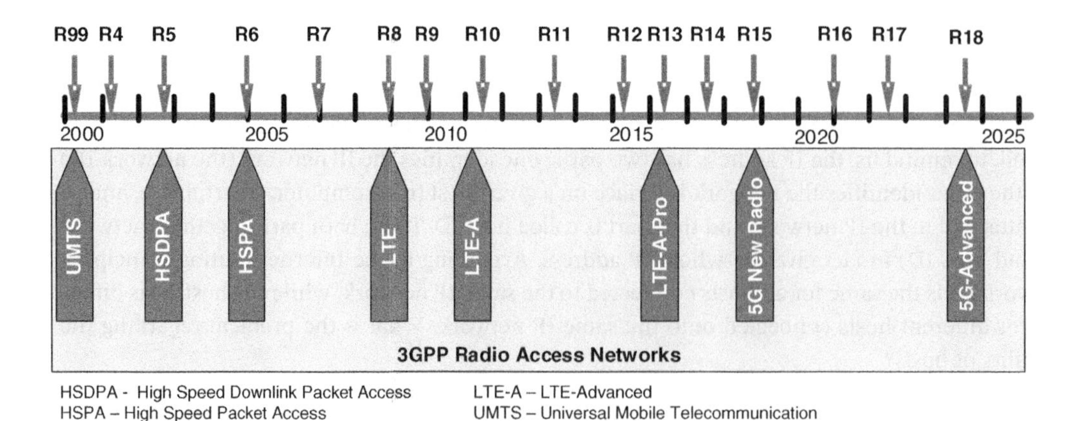

Figure 4.1 Development of 3GPP mobile networks in 21st century: From 3G UMTS to 5G-Advanced.

What contributed to the rise of mobile broadband? Well, Internet connectivity contributed to the emergence and growth of mobile broadband access in the 2000s (with 3G broadband), the 2010s (with 4G broadband), and now in the 2020s (with 5G broadband). Why is that? The reason is that mobile broadband is required for mobile Internet access in the first place (less so for other services), which in the 2020s is largely used for video traffic (e.g. video streaming and video on demand).

On the other hand, mobile networks have also changed the "face" of the Internet by making many services and applications mobile-first, which is quite different from the past (a few decades ago) when the Internet was originally created only for fixed hosts.

4.1 Mobile Broadband Evolution (LTE/LTE Advanced Pro)

The evolution of mobile technologies goes through so-called generations, as shown in Table 4.1, where each successive generation improves bitrates (increases them) and latency (decreases them) as the main performance indicators for most services (of course, in all cases availability, i.e. mobile coverage, is also important). In summary, the main characteristics of mobile generations include:

- 1G (First Generation): It consisted of analog mobile systems, based mainly on Frequency Division Multiple Access (FDMA) in the radio access part, without global roaming, used in the 1980s.
- 2G (Second Generation): This was the first generation of digital mobile systems, based on Time Division Multiple Access (TDMA) and FDMA (for example, GSM is based on FDMA/TDMA). It is Circuit-Switched (CS), with global roaming, where telephony and SMS are main services, started in the 1990s.
- 3G (Third Generation): It is the first generation of mobile systems that included a standard Packet Switching (PS) domain (for Internet access and MMS) in parallel with CS (for voice and SMS). The radio interface was based on WCDMA (broadband code division multiple access) used together with TDMA/FDMA started in the early 2000s. ITU specified the requirements for 3G technologies in umbrella specification called IMT-2000 (IMT stands for International Mobile Communication).
- 4G (Fourth Generation): This is a generation which is by default all-IP in both access and core parts, where the access radio part is based on Orthogonal Frequency Division Multiple Access (OFDMA). The ITU umbrella specification which defined the requirements for 4G technologies is called IMT-Advanced, and by 2012 two technologies were accepted as 4G, LTE Advanced from 3GPP and Mobile WiMAX 2.0 from IEEE [3]. However, by the end of the 2010s, LTE/LTE Advanced took over the whole 4G mobile market, and sending the WiMAX into the history after it was limited to a few niche markets globally.

Table 4.1 Evolution of digital mobile networks, from 2G to 6G.

	2G	3G	4G	5G	6G
Approximate deployment dates	1990–2000	2000–2010	2010–2020	2020–2030	2030–2040
Approximate average individual user speeds	10s to 100s of Kbit/s	Mbit/s to few 10s of Mbit/s	10s to few 100s of Mbit/s	100s of Mbit/s to few Gbit/s	Gbit/s to tens of Gbit/s
Approximate latency in mobile network only	100s of ms	Few 100s of ms	Few 10s of ms	<10 ms	<1 ms

- 5G (Fifth Generation): It is the generation of mobile systems that is massively deployed everywhere in 2020s. The requirements for 5G were specified in ITU's umbrella specification called IMT-2020, which defined increased data rates of 4G by more than 10 times and decreased latency immobile network to only a few milliseconds. Some initial 5G deployments were made in 2019 and 2020, but the official recognition of technologies as 5G was done by ITU-R Working Party 5D in 2021 (for 3GPP 5G SRIT, 3GPP 5G RIT, and 5Gi) and 2022 (for ETSI DECT). The 5G is accompanied by many new spectrum allocations, thus providing opportunities for many new services targeted to legacy and new services as well as to use cases in different verticals (that are initially non-ICT sectors which use mobile networks in their business and/or operations). The 5G decade is the 2020s.
- 6G (Sixth Generation): It refers to the next mobile generation that is expected to appear by 2030 and will mark the 2030s. The requirements for the next generation will be defined by the IMT-2030 specification by ITU which will be known by 2030 according to the timeline [4]. One may expect improvement of performances in terms of latency, data rates, Spectral Efficiency (SE), and capacity with additional new spectrum, mobility, and number of connected smart mobile devices, with omnipresent terrestrial and non-terrestrial (i.e. satellite) connectivity and high level of Artificial Intelligence/Machine Learning (AI/ML) use cases in network operations and service provisioning [5].

4.1.1 E-UTRAN: 4G Radio Access Network from 3GPP

The main 4G technology is LTE/LTE Advanced from 3GPP, which was implemented almost in all countries worldwide by the end of the 2010s. Also, 4G Radio Interface Technology (RIT) is part of the 5G standard from 3GPP called Set of Radio Interface Technologies (SRIT), together with 5G New Radio (5G NR) as the other RIT in the set. The 4G architecture is completely packet-based (i.e. IP-based), however, it evolved from the packet-switching parts of the 3G and 2G mobile systems from 3GPP, so it is called Evolved Packet System (EPS). The EPS network architecture consists of two main parts:

- Evolved Universal Terrestrial Radio Access Network (E-UTRAN) and
- Evolved Packet Core (EPC).

E-UTRAN consists of 4G base stations, which are called eNodeB (or eNB, in short), and interfaces between them and with the core. In E-UTRAN, for the first time (compared to previous mobile generations in respect to 4G), the 3GPP introduces direct interface between base stations called X2 (Figure 4.2). There are no other network entities in LTE Radio Access Network (LTE RAN) called E-UTRAN ("translated", it means it is based on the evolution of UTRAN which is 3G RAN), so eNB is directly connected to LTE core network nodes, that is, to Evolved Packet Core (EPC). Direct connections between eNBs are critical for fast handovers, which were required for real-time services over IP (such as Voice over LTE – VoLTE) in LTE's all-IP environment, including its access and core network parts.

While in previous 3G mobile networks from 3GPP there was a centralized node in the UTRAN, the Radio Network Controller (RNC), with the evolution of the 3GPP mobile network architecture toward 4G architecture, the RNC was removed and radio resource management was fully given to base stations (eNB in 4G). Having only two tiers in the architecture (eNBs on one side and centralized core network nodes on the other) we get a flat architecture. But, why is flat architecture needed for mobile networks?

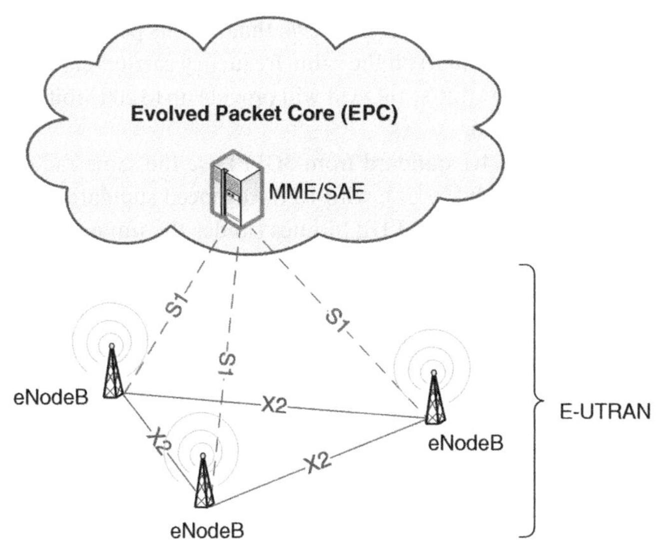

Figure 4.2 4G Evolved UTRAN (E-UTRAN).

Well, because lower number of nodes on the path of IP packets are required to decrease the delay budget (i.e. total delay due to various causes), such as to fit into the end-to-end delay budget for voice when it is carried in a form of VoIP (i.e. VoLTE in 4G LTE/LTE-A mobile networks). Lower delay in the mobile network also contributes to all other services, including real-time and critical as well as non-real-time (e.g. Web services).

As for the radio access technology, both LTE and LTE Advanced use OFDMA in the radio interface in the Downlink (DL), and single carrier FDMA (SC-FDMA) in the uplink (UL).

A generic LTE radio frame for Frequency Division Duplex (FDD) and Time Division Duplex (TDD) has duration of 10 ms and consists of 20 slots with a slot duration of 0.5 ms, where:

- Two adjacent slots form one subframe with a length of 1 ms.
- The so-called Resource Blocks (RBs) comprise either 12 subcarriers with a subcarrier bandwidth of 15 kHz, or 24 subcarriers with a subcarrier bandwidth of 7.5 kHz (each in a time slot with duration of 0.5 ms).

LTE uses different sizes of frequency carriers (1.4, 3, 5, 10, 15, and 20 MHz), and it can operate in both FDD and TDD modes. Most of the deployments are based on FDD in paired spectrum. There, FDD uses separate frequency carriers for DL and UL, typically bots with the same carrier widths, where UL is the one on the lower band of the two in a given pair. However, TDD mode is important for enabling deployments where paired spectrum is unavailable. Also, it allows LTE to be able to operate more efficiently in terms of spectrum usage by allocation of time slots in DL and UL direction upon traffic need in the RAN.

Other innovations in the LTE radio interface (given the 3GPP standardization) to increase the spectral efficiency (that is, providing more bit/s per Hertz) are the introduction of Multiple Input Multiple Output (MIMO) and higher modulation and coding schemes that are resulting in the transmission of more bits per transmitted symbol over the radio interface. For example, 64QAM modulation (64 = 2^6, hence 64QAM transmits 6 bits on one symbol over the radio interface) has approximately 50% more bandwidth than 16QAM modulation (16 = 2^4, that is 4 bits per symbol, hence 6 bits per symbol is 50% more than 4 bits per symbol) on the same frequency carrier (e.g. for 20 MHz frequency carrier with 64QAM results in 300 Mbit/s, 16QAM will provide up to 200 Mbit/s, while QAM modulation provides only up to 75 Mbit/s).

Overall, initial LTE and LTE Advanced (the real 4G standard from 3GPP) use the same radio interface and have the same spectrum efficiency (in bit/s/Hz), with LTE Advanced standardized Carrier Aggregation (CA) providing the required 4G DL and UL bitrates (under the umbrella of IMT-Advanced ITU for 4G).

4.1.2 Evolved Packet Core (EPC)

The 4G core network standardized by 3GPP is EPC [1–3]. It consists of several network nodes, where the control and user data plane are further separated (than in 3G core networks). The user data traffic from E-UTRAN toward the Internet/IP and vice versa traverses through Serving Gateway (S-GW) and Packet Data Network Gateway (PDN-GW, shortly P-GW), similarly to SGSN (Serving GPRS Support Node) and GGSN (Gateway GPRS Support Node) that were standardized for GPRS and UMTS in 2G and 3G, respectively. So, S-GW and P-GW are the two main gateways regarding the user traffic. The EPC network elements are shown in Figure 4.3.

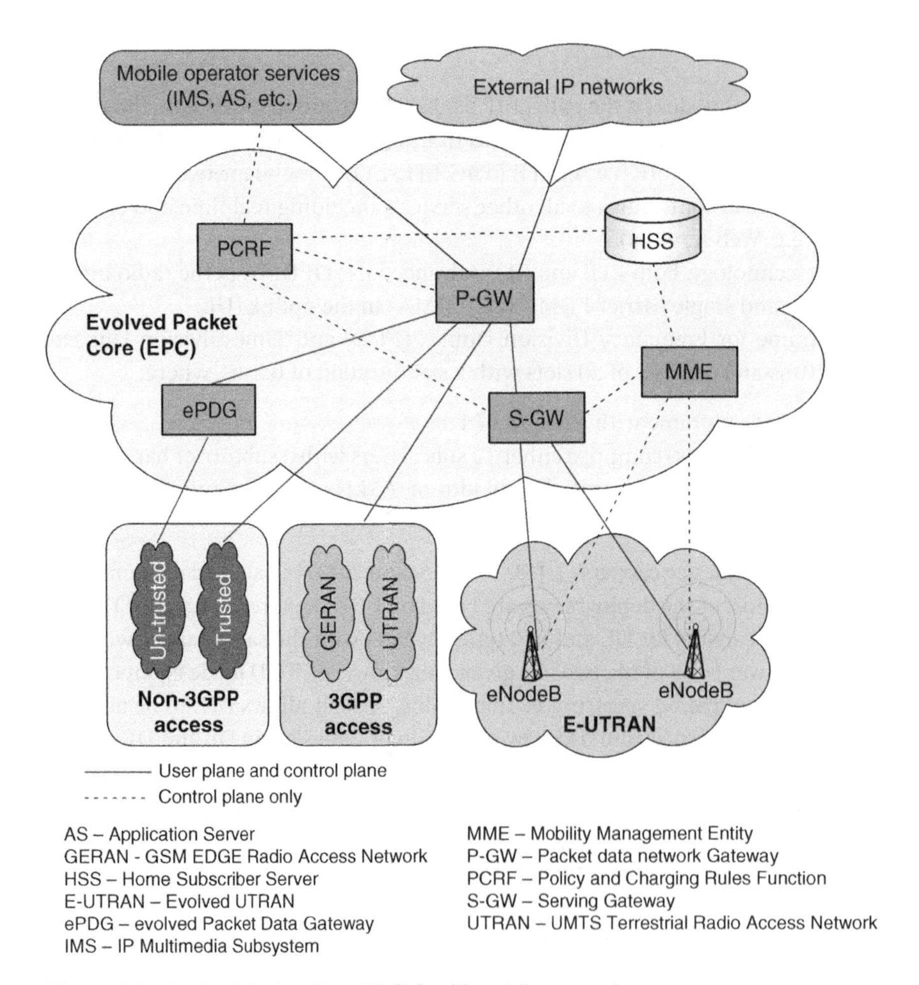

Figure 4.3 Evolved Packet Core (EPC) for 4G mobile networks.

What is the main node regarding the mobility control? It is a Mobility Management Entity (MME), which performs mobility-related signaling. It operates entirely in the Control Plane (CP), so no user traffic passes through the MME. Before 4G, mobility control was implemented in the same network nodes that also managed user traffic, but since 4G the separation of User Plane (UP) and CP has started in a more effective way. Such division continues further in 5G mobile networks and aims to provide better flexibility of mobile Network Functions (NFs) to implement new network features and new services.

Another control node or element in the EPC is the Home Subscriber Server (HSS). It is the main mobile user database, which contains user-related information such as subscriber's profile (e.g. available services to that mobile user), and user authentication information.

However, the main node in EPC in terms of QoS in the mobile network is the Policy and Charging Rules Function (PCRF). It is designed as a software-based network node that is responsible for Policy and Charging Control (PCC). PCRF aims to detect service flows and determine policy rules for them in real time, while implementing an appropriate charging policy for such service in a given mobile network.

For communication between different nodes in EPC, between EPC and E-UTRAN and between eNodeBs (the only nodes in E-UTRAN), appropriate protocols are required. What protocols are used in 4G LTE networks?

All communication between all network nodes is based on SCTP/IP or UDP/IP communication. Stream Control Transmission Protocol (SCTP) is actually a much better version of TCP (typically used in end user hosts, at least before QUIC), because it provides very high reliability (ability to use multiple different paths between two communicating network nodes, with so-called multi-homing and multi-streaming functionalities) and is therefore mainly used for the transmission of CP traffic (i.e. signaling traffic) between network nodes in the EPS (which includes E-UTRAN and EPC) [2]. However, GPRS Tunneling Protocol (GTP) runs over SCTP/IP (for control traffic) or UDP/IP (used for UP traffic, i.e. traffic originating/terminating from/to mobile end terminals), as shown in Figure 4.4.

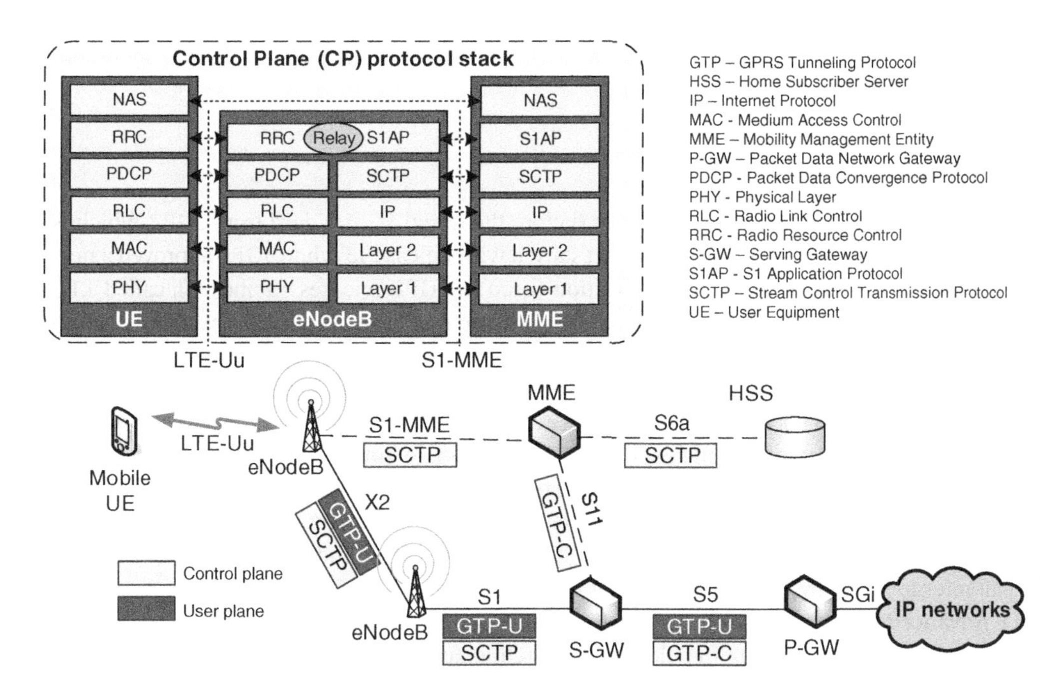

Figure 4.4 Control plane (CP) protocol stack in 4G.

GTP was originally standardized by 3GPP, not the IETF. While the IETF standardizes all the major Internet technologies used today in the telecommunications/ICT world (e.g. IPv4, IPv6, TCP, SCTP, UDP, DNS, and so on) [2], the tunneling protocol used in mobile networks was standardized by 3GPP in the GPRS era (i.e. 2.5G era) and the name has stuck ever since. With GTP, 3GPP actually "keeps" control of how traffic travels between different network nodes in a mobile network (including access and core parts), especially given the fact that end users are mobile and need to quickly switch over established paths in the mobile network to/from a given user device. Also, GTP is used in both UP and CP. So there are two main versions of GTP, one is the GTP-U version (for user traffic) and the other is GTP-C (for control traffic).

What does the use of GTP mean between each pair of nodes in 4G mobile networks? This means that IP packets from different mobile applications do not travel independently in the network, but they are tunneled from one node to another in a well-established hierarchy for the two main planes – CP and UP. So, all traffic is aggregated (e.g. by traffic type, when QoS is provided by class and not by flow) and tunneled into the mobile network through many "tunnels" that are concatenated end to end, where the part that belongs of the 4G mobile network (i.e. EPS, made of EPC and E-UTRAN) is the part that travels through the mobile network.

The same EPC is used for all versions of 4G LTE, including LTE Advanced and LTE Advanced Pro. It is also used for initial 5G deployments that are based on 5G base stations connected to existing EPCs of mobile operators (with new standardized interfaces for such purpose). All this increases the importance of EPC at present and in the near future.

As for services, the evolution of 3G to 4G also resulted in the definition of a common IP Multimedia System (IMS), which was initially standardized by 3GPP in Release 8. IMS is the main part in LTE networks for all services that require signaling (e.g. VoLTE is the main service using IMS). However, IMS was standardized by 3GPP, but was made to be independent access so that the same system is used for telecommunications services over both fixed and mobile access networks.

This evolution (for 4G LTE) has in the long run resulted in reduced CAPEX and OPEX (for mobile telecom operators) compared to previous mobile systems (such as 3G), as all traffic is carried over the same all-IP mobile network (no need for parallel operation of circuit-switched and packet-switched network parts), thus providing higher levels of efficiency in the design, operation, and maintenance of the mobile network.

4.1.3 LTE Advanced Pro

The further development of the 4G standards (before the initial 5G standard from 3GPP, which is in Release 15) was called LTE Advanced Pro. It started with Release 13 when 3GPP approved a new LTE marker for the LTE Advanced specification (used in 3GPP releases 13 and 14), called LTE Advanced Pro. As the name suggests, it is completely based on LTE Advanced, but with some additional features. What does the term "pro" mean in its name?

Well, 3GPP wanted to indicate that the latest version of LTE addresses new markets (new verticals, that is, segments that did not use mobile networks in the past, such as industry, transport, etc.) as well as new functionalities to improve efficiency. So, what are the main innovations in LTE Advanced Pro?

LTE Advanced Pro, besides higher bitrates, also provides additional support for the emerging IoT and enhanced Machine Type Communication (eMTC) via the LTE mobile networks. Also, it provides Narrow Band Internet of Things (NB-IoT) targeted for the low end of the mobile IoT market (we will refer to IoT in more detail in Chapter 6).

LTE Advanced Pro also focuses on device-to-device (D2D) communication, which is targeted for public safety as well as building mobile network architecture for support of emergency services. Also, that contributes to the development of vehicle-to-everything (V2X) communication, which increases its importance starting from LTE Advanced Pro and then continues in 5G specifications and beyond.

Considering the dedication to new verticals, LTE Advanced Pro aims to introduce new delay critical services, which again further continue their development in 5G standards. Hence, the QoS framework developed further in LTE-A Pro on the work completed on LTE Advanced before.

It is also important that LTE has entered a new segment with LTE-A Pro toward the use of the standardized LTE technology in the unlicensed spectrum (on 5 GHz) for mobile data traffic offload in hotspot areas or private use (license-free) of the technology, a process that continued further.

The LTE-A Pro uses the same LTE carriers, 1.4, 3, 5, 10, 15, and 20 MHz (of course, wider carrier gives proportionally higher bitrate). So, LTE Advanced has higher bitrates (than LTE only) due to the possibility for frequency CA on the protocol layer 2 (i.e. below the IP layer) of up to five non-continuators frequency carriers (each with size up to 20 MHz). Then, the LTE Advanced Pro provides further enhancement by allowing CA of up to 32 frequency carriers, i.e. Component Carriers (CCs) (each carrier up to 20 MHz), which results in the maximum use of 640 MHz (= 32 × 20 MHz) carrier bandwidth in total. However, the disadvantage is that many mobile operators may lack enough spectrum to take advantage of the maximum possible aggregation provided by LTE-A Pro.

Overall, 4G LTE/LTE-A/LTE-A Pro gives higher bitrates and lower delays in the evolution from LTE to LTE-A-Pro, and all that results in a better experience when using mobile services than before. The 4G is part of the 5G SRIT standard from 3GPP, and hence it remains important in the 5G era. However, the 5G RIT standard from 3GPP uses novel 5G technologies in both RAN and core network parts.

4.2 5G New Radio

The newest mobile generation is 5G (Fifth Generation). Similar to the development of IMT-2000 for 3G and IMT-Advanced for 4G, the ITU-R (ITU Radiocommunication sector) has also created an umbrella called IMT-2020 for 5G mobile systems [6]. Each such umbrella for each newer mobile generation included stricter requirements than the previous, thus ensuring continuing progress in mobile systems. In that manner, comparison of requirements in IMT-Advanced (for 4G) and IMT-2020 (for 5G) is given in Table 4.2.

Table 4.2 IMT-2020/5G vs. IMT-advanced/4G.

	IMT-advanced	**IMT-2020**
Minimum peak bitrate	Downlink: 1 Gbit/s Uplink: 0.05 Gbit/s	Downlink: 20 Gbit/s Uplink: 10 Gbit/s
Bitrate experienced by individual mobile device	10 Mbit/s	100 Mbit/s
Peak spectral efficiency	Downlink: 15 bit/s/Hz Uplink: 6.75 bit/s/Hz	Downlink: 30 bit/s/Hz Uplink: 15 bit/s/Hz
Mobility	350 km/h	500 km/h
User plane latency	10 ms	1 ms
Connection density	100 thousand devices per square kilometer	1 million devices per square kilometer
Traffic capacity	0.1 Mbit/s/sq. m.	10 Mbit/s/sq. m. in hot spots

Table 4.2 shows that the requirements set in IMT-2020 for 5G systems is aimed at a significant improvement over IMT-Advanced (i.e. 4G mobile systems). For example, IMT-2020 set the requirement for the minimum supported bitrate in the DL to be 20 Gbit/s and in the UL to be 10 Gbit/s, while the bitrate experienced by individual mobile users is 100 Mbit/s or higher. Also, the connection density requirements in IMT-2020 increase tenfold compared to IMT-Advanced, set to support a minimum of 1 million connections per square kilometer in 5G systems (this includes smartphones, IoT devices and all other IMT-2020/5G devices). On the other hand, the traffic capacity is required to be 100 times higher in 5G systems than in 4G systems. Also, an important requirement in IMT-2020 is support for UP latency up to 1 ms, which is required for the intended critical services in IMT-2020 (i.e. 5G) systems.

The main 5G/IMT-2020 use cases include the following:

- Enhanced Mobile Broadband (eMBB) – aims to provide services such as enhanced indoor and outdoor broadband Internet access, including enterprise collaboration, augmented and virtual reality, enhanced Fixed Wireless Access (FWA) services over 5G, mobile TV, etc. This mainly refers to mobile access to Internet services, as well as to mobile voice and TV services.
- Massive Machine-Type Communications (mMTC) – intended for use cases based on massive deployment of IoT devices which do not have strict requirements on QoS parameters such as delay; examples are asset tracking, smart agriculture, smart home, non-critical smart city services, smart utilities, energy monitoring, and remote monitoring.
- UltraReliable Low-Latency Communications (URLLC) – targeted to performance critical use cases such as autonomous vehicles, smart grids, remote patient monitoring and telehealth, and industrial automation.

Some initial 5G deployments were made in 2019 and 2020, but the official recognition of technologies as 5G is done by ITU-R working parties in 2021 (for 3GPP 5G SRIT, 3GPP 5G RIT, and 5Gi) and 2022 (for ETSI DECT).

Although there are four accepted IMT-2020 technologies, the main 5G standards accepted by the industry and telecom operators are in fact from 3GPP, which consists of two separate and independent submissions:

- 5G radio interface technology (5G-RIT): that is, NR submitted as a RIT proposal for IMT-2020.
- 5G set of radio interface technologies (5G-SRIT): in this case, the proposal for IMT-2020 includes a set of two RITs, which include NR (5G RIT) and E-UTRA/LTE (4G RIT).

The 3GPP standardization of 5G started with Release 15 and then continued with Release 16, and ongoing Releases 17, 18, and beyond [7]. The initial 5G deployments in the 2020s are based on 5G-SRIT (based on both 5G RIT and 4G in access networks connected to 4G EPC core), while 5G-RIT is becoming dominant in the mid and late 2020s (with 5G and 4G base stations in radio access network, connected to 5G core). So, one may say that the main difference between 5G RIT and 5G SRIT is core network, and correspondingly interfaces toward base stations.

4.2.1 5G New Radio (NR) Characteristics

5G mobile networks have an NR interface called 5G NR. However, it is not completely new because it evolves from the LTE radio interface. In that way, both LTE and NR use Orthogonal Frequency Division Multiplexing (OFDM) modulation, which divides available time resources into 10 ms frames with 1 ms subframes. In addition, subframes are further divided into slots and symbols, where the combination of one OFDM symbol and one carrier constitutes the smallest physical resource in NR.

While with LTE the symbol duration and Subcarrier Spacing (SCS) are fixed, with NR it is possible to configure different so-called OFDM numerologies on a per-frame basis, i.e. each subframe is independent and can be characterized by a different numerology. Scalable OFDM numerology with carrier spacing scaling includes:

- LTE supports carrier bands up to 20 MHz with mostly fixed OFDM numerology – 15 kHz SCS.
- NR offers scalable OFDM numerology to support a variety of spectrum bands and deployment models. It can operate also in mmWave bands with wide channel widths (hundreds of MHz) and the OFDM SCS should be able to scale appropriately so that the Fast Fourier Transform (FFT) complexity does not increase exponentially for wider bandwidths.

The numerologies provide the possibility to implement different 5G services with a single radio access technology - RAT (e.g. eMBB, mMTC, and URLLC services). For example, a shorter OFDM symbol duration combined with larger SCS can be used for high data rate and Low Latency (LL) traffic (e.g. eMBB or URLLC). On the other side, smaller SCS can be used for narrowband low-frequency communications (e.g. mMTC). Figure 4.5 shows the 5G NR timeslot with different possible SCS.

The finest granularity of 5G NR resource allocation is Resource Element (RE). A single RE uses spectrum of single SCS. Then, 12 such REs form a so-called Physical Resource Block (PRB). For example, SCS = 15 kHz (numerology = 0) gives a spectrum for PRB of 12×15 kHz = 180 kHz, SCS = 30 kHz (numerology = 1) gives a spectrum for PRB of 12×30 kHz = 360 kHz, and so on for

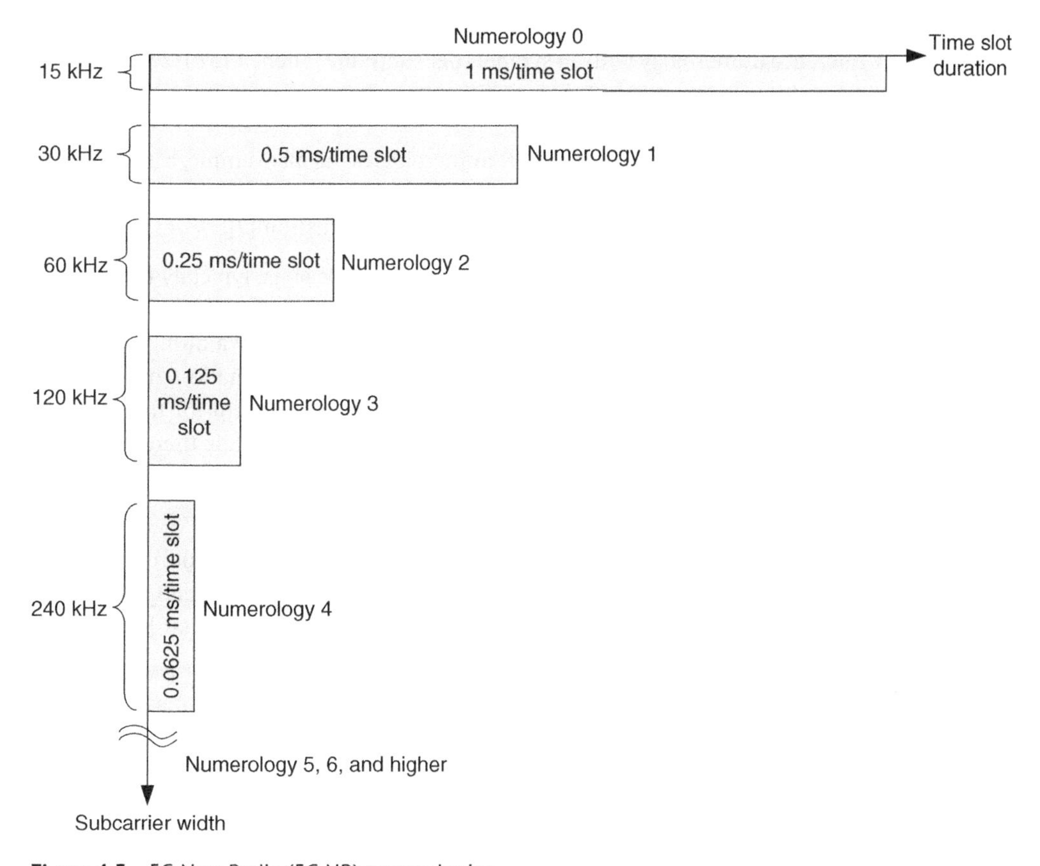

Figure 4.5 5G New Radio (5G NR) numerologies.

Table 4.3 NR numerologies and physical resource blocks (PRBs).

NR numerology (μ)	Distance between subcarriers $\Delta f = 2^{\mu} \cdot 15$ (kHz)	Spectrum used by NR physical resource block (PRB) (kHz)
0	15	12×15 = 180
1	30	12×30 = 360
2	60	12×60 = 720
3	120	12×120 = 1440
4	240	12×240 = 2880
5	480	12×480 = 5760
6	960	12×960 = 11520

other numerologies as given in Table 4.3 (number of numerologies in 5G NR is not fixed, so each new 3GPP release may add new numerologies besides the existing ones at a given time).

In 5G NR, similar to 4G LTE, the radio frame has a time length of 10 ms, consisting of 10 subframes of 1 ms each. However, unlike LTE which has a Fixed Carrier Spacing (FCS) of 15 kHz, NR supports scalable numerology for more flexible deployments covering a wide range of services and carrier frequencies. One should note that SCSs of 15, 30, and 60 kHz apply to carrier frequencies of 6 GHz or lower (below 6 GHz), while SCS values of 60, 120, 240, 480, and 960 kHz are applicable to carrier frequencies above 6 GHz (in practice, that is above 24 GHz). The 1 ms subframe duration is based on a 15 kHz reference numerology with 14 symbols per subframe. Then, a 15 kHz SCS results in a slot duration of 1 ms/slot (Figure 4.5), a 30 kHz SCS results in 0.5 ms/slot, and so on, a 240 kHz SCS results in 0.0625 ms/slot, etc.

However, there are mappings between different NR numerologies. As an example, a 30 kHz SCS has a slot duration of 0.5 ms, which can be mapped to two slots (0.25 ms each) for a 60 kHz SCS. In addition to slots, NR frame structure supports slot and mini-slot aggregation [7]:

- Aggregation of slots refers to the transmission using two or multiple slots, typically for a wider reach (e.g. cells in rural areas with a big radius).
- Mini-slots refer to the special case in which the NR transmission spans only a number of symbols less than the actual number of 14 OFDM symbols per slot. The smallest mini-slot is 1-symbol. This way, it is possible to have more flexible resource management for the cell and contributes to LL considering that smaller resource allocation requires less time than allocation of a whole time slot. In fact, this is targeted to the LL which combined with ultra-reliability gives the new service type in 5G NR, targeted for use for provisioning of URLLC services.

The flexible slot structure is an important component of NR, used for flexible resource management for current deployments, but also necessary for future compatibility for beyond 5G mobile systems. To this end, NR supports up to two DL/UL switching point slots in particular [7]:

- Zero switching point in the slot – considering that NR time slots can carry 14 OFDM symbols, this means that there are either 14 DL symbols or 14 "flexible" symbols (can be dynamically assigned to DL or UL, depending on the UE need) or 14 UL symbols.
- One switching point in a slot – in this case, the slot is starting with zero or more DL OFDM symbols and ending with zero or more UL OFDM symbols. Between DL and UL symbols can be placed so-called "flexible" symbols.

- Two switch points in a single slot – in this case, the first (or second) 7 symbols start with zero or more DL OFDM symbols, while on the other side it ends with at least one UL symbol. Also, there can be zero or several "flexible" symbols in between.

The maximum channel bandwidth supported by NR is 100 MHz for sub-6 GHz and 2000 MHz above 6 GHz. It should be noted that the maximum supported UL/DL channel bandwidth in the same band may be different (similar to LTE different channel widths). The minimum channel bandwidth for NR sub-6 spectrum is 5 MHz. Minimum channel bandwidth above 6 GHz (e.g. in mmWave spectrum) is 50 MHz. As usual, one may expect that new channel widths can be added in NR future releases.

4.2.2 5G Radio Access Network (5G RAN) Architectures

NR is implemented in 5G base stations in a 5G access network (5G AN or 5G RAN). Additionally, 5G AN (or 5G RAN) is also referred to as NG-RAN (Next Generation RAN) [8]. In that manner, one should note that 5G AN, 5G RAN, and NG-RAN (Next Generation RAN) are used in this book interchangeably, and the main nodes in such RAN node are the following:

- gNodeB (gNB), which provides NR radio access or
- Next-Generation eNodeB (ng-eNB), which provides 4G radio access and is connected to 5G core network.

The two deployment options for 5G mobile access networks defined by 3GPP are as follows:

- Non-standAlone (NSA) architecture, where the 5G RAN and its new interface radio (NR) are used together with the existing LTE and EPC infrastructure (i.e. 4G radio and 4G Core), which makes NR technology available without replacing all parts of the network. In this configuration, only 4G services are supported (not new 5G services), but the new 5G radio (lower latency, etc.) has increased capacity and individual bitrates. NSA is also known as "E-UTRA-NR Dual Connectivity (EN-DC)." This corresponds to the 5G-SRIT proposal from 3GPP.
- StandAlone (SA) architecture, where the NR is connected to the 5G core network. Only, in this configuration, all 5G services are supported. This architecture corresponds to the 5G-RIT proposal from 3GPP.

4.3 SDN, NFV, and Network Slicing in 5G

The new 5G targets that started with LTE Advanced Pro are vertical markets, new markets for the telecom/ICT world. So, the question is how to provide heterogeneous services (required for different services) with high scalability of the mobile network, easily scale up or down, and add or delete service or microservice when needed?

Well, it can be done if we make the telecom applications/services in the mobile network similar to the applications we install and uninstall from our laptops, computers, or smartphones. But that's easier said than done. And, the answer is, by adding more flexibility to the network in terms of the allocation of its resources (e.g. bandwidth, memory or processing power of nodes, and so on) with respect to all planes (UP, CP, and management plane). What "tools" are available for this purpose?

The "tools" are Software-Defined Networking (SDN) and Network Functions Virtualization (NFV). They are seen as enablers to reduce operating costs and provide new potential revenue streams by expanding the world of telecommunications in various verticals (e.g. machine-to-machine communications and– IoT, used for various services in various so-called verticals such as smart cities, smart homes for smart industry, critical IoT services, and so on).

On the other hand, telecom operators have a large amount of equipment that does not have SDN or NFV capabilities. So, it is very likely that many telecom operators can take advantage of SDN and NFV in a so-called hybrid implementation that includes legacy (non-SDN and non-NFV) network elements together with SDN and NFV enabled network elements.

In general, it can be expected that 5G mobile networks and all future networks (not only 5G) will be more complex (with many new services in different verticals) and more efficient (i.e. higher utilization of the same network infrastructure for all services). What is the name of such an approach in 5G?

Well, the standardized form of NFV is called network slicing, that is, network slices are defined in 5G mobile networks [9]. However, there are several other earlier terms that are related to the same approach. For example, in a series of recommendations for Next Generation Networks (NGNs) and future networks, ITU-T standardized the so-called Logically Isolated Network Partition (LINP), which is actually now called network slicing – in short.

In general, network slicing is based on network virtualization, and network virtualization is based on SDN, NFV, as well as cloud computing technologies. It contributes to IMT-2020/5G using software approaches for mobile network design, implementation, deployment, management, and maintenance.

With network virtualization, the underlying physical infrastructure is abstracted as either network, storage, or computing resources, and the main logical building blocks in IMT-2020 (5G) are the network parts. Now, let us define the network part, what is it?

By definition, a network part in IMT-2020 (i.e. 5G) is a logical network that provides certain specific capabilities and characteristics for a given mobile service market scenario. In general, there are various possible uses of network slicing in 5G. For example, it can contribute to the creation of dedicated network parts for massive IoT deployments and for improved mobile broadband Internet access. Due to very low latency in 5G (IMT-2020) mobile networks, a dedicated network slice (or slices) can be allocated to ultrareliable and low-latency services. Many different combinations are possible in terms of 5G cutting, which ultimately depends on the mobile operator's 5G offerings and its business strategy.

Further, network slicing provides possibilities for offering customized services with different requirements on QoS (and other NFs) between slices, and within a given network slice. Different slices may be able to provide different QoS for the same type of service (e.g. managed carrier-class VoIP over a dedicated slice and OTT VoIP over mobile Internet access, as two examples).

So, we can have different network slices in 5G. However, with the aim of 5G slicing technology to be truly successful, it will need entire ecosystems to come together for the purpose of solving and standardizing their end-to-end applications. The examples are the automotive industry (many services in 5G are dedicated to vehicle-to-X communication), healthcare (different types of services are possible, from health monitoring to emergency services or surgery assistance at a distance, where the first ones can go over slices with less strict QoS requirements, while the latter ones should go through URLLC network slices in 5G), agriculture (e.g. the mMTC), manufacturing (discrete automation for Industry 4.0 and beyond should be provided via privately managed URLLC network slices, dedicated to a given industry "customer"), and so on. Such varied ecosystems (that exist in different sectors, in different verticals) need to become more and more involved in 5G, and all that with the aim to help drive the potential that slicing can provide for targeted and customized 5G services in different verticals. However, this is a long-run scenario for mobile operators, which may span well beyond 2030.

4.3.1 Network Slicing in IMT-2020

Network slicing is used in IMT-2020 networks to improve network flexibility to accommodate different service scenarios. According to ITU-T [9, 10], the IMT-2020/5G network enables a variety of services, including eMBB services, mMTC-based services, and URLLC on the same infrastructure of network and computing resources.

Network slicing enables IMT-2020/5G network operators to create logically segmented networks providing customized solutions for different market and business scenarios.

What is a network slice?

According to the ITU definition, a network slice is a logical network that provides certain network capabilities for a given service or group of services. The goal is to provide flexible solutions for different market scenarios that have different requirements in terms of functionalities, performance and resource allocation. The operation of the network part is realized through the use of Network Slice Instances (NSIs).

In doing so, a given NSI is composed of a set of Network Function Instances (NFIs). Each NSI represents a logical network that provides specific network capabilities and features. Slice instances are created and managed with slice lifecycle management functions. These functions deal with the orchestration, control, and management of the network part, NFs, and resources during the lifecycle of the given network slice in which NSIs are used.

A single NSI can support one or more application services. Also, the same application service can be supported by multiple NSIs. An NSI can be composed of one or more subnetwork slice instances (sub-NSIs). Similar to an NSI, a sub-NSI is composed of a set of NFIs operating over allocated resources. A sub-NSI can be used by one, two, or more NSIs.

Different NSIs can operate on the basis of different rules and policies. For example, one NSI may be used for Internet access service (in this case, over a mobile network) where network neutrality applies, a second NSI can be used for carrier grade VoIP (which uses managed IP networking), a third NSI can be used for IPTV provisioning (which is telecom service, not over the public Internet), a fourth NSI can be used for IoT services for private 5G network implemented in a smart factory), a fifth NSI can be used for services of driverless vehicles or drones (which requires ultrahigh reliability and quality), and so on. Of course, multiple services can be also provided over a single NSI (e.g. carrier grade VoIP and IPTV services). So, in theory, all combinations are possible regarding network slices and service provisions over them.

Figure 4.6 illustrates the network slicing concept for IMT-2020/5G. The network is required to be flexible to support the extreme variety of requirements in terms of application services. In 5G, we have four service types, which include eMBB, mMTC (i.e. massive IoT), URLLC (also known as critical IoT), and V2X. In general, network virtualization based on SDN, NFV, and cloud computing technologies contribute to the IMT-2020 framework by using software for mobile network design, implementation, deployment, management, and maintenance. With such virtualization, the underlying physical infrastructure is abstracted such as network, storage, or computing resources.

The 5G end-to-end network slices include everything needed to form a full public land mobile network (PLMN), including access and core NFs as well as applications servers (which may be also third party services via network exposure functions – NEFs), as shown in Figure 4.6. The two smartphone network parts illustrate that an operator can deploy multiple network parts with exactly the same system features, capabilities, and services, but dedicated to different business segments and, therefore, each of them may provide a different capacity for the number of UEs and data traffic.

Each of these types of 5G slices can have multiple instances in a single network, such as two or more eMBB network slices. Also, different parts of a 5G network may have different QoS support.

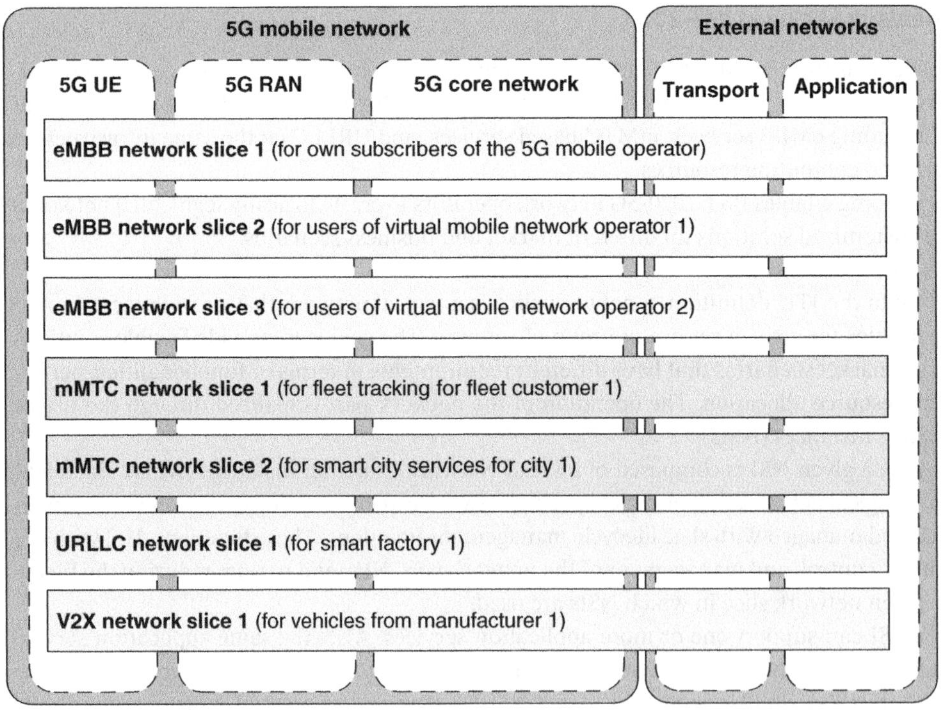

5G mobile network
External networks
5G UE
5G RAN
5G core network
Transport
Application
eMBB network slice 1 (for own subscribers of the 5G mobile operator)
eMBB network slice 2 (for users of virtual mobile network operator 1)
eMBB network slice 3 (for users of virtual mobile network operator 2)
mMTC network slice 1 (for fleet tracking for fleet customer 1)
mMTC network slice 2 (for smart city services for city 1)
URLLC network slice 1 (for smart factory 1)
V2X network slice 1 (for vehicles from manufacturer 1)

eMBB – enhanced Mobile BroadBand
mIoT – massive Internet of Things
RAN – Radio Access Network
UE – User Equipment
URRLC - Ultra-Reliable Low-Latency Communications
V2X – Vehicle to Everything

Figure 4.6 IMT-2020/5G network slicing.

Note that this approach does not threaten net neutrality, which also applies to mobile networks, but it is more aimed at expanding the portfolio of mobile services to other verticals, such as mission-critical services which are not provided over the Internet but through managed networks, such as connected car services, industrial automation, etc. Ultimately, network slicing in 5G and subsequent mobile network like technology enables mobile operators to develop innovative new business models. In the long term, it can be expected that future mobile generations (6G, 7G) will be directed toward mass network slicing, that is, the possibilities of creating and managing a very large number of network slices through the use of a given physical network infrastructure.

4.4 5G Next Generation Core

To support and enable known and unknown future use cases, a new 5G core network was defined. This new core network is called 5G Next Generation Core (NG-Core or NGC) or 5G Core Network (we will use all these terms interchangeably) to go along with 5G NR.

The 5G network architecture consists of two main parts (Figure 4.7), which include the 5G CN and 5G RAN domains.

As previously noted for the gNB in the 5G RAN, there is a separation of functions with control plane functions (CPFs) and user plane functions (UPFs) also in the 5G core, thereby ensuring the separation of control and user planes for greater flexibility in providing new emerging services.

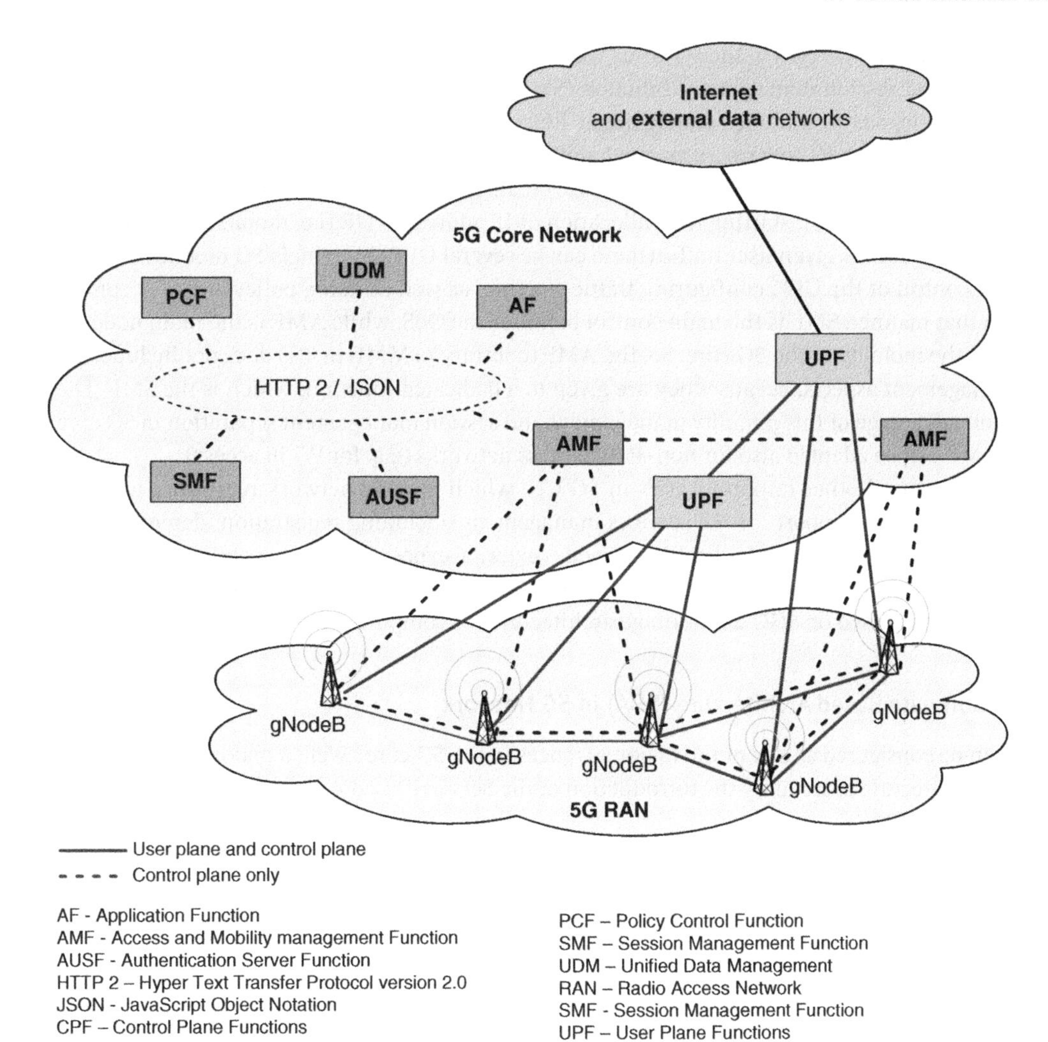

Figure 4.7 5G architecture.

4.4.1 5G Core Network Functions

The 5G CN provides full separation of UP and CP functions, although that process has started in previous generations regarding the mobile core networks. However, the UP–CP separation also exists in the 5G RAN. So, the CP/UP split exists everywhere in the 5G architecture, end to end. How is 5G core organized?

Well, the approach in organization of the 5G core is based on NFV and also NGN principles standardized by the ITU [3]. There are no more "traditional" gateway nodes (like in 4G), but different so-called NFs in the core that can be combined together for provision of a given 5G service. The functions in 5G core can belong either to the UP or to the CP.

Then, which is the main UPF responsible for all user traffic (from/to users or devices)? Well, that is the UPF, as its name indicates. The UPF provides various functionalities for the UP traffic (i.e. traffic from/to end mobile devices) which includes packet routing and forwarding, with appropriate QoS handling.

What are the main CP NFs in the 5G core? Well, they are access and mobility management function (AMF) and session management function (SMF). The AMF provides mobility management functions, analogous to mobility management functions of the MME in 4G CN. However, the SMFs of the MME (in 4G core) are separated and combined with the data plane control functions of the S-GW (in 4G core) and P-GW (in 4G core) to create the SMF in the 5G core. So, the SMF deals with session management, starting with allocation of IP address to UE (i.e. mobile devices), selection of the UPF for the given user traffic (there can be several UPFs in a single 5G mobile network) and then control of the UPF, configuring traffic steering, as well as doing policy and QoS control "job." In that manner, SMF is the main control point for the QoS, while AMF is the main node for control of the mobility in the 5G core. So, the AMF (contrary to MME in 4G) does not include session management aspects, because they are given to a dedicated function, which is the SMF. The important advantage of this mobility management and session management separation in 5G core is that AMF can be adapted also for non-3GPP access networks (e.g. for Wi-Fi access).

There are several other important NFs in 5G CN, which include network repository function (NRF) for provision support for NF services management (including registration, deregistration, authorization, and discovery), NEF which provides external exposure of the capabilities of the NFs from 5G core (can be used for creation of different 5G services), and Unified Data Management (UDM) which is based on 5GC data storage architecture for computer and storage separation.

4.4.2 Software Based Architecture (SBA) in 5G Network

What can be considered as one of the major advances in the 5G core? Well, a major advance in the 5G core architecture is certainly the introduction of the Service-Based Architecture (SBA). What is the purpose of SBA?

In the 5G core, NFs employing the SBA offer and consume services of other NFs. Allowing any other NF to consume services offered by a NF enables direct interactions between NFs. For example, the Policy Control Function (PCF) can directly subscribe to the location change service offered by the AMF rather than having to do that via the SMF.

In general, a 3GPP 5G network is based on building blocks that are defined as NFs. Unlike 4G, where there are network nodes (such as serving gateway or packet data network gateway), in 5G the NGN approach of ITU [3] is implemented where the core network is based on NFs that offer their services through interfaces of a common framework through which a given NF can use one or more other NFs and vice versa. In order for a given function to discover which services are enabled by other NFs in the 5G standard from 3GPP are defined the so-called NRF. Such a network model, by its very nature, accepts principles such as modularity, reuse, and self-maintenance of NFs. Of course, it is made possible by the softwarization and virtualization of network resources with SDN and NFV technologies.

Using a repository of NFs in the 5G network creates the possibility for 5G functionalities to be available to third parties (with a suitable agreement with the mobile operator who owns the 5G network and has implemented the NFs in it), such as providers of services and vertical industries outside the domain of the operator. The availability of 5G NFs to third parties is enabled by the NEF. The interface provided by NEF to third parties can be considered as one of the primary points through which 5G interacts more closely with different vertical industries.

Thereby, 3GPP decided in its standards that the 5G service exposure by NEF should be based on so-called RESTful APIs (Figure 4.8). With the RESTful API the underlying server implementation is hidden by an abstraction layer [11]. The data for a given service is defined as a "resource" that is uniquely identified by a Uniform Resource Identifier (URI), which is a way of addressing resources in the Internet/IP world. As in Internet technologies [2], service consumers use HTTP methods

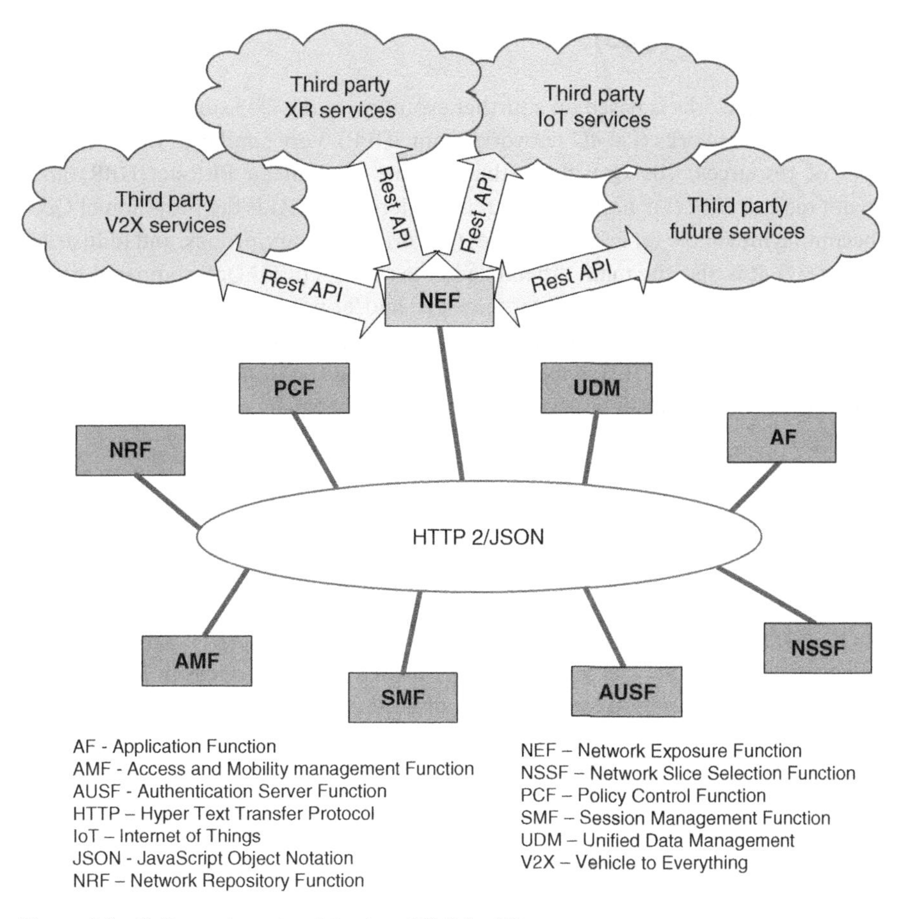

Figure 4.8 Software based architecture (SBA) for 5G.

(HTTP 2, HTTP 3) to access only a representation of resources, where the representation can be delivered over HTTP using different file formats (e.g. JSON, XML, and others). In this approach, the NFs that form the 5G SBA communicate with each other through Service-Based Interfaces (SBIs).

The basic operations for persistent data storage are Create, Read, Update, and Delete (CRUD) [12]. Considering that in the time of 5G specification writing the actual HTTP standard was 2.0, it is noted in 3GPP specifications (one should note that in 2022 it was standardized HTTP 3, covered in Chapter 2 of this book). However, the main HTTP operations from the CRUD are HTTP methods GET, POST, PUT, and DELETE, which are used on resources that are identified by URIs. However, using RESTful API is not mandatory. While HTTP 2 or HTTP 3 are standards, there are also possible services operations to be implemented as custom operations (instead of operations on a resource for which HTTP methods are used) by calling certain NFs from the NF provider (e.g. 5G mobile operator). Also, it is possible to combine the RESTful approach with custom operations within the same API in 5G CN.

So, using the SBA approach, the internal communication in the 5G CN is based on the same principles as the network functional exposure. This allows the technological approach of the 5G system to be aligned with the need to introduce new services as needed for end customers, which is intended for services in different verticals, such as automation of different industries, smart transport, smart cities, smart agriculture, and so on. Both existing and new services have certain QoS requirements that the 5G network needs to handle.

4.5 5G Quality of Service (QoS)

QoS support in 5G mobile networks is based on a further evolution of the QoS support defined for LTE/LTE Advanced mobile networks (i.e. 4G networks from 3GPP). Very similar to LTE, 5G QoS supports two types of resources, viz. QoS flows that require Guaranteed BitRate (GBR) and QoS flows that do not require GBR (i.e. non-GBR). But the difference in 5G is the provision of QoS stream which is becoming inevitable for critical services (e.g. driverless cars, drones, and industrial machine control over 5G). It is also the main difference in the concept of 5G QoS compared to 4G QoS. Of course, QoS per stream is only possible for managed and/or private services, not for OTT services offered over public Internet access. To that end, to identify per-stream QoS in 5G systems, a Quality of Service Flow Identifier (QFI) is used. So traffic having the same QFI gets the same traffic management access. Such QFI can be assigned dynamically or can be equal to standardized 5QI (5G QoS indicators).

4.5.1 5G QoS Indicators (5QIs)

The list of 5G QCIs (5QIs) is given in Tables 4.4–4.6, keeping in mind that it is not a definitive list but is constantly being upgraded as new use cases of 5G networks emerge in newer 3GPP specifications. This list of QoS identifiers includes all QCIs from LTE/LTE Advanced/LTE Advanced Pro (i.e. up to 3GPP Release 14) and adds new QoS indicators related to new services in 5G, such as remote control, Intelligent Transportation Systems (ITS), discrete automation, and AR.

What is new about 5G that will certainly continue to spread in 6G? A new type of resource in 5G is the delay-critical GBR, which has a requirement for very low delays (up to 20 ms) as well as low

Table 4.4 5G QoS indicators (5QIs) for GBR.

5QI (QCI)	Resource type	Priority level	Packet delay budget (ms)	Packet error rate	Maximum data burst volume	Example services
1	GBR	20	100	10^{-2}	—	Voice over IP (VoIP)
2		40	150	10^{-3}	—	Video call and live streaming
3		30	50	10^{-3}	—	Real-time gaming, V2X messages
4		50	300	10^{-6}	—	Buffered video streaming (not conversational)
65		7	75	10^{-2}	—	Mission critical push to talk (MCPTT) voice
66		20	100	10^{-2}	—	Non-mission critical push to talk voice
67		15	100	10^{-3}	—	Mission critical video user plane
75		25	50	10^{-2}	—	Vehicle-to-X (V2X) messages
71		56	150	10^{-6}	—	"Live" uplink streaming
72		56	300	10^{-4}	—	"Live" uplink streaming
73		56	300	10^{-8}	—	"Live" uplink streaming
74		56	500	10^{-8}	—	"Live" uplink streaming
76		56	500	10^{-4}	—	"Live" uplink streaming

Table 4.5 5G QoS indicators (5QIs) for non-GBR.

5QI (QCI)	Resource type	Priority level	Packet delay budget (ms)	Packet error rate	Maximum data burst volume	Example services
5	Non-GBR	10	100	10^{-6}	—	IMS signaling
6		60	100	10^{-3}	—	Buffered video streaming, interactive TCP-based applications (e.g. Web, email, and file sharing)
7		70	100	10^{-3}	—	Voice, live video streaming, interactive gaming
8		80	300	10^{-6}	—	Buffered video streaming, interactive TCP-based applications (e.g. Web, email, and file sharing)
9		90	300	10^{-6}	—	
10		90	1100	10^{-6}	—	Video (buffered streaming)
						TCP-based (e.g. www, e-mail, chat, ftp, p2p file sharing, and progressive video) and any service that can be used over satellite access type with these characteristics
69		50	60	10^{-6}	—	Mission Critical signaling (e.g. MCPTT signaling)
70		55	200	10^{-6}	—	Mission critical data from buffered video streaming, interactive TCP-based applications (e.g. Web, email, and file sharing)
79		65	50	10^{-2}	—	V2X messages
80		66	10	10^{-6}	—	Low latency eMBB augmented reality

Table 4.6 New 5G QoS indicators (5QIs) for delay critical GBR.

5QI (QCI)	Resource type	Priority level	Packet delay budget (ms)	Packet error rate	Maximum data burst volume (bytes)	Example services
82	Delay-critical GBR	19	10	10^{-4}	255	Discrete automation
83		22	10	10^{-4}	1354	Discrete automation
84		24	30	10^{-5}	1354	Intelligent transport systems
85		21	5	10^{-5}	255	Electricity distribution –high voltage
86		18	5	10^{-4}	1354	V2X messages
87		25	5	10^{-3}	500	Interactive service – motion tracking data
88		25	10	10^{-3}	1125	Interactive service – motion tracking data
89		25	15	10^{-4}	17000	Visual content for cloud/edge/split rendering
90		25	20	10^{-4}	63000	Visual content for cloud/edge/split rendering

jitter (up to 20 ms), aimed at critical services such as control of automatic processes or transport vehicles through the mobile network, thus eliminating the need to build a separate network for such late critical services. Of course, this does not mean that all such critical services will be through the same network slice; on the contrary, services with different demands and with different ownership and target groups can (but not always) be provided through different network slices implemented on the same mobile 5G infrastructure. For each QoS flow in 5G there is an associated 5G QoS profile, which must contain the following two parameters:

- 5QI parameter, as defined by 3GPP.
- Allocation and Retention Priority (ARP) parameter – similar to 4G and 3G mobile networks from 3GPP, this parameter defines the priority of the flows that have the same 5QI.

For the GBR only flows, the QoS profile also must include the bitrate parameters:

- Guaranteed flow bitrate (GFBR) for both UL and DL;
- Maximum flow bitrate (MFBR) for both UL and DL.

For GBR streams, the list of QoS parameters can be completed with two additional parameters:

- Notification control, which is the availability of notifications from the 5G RAN when the guaranteed bitrate can no longer be met for a given flow.
- Maximum packet loss rate in the UL and DL, which refer to the maximum error rate that can be tolerated for the QoS flow.

In 5G, in 3GPP Release 17 [13], there are new 5QIs, given in Table 4.4 for GBR, in Table 4.5 for non-GBR, and in Table 4.6 for delay critical GBR [14]. Some of the QCIs in the tables (QCIs with numbers 1, 2, . . ., 9) are actually defined in LTE/LTE Advanced (in the basic QoS version), while some are QCIs added to LTE Advanced Pro (QCIs with the numbers 65, 66, 69, 70, 75, 79). Thus, the resource types and QCI in 3GPP mobile networks started for 4G continued in 5G mobile networks. Also, the priority values that existed in the 4G QoS frame in 5G are the same values multiplied by 10. For example, IMS signaling had a priority of 1 in the 4G QoS frame and in 5G it has the same priority, but the value is multiplied by 10, which is a priority of 10. This was done to avoid decimal values, such as 6.5 (which is now priority level 65) and so on, for all priority level values.

When QCIs refer to 5G, they are designated as 5G Quality Identifiers (5QIs), which still includes all previously defined QCIs in 4G and adds new 5QIs for new applications expected in future 5G mobile networks. Thus, 5G introduces a new type of resource, called delay critical GBR. As the name suggests, it is intended to be used for 5G services that have the strictest latency requirements. For example, power distribution or motion tracking (e.g. in industrial machinery) requires delays of less than 5 ms, while ITS can have delays of up to 30 ms.

For a non-GBR flow, the QoS profile may also include the reflective QoS attribute (RQA) parameter. Reflective QoS in 5G refers to the ability to map UP traffic to UL QoS flows without having QoS rules from SMF by deriving QoS rules from DL traffic (hence, there is reflection of DL QoS rules in follow-up). This applies to both the IP protocol data unit (PDU) session and the Ethernet PDU session. The derived QoS rule contains several parameters including a packet filter, QFI, and precedence value.

The packet filter set is used to identify one or more IP or Ethernet streams. The IP packet filter set is based on a combination of the following parameters from IPv4 or IPv6 packets [1]:

- IP address of source and destination (IPv4 or IPv6);
- Source and destination port numbers (e.g. TCP port and UDP port);

- Protocol Identifier (ID) (in IPv4) or Next Header (in IPv6) that gives the protocol above the IP layer in the stack (e.g. TCP and UDP);
- Type of Service (ToS) field in the IPv4 header or Traffic Class in the IPv6 header for the differentiation of traffic on the IP layer based on defined classes (e.g. with DiffServ);
- Flow Label, which is only available in IPv6 headers (flow identification in IPv4 uses a quintuple, consisting of source and destination IP addresses, source and destination ports, and transport protocol over IP);
- Security parameter index, which is used in case of IPsec protected packets;
- Filter direction, which can be either downward or upward.

The priority value of a QoS rule determines the evaluation order, so a lower priority value gives higher priority in the evaluation of QoS rules and vice versa.

4.5.2 QoS Functions in 5G Network

In general, QoS needs to be performed in both directions, downstream and upstream. While the DL direction is easier to control, considering that it starts from the 5Γ base station (gNB), which is connected to 5G core NFs, the upstream direction may be more challenging considering the limited capabilities of UE when compared to the base stations. The classification and labeling of user data traffic in the upstream direction is performed by the UE (i.e. the mobile terminal), based on QoS rules that can be explicitly communicated to the user terminal (through session establishment or modification of PDU), preconfigured or implicitly derived using reflective QoS (i.e. deriving QoS rules from the downstream direction and applying them in the upstream).

The QoS flow mapping in 5G is shown in Figure 4.9. Several NFs in 5G core are used for flow mapping. In that way, the SMF performs the connection of service data streams (SDFs) for QoS flows, through interaction with the mobile terminal, 5G RAN and UPF.

The function of the SMF is, among other things, to determine the authorized QoS of a given QoS flow based on the PCC rules associated with the flow, and to notify the PCF when the QoS targets cannot be met.

For QoS control at the PDU session level, PCF provides a combination of 5QI/ARP for the PDU session. The association of the PCC rule with a QoS flow in a PDU session is called a QoS flow binding. Binding a PCC rule to a QoS flow causes SDF DL transmission to be directed to the QoS flow. For the UL direction, the QoS rule associated with the QoS flow instructs the mobile terminal (i.e. UE) to direct the SDF in the UL direction to the QoS flow in the established association. If the authorized QoS of a given PCC rule is changed, the bindings are reevaluated and then the given SDF can be bound to another QoS flow. So we have three levels of QoS control in 5G: (i) QoS control at the SDF level; (ii) QoS flow level QoS control; and (iii) QoS control at the PDU session level.

Figure 4.9 illustrates the mappings between the three QoS levels in 5G. Note that one or more SDFs can be mapped to a single QoS flow. Also, one or several QoS flows can be mapped to a given PDU session. In the DL direction, data packets (in the UP) are classified by the UPF based on the packet filter sets of the DL Packet Detection Rules (PDRs), applying them in order of their precedence.

UPF conveys the classification of user data traffic belonging to a QoS flow over the interface between UPF and 5G RAN based on marking the UP with QFI. Furthermore, AN binds QoS flows to AN resources, which are Data Radio Bearers (DRBs) in the RAN. However, the relationship between QoS flows and AN resources is not strictly one to one. Furthermore, the AN establishes and releases the AN resources, and in the case of QoS flows indicates this to the SMF.

In summary, the main innovation in 5G QoS is the ability to differentiate data flows, which provide a means to differentiate traffic from different applications/services with different QoS requirements, while at the same time maximizing resource utilization in the AN. Also, 5G is

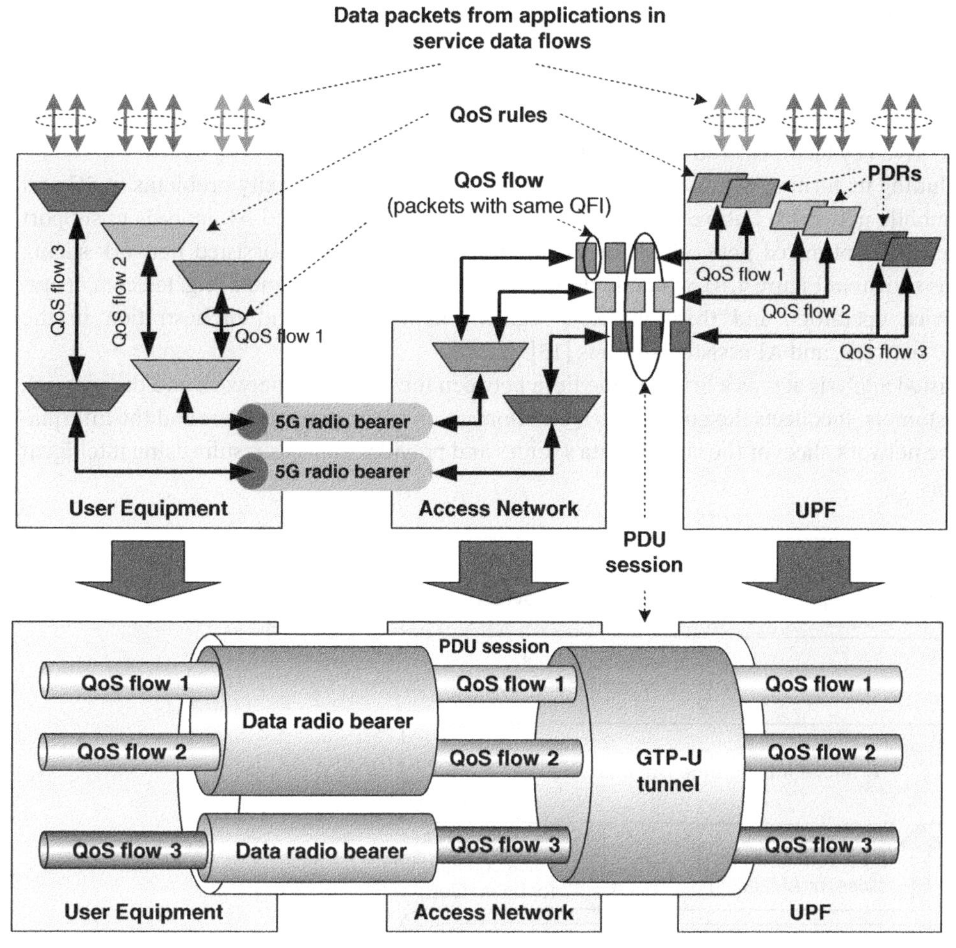

Figure 4.9 QoS flow mapping in 5G mobile network.

designed to support different types of ANs, including fixed ANs. QoS provisioning is symmetrical because it is provided in the same way in both directions, DL and UL, which is possible with the mentioned reflective QoS approach (copying the QoS rules for the DL to the UL). Finally, while in UMTS and LTE/LTE Advanced, QoS support is based on carriers, the 5G cellular core is designed to have QoS based on flow rather than carrier.

4.5.3 5G QoE Analysis with Artificial Intelligence (AI) Assistance

As already noted in this chapter, the IMT-2020/5G network introduces network slicing to provide customized and dedicated network services for many different industries running on a shared physical infrastructure, which is directly related to QoS provisioning considering that different services over different network slices may have different requirements. Compared to traditional networking, network slicing based on SDN and NFV provides flexibility, agility, and variety of network service offerings on the IMT-2020 network. This increases complexity in QoS provisioning. On the other hand, guaranteeing a network slice based service level agreement (SLA) is a key business aspect for both service providers and mobile operators nowadays. This results in the need for high dynamics of establishing and releasing network parts and services through them, depending on business needs. All this increases the risk of unexpected problems on the mobile network, driven by technical and service innovations.

AI including its form of ML is a leading candidate for solving complexity problems in 5G and beyond mobile networks. Future mobile networks (beyond 5G) will need AI analysis to support automated and optimized network slicing. Conceptual overview of AI-assisted network slicing analysis is shown in Figure 4.10, which includes: network segment customers (e.g. telecom operators, service operators, and third parties), segment management and orchestration in the IMT-2020 network, and AI-assisted analysis [15].

AI-assisted analysis acts as a bridging medium between the operator's network and the network part's customers. It collects the customer's QoE information of the network slice and the information of the network slices of the slice as data sources and provides analysis results using intelligent algorithms.

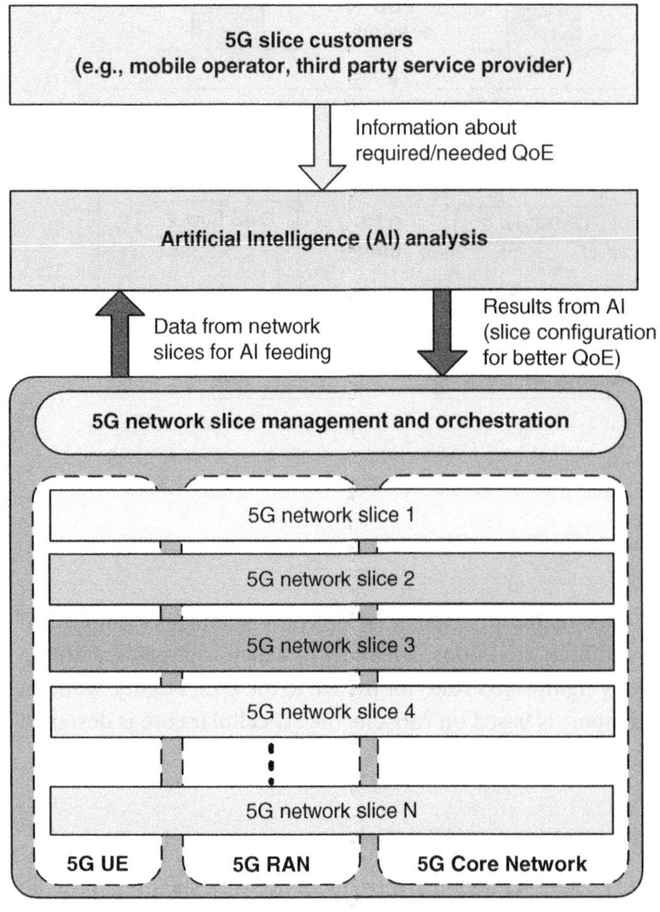

Figure 4.10 QoE analyses with assistance of artificial intelligence.

For example, for resource optimization, AI-assisted analysis calculates can be used for network resource allocation plan such as usage of logical/physical resources, allocated bandwidth portions, and scheduling algorithm (which is usually not standardized by the SDO, because it highly depends on a given network and traffic scenario, which is always dynamic in a mobile network such as 5G or 6G).

4.6 Spectrum Management for International Mobile Telecommunications (IMT)

In terms of spectrum bands and spectrum management for deployment of 5G, they can be subdivided in three macro categories: sub-1 GHz, 1–6 GHz, and above 6 GHz.

Sub-1 GHz bands are suitable for supporting mobile broadband coverage in suburban and rural areas as well as for various IoT services. This is because the propagation properties of radio signals at these frequencies allow the 5G mobile network to create very large coverage areas and also deeper building penetration (for indoor coverage in both rural and urban areas).

The bands originally intended for capacity in urban areas are in the 1–6 GHz range, where there is a balance of coverage (not as good as below 1 GHz) and capacity for mobile services (due to more frequency band available relative to below the 1 GHz area). Meanwhile, some of these frequency areas are occupied by various other technologies that appeared before mobile broadband technologies, such as radio (around 100 MHz) and television (which naturally occupied the spectrum above it in the sub-1 GHz area). The digitalization of television in the first decade of the 21st century (somewhere in the first half of the second decade of this century) enabled the multiplexing of more TV channels in the same band, which enabled the gradual release of the 800 and 700 MHz bands, and below for mobile technologies, such as 4G first in the last decade, 5G in this decade (2020s), and 6G in the next decade (2030s). Initial 5G deployments are actually based on a reasonably balanced use of the low band (below 1 GHz) and mid-frequency spectrum (between 1 and 6 GHz) to provide coverage (outdoor and indoor) and capacity (bitrate).

Spectral bands above 6 GHz provide significant capacity thanks to the very large bandwidth that can be allocated to mobile communications, and thus enable improved mobile broadband applications as well as mission-critical services due to the fewer delays possible.

So, in summary, bands below 1 GHz are used mainly for coverage, 1–6 GHz bands are used for a mix of both coverage and capacity, and bands above 6 GHz are used primarily for capacity in 5G mobile networks and beyond (as shown in Figure 4.11).

There is also a negative side to the high bands in the millimeter wave (mmWave) area (above 24 GHz in 5G). It is the greatly reduced coverage size of a cell operating on that band and its susceptibility to blocking by obstacles in the way such as walls. But it should be emphasized that 5G is the first mobile generation to go into mmWave spectrum territory where no previous mobile generation (including 4G) has gone before. Why is that? Because 5G initially had very stringent requirements in terms of capacity (both per user and per square meter of area, as well as for delays in the mobile network), which could only be achieved by including large new frequency bands, and such are available at high frequencies. Thus, the diversity of spectrum requirements and needs has resulted in the definition and standardization of many 5G deployment options using different spectrum bands for different use cases. Of course, this is followed by appropriate regulatory activities. What does that mean?

Well, in order for the same mobile devices to be used in different countries of the world (which is already well known with roaming from 2G onward), it is necessary to have harmonization in the

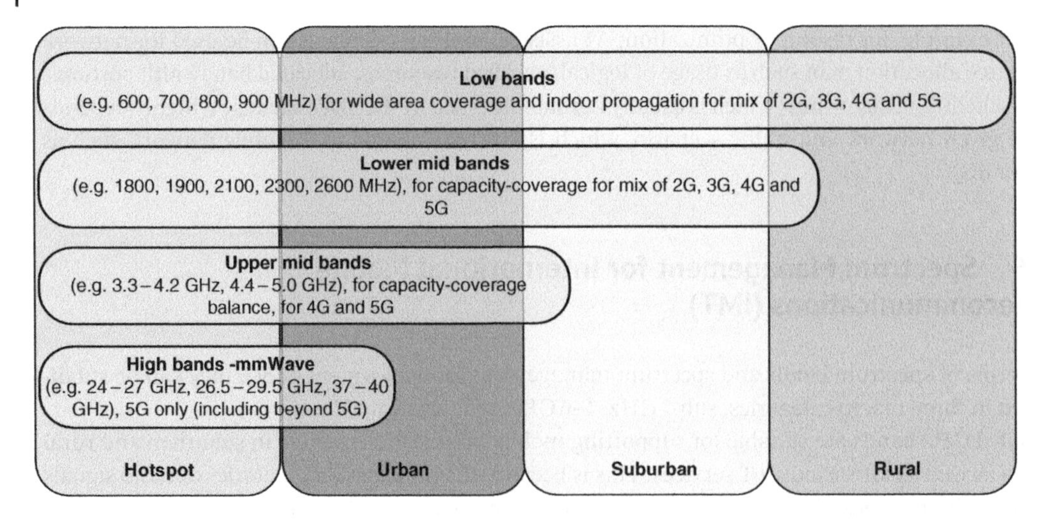

Figure 4.11 Spectrum considerations for mobile networks in rural, suburban, urban, and hotspot areas.

allocation of frequency bands in different regions and in all countries of the world. Such harmonization is realized through ITU Radiocommunication Sector (ITU-R) and World Radiocommunication Conferences (WRCs) held every 3–4 years. Thus, ITU WRC 2019 was the first to define the frequency bands for 5G for all three regions of the world (according to the ITU region classification), while WRC 2023 specifies the bands for 5G-Advanced, as the next stage of development of 5G technologies.

4.6.1 5G Frequency Bands

In terms of 5G spectrum management, two separate radio performance requirements are specified for two frequency bands (FRs), namely FR1 and FR2 [12]. FR1 refers to sub-7 GHz bands (410–7125 MHz) [16], while FR2 refers to so-called mm Wave bands (above 24.25 GHz), as given in Table 4.7.

There are four types of bands specified for NR (one should note that all NR bands are defined with the prefix "*n*" to distinguish them from bands for other Radio Access Technologies – RATs):

- LTE "refarming" band: The bands have the corresponding LTE bands. For example, NR band n7 corresponds to LTE band 7. Hence, the bands are likely to be used by "refarming" (i.e. reusing) the existing 4G LTE bands.
- NR new bands in FR1: Completely new frequency bands for 5G NR in FR1 whose corresponding LTE bands do not exist.
- NR new bands in FR2: New frequency band for 5G NR in FR2.
- Additional Uplink (SUL)/Downlink (SDL) band: These bands have only UL/DL frequency and they can be deployed with other type of NR bands.

Table 4.7 Definition of frequency ranges for 5G in 3GPP standards.

Frequency range name		Frequency range (MHz)
FR1		410–7125
FR2	FR2-1	24250–52600
	FR2-2	52600–71000

Table 4.8 5G new radio (NR) and frequency ranges FR1 and FR2.

Numerology (μ)	Subcarrier spacing (SCS) (kHz)	Frequency range
0	15	FR1
1	30	FR1
2	60	FR1/FR2
3	120	FR2-1/FR2-2
4	240	FR2-2
5	480	FR2-2
6	960	FR2-2
...	...	...
μ_N	$SCS(\mu_N) = 2 * SCS(\mu_{N-1})$	FR2-X

In addition, it should be noted that the ranges for the frequency ranges {n65–n256} and {n257–n512} are reserved as the range number for new ranges in FR1 and FR2, respectively. A band number will be assigned to a new frequency band on a first come, first served basis from the reserved frequency band.

5G NR design is directly dependent upon the spectrum range, FR1 or FR2, as shown in Table 4.8. The SCS is 15, 30, or 60 kHz for bands in FR1, while it is 60, 120, or 240 kHz for bands in FR2. So, 5G uses different numerologies in different frequency ranges (Table 4.8), while the previous generation LTE/LTE Advanced uses only numerology $\mu = 0$, with SCS = 15 kHz. The SCS of LTE (15 kHz) is used as initial one, and each next numerology has doubled SCS width compared to its predecessor, that is, $SCS(\mu_N) = 2 \times SCS(\mu_{N-1})$. All higher numerologies are for operating bands in FR2 which expands over time (with each new release of 3GPP and new spectrum allocations in the ITU WRC). Such trend is indicated in Table 4.8 as FR2+ and for the future provisionally noted as FR-X in Table 4.8, for correspondingly higher ranges toward frequencies (above 100 GHz and toward THz in the future).

Table 4.9 outlines the FR1 bands in the sub-1 GHz spectrum, which are aimed for deployment of larger cells (e.g. up to 20–30 km in radius) for coverage purposes on a national level by mobile operators. This low band spectrum is primarily important for rural areas where the download and upload bitrates are impacted by the availability of such spectrum. In that manner, the trend in 5G and beyond is to provide 600 MHz spectrum in low band (i.e. bands between 410 MHz and 1 GHz), which are aimed to increase mobile broadband speeds in rural areas. NR bands sub-1 GHz in FR1 use FDD and have frequency carriers with a smaller width due to the smaller amount of spectrum.

The most important bands for balanced capacity and coverage aspects in urban areas are the mid-range (that is FR1 bands in range 1–6 GHz). They split between the low mid bands (1–3 GHz) and upper mid bands (3–6 GHz), although such splitting is only provisional [10]. The new 5G bands for licensed spectrum include bands at 3.5 GHz (n77 and n78), as well as n79 (between 4 and 5 GHz). Also, very important are bands at 6–7 GHz, which are targeted to be used as licensed or unlicensed, depending upon the country. Therefore, there are two 5G bands at 6–7 GHz; they are noted as n102 and n104 in Table 4.10. These two 5G bands are defined to be used with shared spectrum channel access [16] depending upon the country regulations, whether the whole band 5925–7125 is defined as unlicensed (in such case, also Wi-Fi technologies, such as Wi-Fi 6th and 7th generation can operate in these bands) or part of it is dedicated for 5G use only (in a licensed manner).

Table 4.9 5G bands in sub-1 GHz spectrum.

NR band	Uplink (UL) bands (MHz)	Downlink (DL) bands (MHz)	Duplex mode
n5	824–849	869–894	FDD
n8	880–915	925–960	FDD
n12	699–716	729–746	FDD
n13	777–787	746–756	FDD
n14	788–798	758–768	FDD
n18	815–830	860–875	FDD
n20	832–862	791–821	FDD
n26	814–849	859–894	FDD
n28	703–748	758–803	FDD
n29	—	717–728	SDL
n67	—	738–758	SDL
n71	663–698	617–652	FDD
n81	880–915	—	SUL
n82	832–862	—	SUL
n83	703–748	—	SUL
n85	698–716	728–746	FDD
n89	824–849	—	SUL
n100	874.4–880	919.4–925	FDD
n105	663–703	612–652	FDD

Table 4.10 New 5G NR bands in upper mid bands (3–6 GHz).

NR band name	Uplink (UL) band (MHz)	Downlink (DL) band (MHz)	Duplex mode
n77	3300–4200	3300–4200	TDD
n78	3300–3800	3300–3800	TDD
n79	4400–5000	4400–5000	TDD
n102	5925–6425	5925–6425	TDD
n104	6425–7125	6425–7125	TDD

The lower middle bands, primarily the 1800 and 1900 MHz bands (used for 2G GSM), 2100 MHz (used for 3G UMTS), 2300 and 2600 MHz (used for 4G in certain regions) are already in use from the mobile generations and before the emergence of 5G. With the gradual reduction of the participation in the mobile market of 2G and 3G terminals, part of their ranges are reformed for 5G. Toward 2030 and beyond, the same will happen with the 4G bands for which licenses have been granted to national mobile operators in the past decade, 2010–2020; namely, the lower middle bands intended to provide high-speed Internet traffic (i.e. data) through 4G networks, with such bands being assigned as FDD in most countries. However, in certain countries (e.g. the USA,

China, and others), 5G is also being deployed in the lower mid-band bands such as 2600 MHz using TDD. In contrast to the lower 5G bands (below 3 GHz, where FDD is still mainly used), TDD is used in the upper mid band bands as well as in the high bands in mmWave areas. In general, FDD enables symmetrical resource allocation in UL and DL and, therefore, simpler planning of RAN resources, because the two directions are physically separated due to the use of separate frequency bands in each direction. In each FDD pair, the lower band is usually in the UL direction and the upper band is in the DL direction, because mobile terminals transmit signals in the UL and they have much lower capabilities than base stations that transmit in the DL direction. In short, NR mid bands 1–6 GHz in FR1 use TDD (as duplex mode) unlike bands in sub-1 GHz that use FDD.

High 5G bands given in Table 4.11 include NR bands in FR2-1 and FR2-2 [12]. Initial 5G bands in mmWave spectrum licensed in many different countries in the late 2010s or early 2020s include n257 and n258 bands, as well as n260 is some countries (e.g. in North America) [17]. High bands use TDD, which means that on the same frequency carrier are multiplexed time slots in the DL and UL directions. A typical DL:UL ratio is 3 : 1 considering that most users still consume more content than they generate (e.g. via social networks content uploads, file sharing via clouds, security camera streaming via mobile network access, and AR/VR services). However, this aspect ratio may change over time with innovation of new services. Also, it depends on the types of services that are implemented in a given area of the mobile network (e.g. in the case of many surveillance cameras for CCTV or smart home/building services, due to affordable pricing schemes for the UL traffic or due to UL bandwidth demanding XR services).

4.6.2 Analysis of 5G Frequency Carriers in FR1 and FR2

With the aim to understand 5G spectrum management, one needs to know also frequency carrier widths for FR1 and FR2 ranges. While the maximum carrier width in 2G was 0.2 MHz (for GSM), in 3G was 5 MHz (for UMTS), in 4G was up to 20 MHz (although there were possible carrier widths of 1.3, 3, 5, 10, 15, and 20 MHz), in 5G we have higher diversity of frequency carrier widths. They are normally smaller for FR1 (given in Table 4.12), ranging from minimum of 5 MHz carrier width up to 100 MHz, where smaller SCS is used for smaller carrier width while larger SCS is used for larger carrier widths. For example, carriers in FR1 that are wider than 50 MHz cannot use a 15 kHz SCS. The main resource unit in 5G is RB. Each 5G frequency carrier has a number of RBs denoted with N_{RB}. The number N_{RB} is proportional to the width of frequency carrier and inversely proportional to the SCS.

Table 4.11 NR bands in FR2.

NR band	Uplink (UL) and downlink (DL) bands	Duplex mode
n257	26500–29500	TDD
n258	24250–27500	TDD
n259	39500–43500	TDD
n260	37000–40000	TDD
n261	27500–28350	TDD
n262	47200–48200	TDD
n263	57000–71000	TDD

Table 4.12 FR1 frequency carrier widths vs. sub-carrier spacing (SCS) vs. number of resource blocks (RBs).

		N_{RB} for FR1		
	SCS (kHz)	**15**	**30**	**60**
FR1 frequency carrier widths	5 MHz	25	11	—
	10 MHz	52	24	11
	15 MHz	79	38	18
	20 MHz	106	51	24
	25 MHz	133	65	31
	30 MHz	160	78	38
	35 MHz	188	92	44
	40 MHz	216	106	51
	45 MHz	242	119	58
	50 MHz	270	133	65
	60 MHz	—	162	79
	70 MHz	—	189	93
	80 MHz	—	217	107
	90 MHz	—	245	121
	100 MHz	—	273	135

The number of RBs for each numerology μ and carrier width (i.e. channel bandwidth) is calculated by using the following (assuming symmetrical guard bands, with values as defined in [18]):

$$N_{RB} = \frac{\text{Channel\_Bandwidth} - 2 \times \text{Guard\_Band}}{\text{One\_Resource\_Block\_Bandwidth}} = \frac{\text{Channel\_Bandwidth} - 2 \times \text{Guard\_Band}}{12 \times \text{SCS}} = \frac{\text{Channel\_Bandwidth} - 2 \times \text{Guard\_Band}}{12 \times 2^{\mu} \times 15\,\text{kHz}} \tag{4.1}$$

For example, by using Eq. (4.1) for numerology $\mu = 0$, and Guard Band of 692.5 kHz, for carrier width of 50 MHz = 50000 kHz (Table 4.12), the number of RBs N_{RB} is 270 RBs, according to the following calculation:

$$N_{RB} = \frac{\text{Channel\_Bandwidth} - 2 \times \text{Guard\_Band}}{12 \times 2^{0} \times 15\,\text{kHz}} = \left\lfloor \frac{50000\,\text{kHz} - 2 \times 692.5\,\text{kHz}}{180\,\text{kHz}} \right\rfloor = 270.\text{Resource\_Blocks} \tag{4.2}$$

For the calculation in the opposite direction, if one knows the number of RBs, assuming that the guard bands are symmetrical, then the minimum guard band can be calculated by using the following equation (derived from Eq. 4.1):

$$\text{Guard\_Band} = \frac{\text{Channel}_{\text{Bandwidth}} - N_{RB} \times 12 \times 2^{\mu} \times 15\,\text{kHz}}{2} \tag{4.3}$$

Similarly, the number of resources blocks (Eq. 4.2) or guard bands (Eq. 4.3) for different combinations of numerologies and 5G carriers can also be calculated.

Table 4.13 gives a similar analysis for FR2 bands on the number of RBs and SCS used for different frequency carrier widths.

What are the use cases of frequency bands of 5G?

Well, to see their possible business cases (at low, medium, and high 5G bands) one needs to consider their coverage, capacity, and latency characteristics. Thus, the widest coverage is obtained with the low bands, so they should be used for nationwide 5G coverage. However, low bands have a moderate capacity (spectrum is limited) and higher latencies than other bands (e.g. high bands), so such low bands are, for example, not suitable for URLLC services with very strict delay requirements. However, low bands are perfectly fine for eMBB and mMTC services.

Going to higher frequencies to high bands, the radio coverage is limited (due to the characteristics of radio waves at higher frequencies), but the latency decreases and becomes very low, in the order of 1 ms (in mmWave bands, above 24 GHz). The high bands also have more capacity because there is a lot of spectrum, which is usually measured in GHz. Besides that, the high bands are suitable for use in hotspots and dense areas of the mobile network. Going to higher frequencies, the bandwidth of the channel increases because the frequency carriers can be much larger (due to the large availability of the spectrum). For example, 5G NR can operate on channels of 50, 100, 200, 400, 800, 1600, and 2000 MHz, in bands above 24 GHz (Table 4.14). However, not all combinations of SCS, NR carrier width, and FR2 band, are possible. For example, according to Table 4.14, carrier width over 400 MHz, or SCS values of 480 kHz or 960 kHz, is possible for n263 (and will be possible for new bands that will appear later on a higher spectrum than this).

What are/were the initial 5G deployment bands? Well, the globally targeted bands for the first 5G "wave" were 700 MHz (also 600 MHz), 3.6 GHz (also 2.5 GHz), and 26 GHz (also higher bands, typically 40 GHz in some countries). Actually, fully coordinated 5G networks include all three types of bands, low band, mid band, and high band.

Low band helps to increase coverage for other bands as well (e.g. mid band) by using CA (in 5G, as before in 4G, it is possible to aggregate carriers from different bands). However, the "sweet" band is definitely the mid band that increases capacity using MIMO and can increase its coverage by CA with low bandwidth carriers.

Expected huge peak bitrates (compared to previous mobile generations) can be provided by also using high bands (in mmWave spectrum) in hotspot areas (otherwise, the bitrates will be more or

Table 4.13 FR2 frequency carrier widths vs. subcarrier spacing (SCS) vs. number of resource blocks (RBs).

		N_{RB} for FR2-1		N_{RB} for FR2-2		
	SCS (kHz)	**60**	**120**	**120**	**480**	**960**
FR2 frequency carrier widths	5 MHz	66	32	—	—	—
	100 MHz	132	66	66	—	—
	200 MHz	264	132	—	—	—
	400 MHz	—	264	264	66	33
	800 MHz	—	—	—	124	62
	1600 MHz	—	—	—	248	124
	2000 MHz	—	—	—	—	124

Table 4.14 Carrier width and SCS for FR2 bands.

		NR carrier width (MHz)						
NR band	**SCS (kHz)**	**50**	**100**	**200**	**400**	**800**	**1600**	**2000**
n257	60	Yes	Yes	Yes				
	120	Yes	Yes	Yes	Yes			
n258	60	Yes	Yes	Yes				
	120	Yes	Yes	Yes	Yes			
n259	60	Yes	Yes	Yes				
	120	Yes	Yes	Yes	Yes			
n260	60	Yes	Yes	Yes				
	120	Yes	Yes	Yes	Yes			
n261	60	Yes	Yes	Yes				
	120	Yes	Yes	Yes	Yes			
n262	60	Yes	Yes	Yes				
	120	Yes	Yes	Yes	Yes			
n263	120		Yes		Yes			
	480				Yes	Yes	Yes	
	960				Yes	Yes	Yes	Yes

less as in 4G network). The mmWave spectrum is new to 5G and the mobile world in general, so it may be last on the list of initial deployments for 5G mobile networks, but it is certainly needed in order to exploit the full potential of 5G in terms of very high bitrates and very low delays. The mmWave spectrum is currently defined (in 3GPP specifications) in a range from the n257 band to the n263 band (in FR2), which is country-specific in terms of their current availability.

However, specifying the frequency ranges sometimes is the easier part, from the regulators' and operators' points of view. The hard part is to set balanced prices of 5G spectrum for different bands (e.g. wider range should result in lower cost per MHz) and to motivate the mobile operator to invest in them, especially for the high spectrum bands. Of course, it will happen eventually everywhere, depending on the drivers of the 5G business at a given time and place (e.g. broadband needs and mobile services in various verticals).

4.6.3 Carrier Aggregation and Bandwidth Adaptation

5G RAN similar to 4G uses CA, where two or more CCs are aggregated and form a single bandwidth "pipe" on network layer (i.e. network layer which is, in fact, the IP layer). In doing so, the mobile terminal can simultaneously receive or transmit by using one or more CCs depending on its capabilities:

- Mobile terminals capable of a single CA timing advance can simultaneously receive and/or transmit to multiple CCs corresponding to multiple serving cells sharing the same time advance [multiple serving cells grouped into one timing advance group (TAG)];

- Mobile terminals with multiple timing advance capability for CA can simultaneously receive and/or transmit to multiple CCs corresponding to multiple serving cells with different time advances (multiple serving cells grouped into multiple TAGs). In doing so, the 5G RAN should be configured so that each TAG contains at least one serving cell;
- A mobile terminal that is not CA-capable can only receive on one CC and also transmit on only one CC corresponding to only one serving cell (that is, one serving cell in one TAG).

In terms of placement of CC in the spectrum, 5G CA enables the aggregation of adjacent carriers, but also the aggregation of carriers that are in different frequency areas. As of 3GPP Release 16, the maximum number of configured CCs for one mobile terminal (mobile interface) is 16 for DL and also 16 for UL. Such CA provided the possibility for 5G NR, in both submitted proposals to ITU-R – 5G RIT and 5G SRIT, to satisfy the ITU's IMT-2020 requirements to be labeled as 5G.

Single NR RB can carry multiple IP packets, and vice versa – single IP packet (protocol layer 3) can be transferred over multiple NR RBs.

Bandwidth adjustment (BA) adjusts the bandwidth of the UE. The width can be ordered to change, such as being reduced during periods of low activity to save power. The location can be moved in the frequency domain, e.g. to increase scheduling flexibility. Also, it can be ordered to change the SCS with the aim to allow different services [6].

For the BA is used configuration of BWP (bandwidth part) as a subset of the total bandwidth available in a given cell. Then, from multiple configures BWP in the cell, BA is realized by providing signaling information to the mobile terminal about which of the BWPs is active. Figure 4.12 shows an example of existence of multiple BWPs configured in a given cell:

- BWP-1 with a width of 50 MHz and SCS of 15 kHz;
- BWP-2 with a width of 20 MHz and SCS of 15 kHz;
- BWP-3 with a width of 40 MHz and SCS of 60 kHz.

Each cell in the 5G network configures an initial bandwidth part (initial BWP) which is used when mobile terminal establishes a connection (noted as connected mode in Figure 4.12). Then,

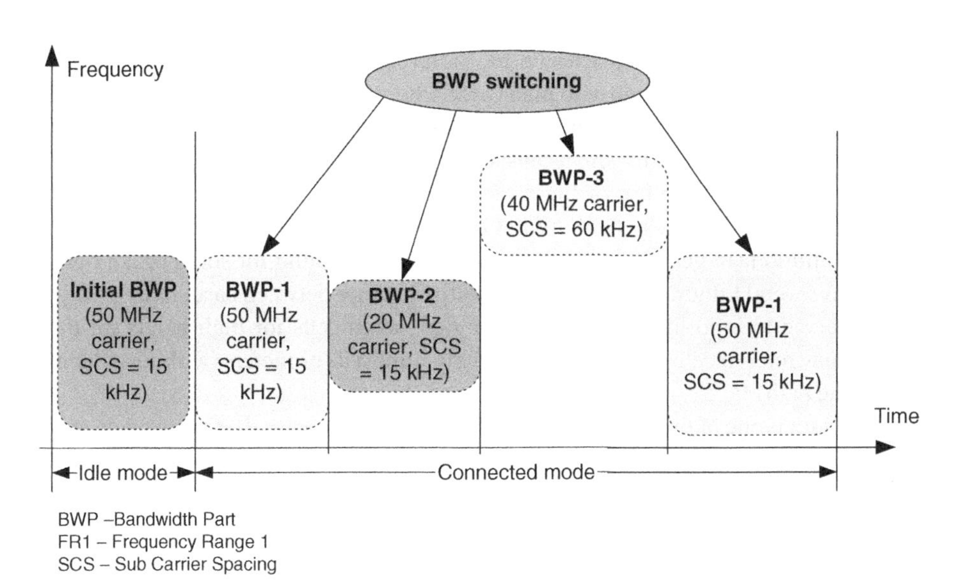

Figure 4.12 Bandwidth adaptation (BA) in 5G.

during the connection it may change BWP, as shown in Figure 4.12, where mobile terminal is switching from BWP-1 to BWP-2, then to BWP-3, and again BWP-1, for the duration of the connected mode.

4.6.4 Discussion on 5G Capacity and User Traffic Versus 5G Spectrum

Even with 5G, most of the traffic will be consumed for Internet access services (including DL and UL direction), noted also as "mobile data" traffic. From the total traffic, the highest percentage comes from the video content, which is over 71% in 2023 and expected to rise to 80% by 2028 [19]. Besides the rise of video traffic due to higher resolutions of video contents, better capabilities of mobile terminals (smartphones) over time and larger mobile data caps (e.g. unlimited data caps), the data traffic generated per minute of usage is expected to increase in future in line with the expected uptake of gaming, XR, and various video-based applications (e.g. social networking).

The real traffic capacity in a given cell depends upon the number of carriers used and their widths, as well as SE. The SE depends upon the modulation and coding scheme in use as well as the MIMO. For example, $N \times N$ MIMO increases the capacity about N times than single-input single-output (SISO). For example, if SISO over a given carrier gives throughput of 1 Gbit/s, then 8×8 MIMO gives over the same spectrum throughput of 8 Gbit/s, of course, if that MIMO is supported on both ends of the radio link (e.g. base station, gNB in 5G, and mobile terminal, e.g. smartphone). The other technology that influences directly the SE is modulation and coding scheme, which can be 64 $(= 2^6)$ QAM, 128 $(= 2^7)$ QAM, 256 $(= 2^8)$ QAM, 512 $(= 2^9)$ QAM, 1024 $(= 2^{10})$ QAM, and so on with step of $2m$ for QAM in future standards (e.g. in the 2030s). For comparison, 16 QAM transmits 4 bits per symbol, 256 QAM transmits 8 bits per symbol, so 256 QAM has $8/4 = 2$ times higher capacity (bitrates) on the same frequency carrier (with a given width). The modulation scheme is directly influenced by the signal-to-noise ratio (SNR) on one side and the support for the given modulation and coding scheme at both ends of the radio link. Higher QAM requires better SNR. As SNR decreases the modulation scheme used should also be at a lower level (e.g. 16 QAM, not 256 QAM), and vice versa. When can we have better SNR?

Well, on average, better SNR is when there is a shorter distance between the mobile terminal and base station, and when there are fewer obstacles on the path (e.g. walls). Shorter distances mean that we need smaller cells with the aim to go with high QAM schemes (e.g. 256 QAM, 512 QAM, and 1024 QAM), and vice versa. Considering that increase of SE is required to achieve high capacity in 5G and beyond RANs, it gives the need for small cells, especially in hotspot and urban areas. Also, smaller cells provide more frequent reuse of the given spectrum resources at shorter distances, which also increases the capacity in the network. What is a small cell in urban environments?

There is no strict definition (in general, various definitions can be found for small cells); however, typically, it means several hundreds of meters (e.g. 400 m distance between base stations [10]), which are typically served by mid bands. On the other side, small cells for high bands (in the mmWave spectrum) may refer to cells of tens of meters up to hundreds of meters (without evident "hard" obstacles on the path).

Overall, high bandwidth is one of the key drivers for 5G. However, even with the first two releases, 15 and 16, 5G NR managed to achieve aggregate DL and UL bitrates, as shown in Tables 4.15 and 4.16, which was confirmed in an evaluation by ITU-R side of the 3GPP proposals for IMT-2020/5G in 2020 [20]. Such bitrates are obtained by aggregations of 16 CCs, where each such CC has a different bandwidth (in MHz) and a different SCS is used depending on whether we use FR1 or FR2 frequency band carriers for 5G.

Table 4.15 Peak data rates for NR DL cases.

Duplexing	SCS (kHz)		Per CC BW (MHz)	Peak data rate per CC (Gbit/s)	Aggregated peak data rate over 16 CCs (Gbit/s)	Required DL bandwidth to meet the requirement (MHz)	IMT-2020 Req. (Gbit/s)
FDD	FR1	15	50	2.4016	38.4256	417	20
		30	100	4.8600	77.76	412	
		60	100	4.7646	76.2336	420	
TDD	FR1	15	50	1.8249	29.1984	548	
		30	100	3.6936	59.0976	542	
		60	100	3.6182	57.8912	553	
	FR2	60	200	5.453	87.248	734	
		120	400	10.923	174.768	733	

Table 4.16 Peak data rates for NR UL cases.

Duplexing	SCS (kHz)		Per CC BW (MHz)	Peak data rate per CC (Gbit/s)	Aggregated peak data rate over 16 CCs (Gbit/s)	Required UL spectrum to meet the IMT-2020 requirement (MHz)	IMT-2020 Req. (Gbit/s)
FDD	FR1	15	50	1.2379	19.8064	404	10
		30	100	2.5032	40.0512	400	
		60	100	2.4750	39.6	405	
TDD[a]	FR1	15	50	0.2671	4.2736	1872	
		30	100	0.54	8.64	1852	
		60	100	0.534	8.544	1873	
	FR2	60	200	1.0059	16.095	1988	
		120	400	2.0118	32.189	1988	

[a] Note: For TDD case, the performance can meet ITU-R requirement with sufficient bandwidth support or adopting full uplink frame structure.

Table 4.17 Peak data rates for LTE Advanced Pro DL cases.

Duplexing	Per CC BW (MHz)	Peak data rate per CC (Gbit/s)	Aggregated peak data rate over 32 CCs (Gbit/s)	Req. (Gbit/s)
FDD	20	0.8876	28.4	20
TDD	20	0.674	21.568	

Table 4.18 Peak data rates for LTE advanced Pro UL cases.

Duplexing	Per CC BW (MHz)	Peak data rate per CC (Gbit/s)	Aggregated peak data rate over 32 CCs (Gbit/s)	Req. (Gbit/s)
FDD	20	0.4246	13.5872	10
TDD[a]	20	0.084	2.688	

[a] Note: For TDD case, the performance can meet IMT-2020 requirement with sufficient bandwidth support or adopting full uplink frame structure.

In addition to 5G, the deployment of 5G SRIT also depends on the characteristics of 4G. According to Table 4.17 (for DL) and Table 4.18 (for UL), also LTE Advanced Pro with aggregation over 32 CC achieves bitrates above the IMT-2020 requirements for 5G, which are 20 Gbit/s in the DL and 10 Gbit/s in UL. Of course, for the possible bitrates (which are provided by CA in the RAN) to be effective, it is necessary to provide adequate fronthaul, midhaul, and backhaul capacity for the 5G network (and the necessary CAPEX).

Besides 5G, the NSA deployment is dependent also on 4G characteristics. According to Table 4.17 (for DL) and Table 4.18 (for UL), also the LTE-A Pro with aggregation over 32 CCs achieves bitrates above the IMT-2020 requirements for 5G, which are 20 Gbit/s in DL and 10 Gbit/s in UL. Then, the aggregate bitrates (achieved with the RAN spectrum configuration per cell or site) directly influence the required capacity of fronthaul, midhaul, and backhaul for the 5G network (and needed CAPEX), including both deployment scenarios, NSA or SA.

4.7 Mobile Access in Unlicensed Bands

Besides licensed spectrum, there are three bands of unlicensed spectrum positioned at 2.4, 5–7, and 60 GHz. These bands are used by many technologies, from which Wi-Fi are the most widespread and mobile technologies are the most emerging ones (4G and 5G operations in the unlicensed bands).

Unlicensed spectrum is likely to play an important role for numerous verticals by allowing them to build their private networks with advanced capabilities, or outsource to third-party providers. Wi-Fi provides high-speed connectivity and the recent standard Wi-Fi 6 (IEEE 802.11ax) can support growing numbers of users and traffic growth (one may recall the huge traffic offload from mobile devices via Wi-Fi networks, usually connected through fixed broadband access).

At the same time, unlicensed IoT technologies also play a role for verticals such as low power wide area services for utilities (e.g. LoRa and SigFox) or local connectivity for smart devices (e.g. Zigbee and Z-Wave).

However, mobile technologies have evolved to make use of unlicensed spectrum to allow verticals to build their own private networks. 4G LTE networks (in fact, the LTE Advanced Pro, standardized with 3GPP Release 13 and 14) as well as 5G NR can be deployed entirely in unlicensed spectrum (at 5–6 GHz) to support a wide variety of use cases ranging from high-speed broadband to low power IoT connectivity. For instance, private LTE networks in the unlicensed spectrum are being used to successfully automate warehouses.

4.7.1 4G LTE and 5G NR in Unlicensed Bands

LTE mobile networks were the first to enter the unlicensed frequency bands (in addition to the licensed bands where only mobile systems originally operated, such as 2G and 3G). But from 3GPP Release 12, each subsequent release further specifies the use of LTE in unlicensed bands with different uses, so we have the following technologies that emerged during the 4G decade (2010s) [1]:

- LTE in unlicensed spectrum/licensed assisted access (LTE-U/LAA): It was introduced in 3GPP Release 13, as part of LTE Advanced Pro, and works in 5 GHz unlicensed band. It provides aggregated simultaneous use of LTE in both licensed and unlicensed spectrum bands in the DL direction to ensure the most efficient use of both available bands. The later enhanced LAA (eLAA) of 3GPP Release 14 allows such aggregation in the UL as well in addition to the DL aggregation of LAA. The LAA is aimed to deliver higher bitrates in the Gbit/s range with a single 20 MHz LTE carrier and an unlicensed 5 GHz band.
- LTE Wi-Fi link aggregation (LWA) enables the aggregation of LTE in a licensed band and a Wi-Fi network that by default works in an unlicensed band. Unlike the previous two uses of LTE in unlicensed bands, LWA does not use LTE in unlicensed bands.
- LTE in Citizens Broadband Radio Service (CBRS) [21]: It is tiered access in the spectrum at 3.5 GHz (3550–3700 MHz band), introduced in the United States in 2015 by the Federal Communications Commission (FCC) based on the three tiers (incumbent access, priority access, and general authorized access), based on the use of approved (by FCC) spectrum access system (SAS) and environmental sensing capability (ESC) which should detect transmissions from radar systems and then to transmit such information to the SAS.
- MulteFire: This is a fully operational use of LTE in unlicensed spectrum at 5 GHz, specified by MulteFire [22] which is a standalone deployment of LTE in unlicensed bands (e.g. independent from mobile operators). MulteFire provides improved local broadband radio access without a SIM card (similar to Wi-Fi access) and opens new possibilities for IoT verticals.

Similar to 4G LTE, the 5G NR is also defined in various combinations for use in unlicensed bands (as shown in Figure 4.13). However, the spectrum of 5G goes also above 6 GHz, which is not the case with 4G LTE. For example, 4G LTE uses unlicensed bands in the range 5150–5925 MHz (referred also as 5 GHz unlicensed band), which was accepted at ITU WRC 2019 (held in 2019) to be unlicensed bands for radio local area networks (RLANs). The RLAN can be Wi-Fi technology or mobile technology, such as 4G and 5G. 5G in the unlicensed spectrum (5G-U) is standardized with 3GPP Release 16 (not with initial 3GPP Release 15 which introduced 5G NR) in both DL and UL for unlicensed spectrum at 5 GHz and also in 6 GHz (that is, spectrum 5925–7125 MHz). There are several main implementations of 5G-U (similar to 4G in unlicensed bands) that include the following:

- NR based LAA: It provides aggregated concurrent use of NR in licensed and unlicensed spectrum bands for both directions, DL and UL.

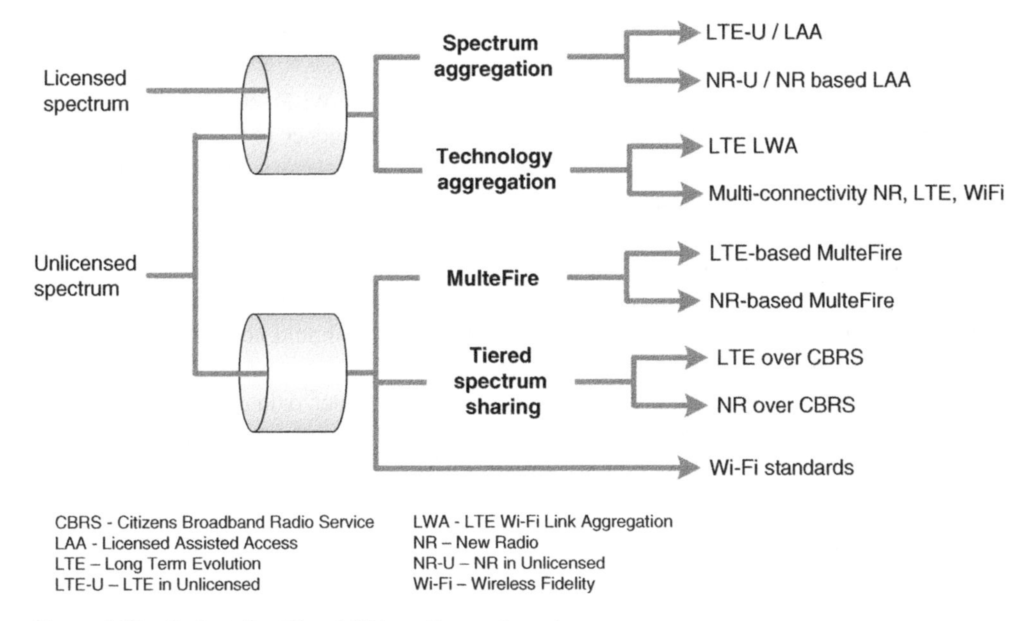

Figure 4.13 Options for 4G and 5G in unlicensed spectrum.

- Multi-connectivity NR-LTE-Wi-Fi: enables the aggregation of NR and LTE in a licensed band and a Wi-Fi network in unlicensed band.
- NR-based tiered sharing: This is based on use of CBRS at 3.5 GHz (e.g. in the US), similar to its use for LTE, based on the noted three tiers. Tier-1 (incumbent access) refers to fixed satellite service (FSS), tier-2 (priority access) refers to authorized federal users in the US, and tier-3 is generalized authorized access (GAA) for open flexible access to the widest group of citizens.
- NR-based MulteFire: This refers to standalone private implementations of 5G NR mainly in unlicensed bands, specified by MulteFire alliance [22].

Considering that 5G NR supports mmWave bands, which with FR2-2 range include also 60 GHz unlicensed spectrum, the next step in NR-U is its use at 60 GHz unlicensed spectrum, which was initially used by Wi-Fi standard IEEE 802.11ad, and then by its successor IEEE 802.11ay.

But, what are the possible use cases of NR-U? In general, unlicensed spectrum has lower QoS than licensed spectrum (due to random access to it by different independent users) and hence it is suitable for eMBB and mMTC services (similar to Wi-Fi), but less suitable for URLLC services. However, in controlled environments, without any interference from other networks that may operate in the given unlicensed spectrum, the QoS becomes more predictable [23] and can be used for support of URLLC services in private 5G-U networks (e.g. in a smart factory).

4.7.2 Access Traffic Steering, Switching, and Splitting for 5G-WLAN

Furthermore, 3GPP includes standards with each new release that enable interoperability between non-3GPP technologies (which is typically Wi-Fi) and 3GPP cellular technologies. As for 5G, its first release, 3GPP Release 15, included optional 3GPP accesses for native 5G services over such non-3GPP access networks.

Later, Release 16 introduced Access Traffic Management, Splitting, and Switching (ATSSS), enabling both 3GPP and non-3GPP connectivity to multiple access networks.

Wireless local area networks (WLANs) are primarily Wi-Fi networks, which can be used with 5G in either trusted or untrusted mode.

Untrusted WLANs include public hotspots, home Wi-Fi, corporate Wi-Fi, etc. that are not controlled by the mobile network operator. However, by enabling convergence to a single 5G CN providing various IP-based services untrusted non-3GPP access, i.e. WLAN can supplement the 3GPP access networks to reduce data congestion and backhaul costs via traffic offloading, provide better coverage and connectivity in hotspot dense areas, enable new business opportunities, and reduce operator capital and operational costs.

ATSSS is standardized as part of 3GPP Release 16. It manages traffic over multiple accesses, which in this case include 5G and WLAN (i.e. Wi-Fi). However, it should be noted that it is an optional feature for both mobile terminals and 5G network.

ATSSS introduces the notion of a multi-access PDU session, for which data traffic can be served over one or more concurrent accesses including a 3GPP access network such as 5G, as well as trusted or untrusted non-3GPP access such as Wi-Fi [24]. The three parts of ATSSS refer to the following:

- Steering: It refers to choosing the best network between 5G or Wi-Fi, based on the information about which radio access interface is better at the given time on a given location.
- Switching: It means changing the access network from 5G to WLAN (offload from mobile network to Wi-Fi) or vice versa (from Wi-Fi to 5G).
- Splitting: It refers to parallel use of 5G and WLAN, and in this case the mobile terminal is connected to both access networks by using aggregation of 5G and WLAN (i.e. Wi-Fi).

In order to support the ATSSS feature, the 5G System architecture is extended as shown in Figure 4.14.

The UE supports one or more of the steering functionalities, e.g. multi-path TCP (MPTCP) functionality and/or access traffic management, splitting, and switching low layer (ATSSS-LL) functionality. The 5G mobile network provides ATSSS rules for enabling for the mobile terminal traffic steering, switching, and splitting across 3GPP access (5G) and non-3GPP access (Wi-Fi).

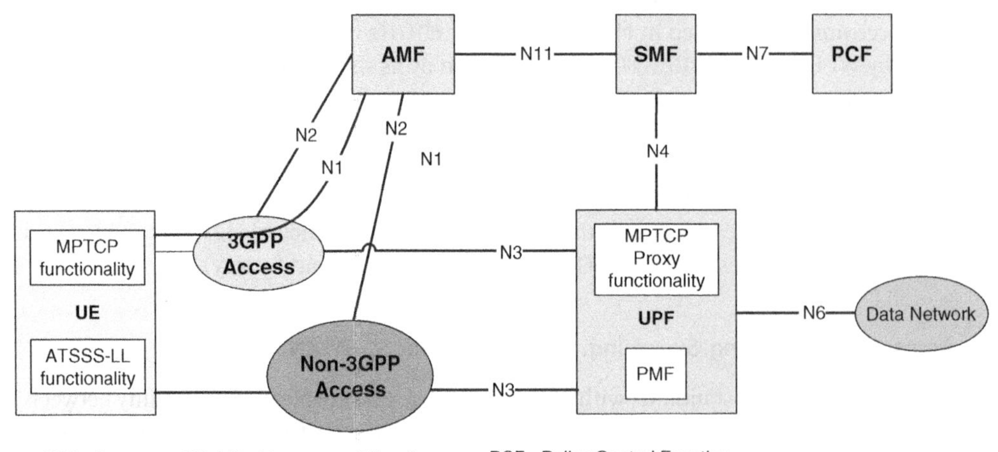

Figure 4.14 ATSSS architecture for 5G.

4.7.3 5G Mobile Technologies in 6 GHz Band

The new 6 GHz bands attract both 3GPP mobile networks and Wi-Fi latest generations (6th and 7th). Discussions and decisions by governments on the use of the 6 GHz band (5925–7125 MHz) are ongoing.

As shown in Figure 4.15, approaches and considerations have ranged from allocating the entire spectrum for unlicensed use (e.g. for Wi-Fi and 5G) to considering the entire spectrum for licensed use (for 5G), or a hybrid option where the lower part (5925–6425 MHz) is intended for unlicensed use and the upper part (6425–7125 MHz) is licensed.

The role of IMT ecosystem development is a topic which cannot be ignored as 6 GHz spectrum management decisions are being made. Where robust 6 GHz IMT device and network infrastructure ecosystems are expected, decisions enabling IMT usage for the 6 GHz band will be justified. At the same time, slow or lacking ecosystem development would complicate support for 6 GHz IMT allocations [25].

With the harmonization of the use of the 6 GHz area by region, an increased use of these bands is expected both for unlicensed 5G networks (private 5G networks) or Wi-Fi networks from the newer Wi-Fi generations (which appear in the 2020s), and for licensed 5G uses that will be in certain regions and countries (e.g. in Europe), driven by ITU WRC-23 and national regulatory decisions.

The 6 GHz band represents the largest remaining contiguous block of mid band (sub-7 GHz) that can be used for licensed mobile services for the foreseeable future. On the other hand, private entities may want to implement private 5G networks for various purposes, such as warehousing, logistics, etc.

Mobile customers (with smartphones) use more data on 5G than 4G, and less traffic can be carried over Wi-Fi as mobile network speeds increase and data allowances (data caps) increase along with mobile network capacity. However, this also depends on the availability and bitrates provided by fixed broadband solutions, given that Wi-Fi is primarily used as a wireless extension to fixed broadband access (e.g. fiber access networks). Also, this can be a valid statement for smartphones, while for a laptop Wi-Fi remains the main approach to wireless connectivity in the home, office, or public places. On the other hand, national mobile operators encourage digital development by providing nationwide connectivity, and they will certainly want to use 6 GHz in a licensed manner. But Wi-Fi 6 and Wi-Fi 7 are already defined to operate in the 6 GHz band, so again it will depend on the region (or country) whether the 1200 MHz in the 6 GHz spectrum will be used for unlicensed, licensed only (5G), or hybrid.

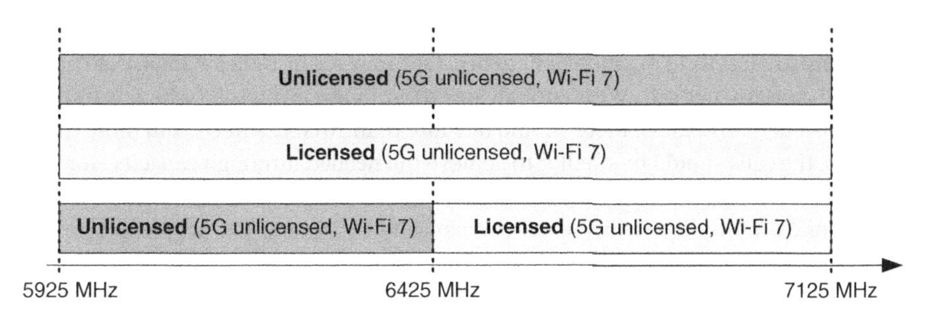

Figure 4.15 Licensed vs. unlicensed possible use cases for 6–7 GHz spectrum.

4.8 Business and Regulatory Aspects of Mobile Broadband

Regarding the regulatory part, licensing and spectrum policy have an initial impact on the 5G roll-outs. It is very important to avoid 5G spectrum and technology fragmentation as well as artificially created high spectrum prices [26]. This refers to all types of 5G bands, including low band, mid band, and high band.

As usual, the availability of the new 5G spectrum is dependent upon national regulators, which are the ones that should free up spectrum for use by national mobile operators, based on WRC harmonization by ITU and regional regulatory harmonization such as the Body of European Regulators for Electronic Communications (BEREC) for the Europe region, as one of many possible examples. This is especially important for spectrum allocations for mmWave because they are very wide, such as several hundreds of MHz or more (e.g. GHz) per operator, considering that mmWave carriers can be up to 2000 MHz wide in 5G, in FR2-2 range.

Another important aspect of 5G regulation is technology neutral spectrum allocation. In fact, 5G has facilitated the decision to adopt technology-neutral spectrum licenses, because it can work in all existing IMT bands (in FR1) and in new bands (in FR2). So, the idea is to re-farm the existing bands (e.g. bands allocated to 2G, 3G, and 4G), so they are used simultaneously for several mobile technologies, including the main two in this decade (until 2030), 4G and 5G. Also, spectrum neutrality provides the possibility for mobile operators to introduce 5G in phases in line with the demand for eMBB while at the same time continuing the support for legacy users.

Let's look to 5G RAN practical deployment. First, there is required a spectrum for 5G NR, and that should be made available by the NRA. Then, when spectrum is available, radio network planning is required, which includes calculations of the required CAPEX (i.e. investments in 5G) and OPEX (operational costs for running 5G by a telecom operator), as well as return on investment (ROI) in order to decide when the time is right to deploy 5G in certain bands. And, the time to deploy 5G RAN in certain bands is different in different countries.

Let's move to the technical part regarding the 5G RAN design. We know that 5G base stations can work on different frequency bands, and that directly influences the positioning of the base stations, such as indoor coverage with mmWave bands and outdoor coverage with sub-6 GHz bands. However, with multi-user MIMO, mmWave can be used to connect small cells.

Overall, for 5G RAN, there is a need for design of high speed, low latency, and high reliability transport network that connects the base stations (gNBs) or, more specifically, connects the remote radio units (RRUs) [27].

In the organization of the 5G RAN (as in other mobile networks) we have backhaul, midhaul, and fronthaul. However, the transport networks should be fiber networks where possible. Why?

Well, that is needed due to the enormous capacity needed for 5G RAN, especially in dense urban areas with many small cells in the mobile network. One may recall Tables 4.15–4.18 to have a view on the needed capacity per cell. With enough spectrum in the given cell (which is not an easy task to accomplish in many cases), there should be more than 20 Gbit/s in DL and more than 10 Gbit/s in UL. Also, the cells should be smaller to achieve the needed ultrahigh capacity (for 5G and beyond), such as a base station every several hundred meters in urban areas and multiple cells on each one, which multiplied with the 5G cell capacity in DL and UL may lead to over Tbit/s capacity in DL (e.g. for 500 cells with maximum capacity of 20 Gbit/s in DL) and 500 Gbit/s in the UL.

But what is the maximum capacity in the 5G RAN? Well, during the evaluation of the 5G proposals from 3GPP, SRI, and RIT (initially made in February 2020), the maximum data rates of 5G NR were tested for both FDD (in FR1) and TDD (in FR1 and FR2), but also the maximum data rates for LTE Advanced Pro. The 5G NR goes with maximum aggregation of 16 CCs, while LTE-A Pro uses up to 32 CCs. Also, the obtained data showed that 5G NR satisfies 20 Gbit/s in the DL and 10 Gbit/s aggregate (peak) rates specified in IMT-2020 (ITU's specification for 5G), but also does LTE Advanced Pro. Thus, LTE Advanced Pro with aggregation with up to 32 CCs can also be referred to as 5G technology. So, one may need tens (or hundreds in future) of Gbps per gNB. However, the "real" capacity in the 5G RAN depends upon the available spectrum that can be used for 5G NR and upon cell size (smaller cells have higher SE than larger cells – which applies to 5G as well as all other mobile technologies). Then, such projected "real" 5G RAN capacity should be used for design of the 5G transport network, i.e. for fronthaul, midhaul, and backhaul. Of course, when designing an optical transport network, transport resources should also be projected for future increases in 5G RAN capacity – as the number of customers and services will increase over time. The number of mobile devices connected via 5G NR will increase exponentially in the following several years until 2030 (when the new generation will appear, 6G).

So, we have to consider different aspects regarding the 5G RAN practical implementations, which include regulatory aspects (e.g. spectrum allocations for 5G), business aspects (decision on CAPEX and projected 5G deployments based also on return of investment calculations), and finally technical aspects which refer to proper design of the 5G RAN, including the transport networks for provision of high-quality services offered through the 5G.

What is new for businesses in 5G? The 5G is targeted to reduce the cost of operating the networks and, with that cost per GB of traffic, thus becoming more cost efficient than 4G mobile networks and earlier mobile generations. With new software-based network architecture and network slicing approach, 5G supports a wider set of use cases with different requirements in terms of speed, latency, density of connections as well as mobility, and all these mean more monetization opportunities. But, future 5G services in different verticals are aimed to transform the whole society and the whole economy by adding more safety, more automation, higher efficiency, and higher quality of life.

Regarding the telecom market, one may expect increased competition with 5G deployments as well as cooperation between the operators and other players in the 5G wide ecosystems. All that will bring new models for ownership of the infrastructure as well as new models for business partnerships during the 5G era, including both private and public 5G networks.

Initial 5G deployments are in 5G NSA mode, that is 5G base stations connected to 4G core, which does not bring all benefits from 5G (regarding the possible services and their performances) that can be obtained through 5G SA with 5G core in operation [28]. Also, almost everywhere there is a need to provide more spectrum to mobile operators. The expectation is that 5G will provide economic and social benefits for all – and that is indeed the long-term expectation of the 2020s mobile generation.

Finally, one may say that the largest challenge to both business and regulatory aspects will be finding the right balance and model that allows innovation to continue to develop with 5G, while at the same time ensuring that telecom operators will continue to receive sufficient economic returns on their investments to build and maintain the mobile networks that will be required to support future products and services.

Thanks to technology advances of IoT, clouds, and AI in many different fields, 5G networks will be at the center of an ecosystem that pushes society's continued digital transformation. In that manner, 5G goes from the well-known "ground" of mobile broadband toward new opportunities in many vertical markets.

References

1 Toni Janevski, "QoS for Fixed and Mobile Ultra-Broadband", John Wiley & Sons, USA, April 2019.
2 Toni Janevski, "Internet Technologies for Fixed and Mobile Networks", Artech House, USA, November 2015.
3 Toni Janevski, "NGN Architectures, Protocols and Services", John Wiley & Sons, UK, April 2014.
4 ITU-R, "Overview Timeline for IMT Towards the Year 2030 and Beyond", August 2022.
5 Kitanov, S., Petrov, I., and Janevski, T. (2021). 6G mobile networks: research trends, challenges and potential solutions. *Journal of Electrical Engineering and Information Technologies* 6 (2).
6 ITU-R Rec. M.2150-1, "Detailed Specifications of the Terrestrial Radio Interfaces of International Mobile Telecommunications-2020 (IMT-2020)", February 2022.
7 3GPP releases, https://www.3gpp.org/specifications-technologies/releases, last accessed in May 2023.
8 3GPP TS 38.401 V17.2.0, "NG-RAN; Architecture Description", September 2022.
9 ITU-T Rec. Y.3112, "Framework for the Support of Network Slicing in the IMT-2020 Network", December 2018.
10 GSMA, "Estimating the Mid-Band Spectrum Needs in the 2025–2030 Time Frame", July 2021.
11 3GPP, "RESTful APIs for the 5G Service Based Architecture", 2018.
12 3GPP TS 29.501 V18.2.0, "5G System; Principles and Guidelines for Services Definition; Stage 3 (Release 18)", June 2023.
13 3GPP TR 21.917, V17.0.1, "Release 17 Description; Summary of Rel-17 Work Items (Release 17)", January 2023.
14 3GPP TS 23.501 V17.6.0, "System Architecture for the 5G System (5GS); Stage 2", September 2022.
15 ITU-T Rec. Y.3156, "Framework of Network Slicing with AI-Assisted Analysis in IMT-2020 Networks", September 2020.
16 3GPP TS 38.104, V17.3.0, "Base Station (BS) Radio Transmission and Reception (Release 17)", September 2022.
17 Qualcomm, "Global Update on 5G Spectrum", November 2019.
18 3GPP TS 38.101, "User Equipment (UE) Radio Transmission and Reception; Part 1: Range 1 Standalone (Release 18)", June 2023.
19 Ericsson, "Ericsson Mobility Report", June 2023.
20 ITU-R, "Final Evaluation Report on the IMT-2020 from 3GPP", February 2020.
21 FCC, "3.5 GHz Band Overview", https://www.fcc.gov/wireless/bureau-divisions/mobility-division/35-ghz-band/35-ghz-band-overview, last accessed in July 2023.
22 MulteFire Alliance, (MFA), https://www.mfa-tech.org/technology/, accessed in July 2023.
23 Qualcomm, "How Does Unlicensed Spectrum with NR-U Transform What 5G Can Do For You?", June 2020.
24 3GPP TR 38.832 V17.0.0, "Study on Enhancement of Radio Access Network (RAN) Slicing (Release 17)", June 2021.
25 GSMA, "6 GHz in the 5G Era, Global Insights on 5925–7125 MHz", July 2022.
26 GSMA, "5G Spectrum – Public Policy Position", March 2021.
27 ITU-T G-Series Supplement 67, "Application of Optical Transport Network Recommendations to 5G Transport", July 2019.
28 ITU Report, "Setting the Scene for 5G: Opportunities & Challenges", 2018.

5

Future Mobile and Wireless Broadband

5.1 5G-Advanced

The development of 5G technology progresses and evolves over the course of various 3GPP releases. While the initial Release 15 introduced the new 5G radio and ushered in the 5G era, which is commercially based on 3GPP systems, the evolution continues as in previous cellular generations. Just to remind you, in 4G there was LTE (Releases 8 and 9), then LTE Advanced (Releases 10–12) and later LTE Advanced Pro as a new label that marked the last 4G release with 3GPP (Releases 13 and 14) [1]. Similarly, in 5G we also have a new designation from 3GPP Release 18 onward and that is 5G-Advanced.

The initial 5G installations in the period 2020–2025 are intended mainly to deliver Enhanced Mobile Broadband (eMBB) thanks to its support for high bandwidth in the new 5G spectrum and low latency, due to new radio and core network architectures. Furthermore, Release 16 and Release 17 bring support for new features and additional frequency bands that further expand the capabilities of eMBB on 5G. Together, these capabilities provide eMBB support for a variety of services, including entertainment, education, and various media. However, the main type of traffic realized is video, which can be viewed in the 5G era on mobile smartphones due to better devices (4K screens), longer battery life (bigger battery and Radio Access Network [RAN] efficiency), and higher capacity resulting in higher average user data rates (assuming the 5G network is well planned for the services offered to end users), as well as higher network data caps that remove all obstacles to consuming any content in any format at any time (e.g. there are no price caps when the same flat price covers all bandwidth consumption for mobile broadband). However, as with previous mobile generations, in the first part of the decade, 5G is differentiated from 4G and previous mobile generations (which exist in the field) and is sold to mobile users for an additional cost or as a separate bundle offer. However, with further evolution of 5G it can be expected (e.g. similar to the development of 4G a decade ago [2, 3]) that such differentiation will diminish, especially considering new features in newer equipment and mobile devices, such as LTE-NR dual connectivity. But the development of mobile networks and devices never stops and moves on, so the wheel keeps rolling and it's good for all parties in the mobile business value chain — vendors produce new equipment, including hardware and software, telecom operators buy and deploy new equipment, end users and businesses buy new devices and new services and move forward in the same way in the future. What is the future?

Well, the near future (e.g. 2024–2029) in the mobile world will be devoted to the development and then the deployment of 5G-Advanced, the new designation for 5G. Also, with 5G-Advanced we are already approaching 6G (which can be expected around 2030), as shown in Figure 5.1.

Future Fixed and Mobile Broadband Internet, Clouds, and IoT/AI, First Edition. Toni Janevski.

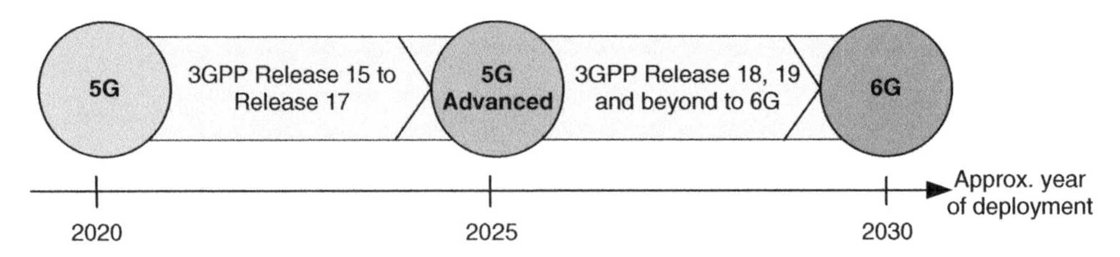

Figure 5.1 Developments from 5G via 5G-advanced to 6G.

5.1.1 Main Characteristics of 5G-Advanced

From the outset, 5G addresses three main usage scenarios: eMBB, massive machine-type communication (mMTC), and Ultrareliable and Low-Latency Communications (URLLC) in accordance with ITU IMT-2020 requirements. The new spectrum for 5G certainly provides the basis for eMBB, which is higher bit rates and capacity than before. 5G massive MTC is based on the use of LTE-M (i.e. eMTC – enhanced machine type communication) and NB-IoT (Narrowband-Internet of Things) which were established with LTE Advanced Pro (and we covered them in previous modules). Releases 16 and 17, which are a further evolution of 5G from the initial 5G standard (provided by 3GPP Release 15), focused in multiple aspects of URLLC services such as industrial IoT as well as vehicle-to-everything (V2X) communication. Along with improvements targeting existing use cases such as mobile broadband, industrial automation, and vehicle-to-everything (V2X), the 3GPP 17 release introduces support for new ones, including public safety, Non-terrestrial Networks (NTNs), and Non-public Networks (NPNs).

Furthermore, 5G-Advanced (3GPP Release 18 and beyond) shows that it will significantly develop 5G in the areas of artificial intelligence and augmented reality. Thus, 5G-Advanced will be developed to provide all kinds of smart connectivity, including smart transportation, smart city applications, and entertainment services with enhanced quality and user experience.

5G-Advanced introduces a range of improvements to 5G systems including (but not limited to):

- Mobile Integrated Access and Backhaul (IAB): It refers to the extension of IAB capability on cars/trains to extend 5G coverage using sub-7 GHz and using mmWave for in-vehicle and out-of-vehicle users, as well as supporting repeaters with traffic awareness and enhanced beamforming capabilities.
- Advanced MIMO in both directions, downlink and uplink: It is intended to provide improved throughput, coverage, as well as more efficient energy consumption.
- Enhanced mobility: It is realized through lower handover latency in 5G-Advanced and improved robustness through lower protocol layer mechanisms (that is inter-cell mobility based on layers 1 and 2 in 3GPP releases) and more seamless carrier aggregation and dual connection operations.
- Using AI/ML data-driven network and application design: It refers to extending the AI/ML framework to optimize mobile network energy savings, traffic balancing and mobility with improved data collection and signaling support, as well as using AI/ML to automatically support air interface functions (e.g. positioning).
- Green mobile networks: It refers to less power consumption per bit per second in mobile networks by defining base station energy consumption models, including the evaluation methodology, Key Performance Indicators (KPIs) and energy reduction techniques specifically for sub-7 GHz in urban and rural areas, as well as use of mmWave bands for dual-link deployments.

Another set of 5G-Advanced targets is spreading 5G to virtually all devices and all possible use cases regarding the provision of services to end users.

One such enhancement is RedCap – Reduced Capability (as NR-Light Evolution), which is defined as extension of the NR-Light services, both aimed to be used for massive IoT over 5G NR, tailored to reduced capability devices which require lower bitrates and hence need less spectrum. RedCap therefore further reduces the bandwidth possibly down to 5 MHz and provides low power modes while being consistent with other NR devices.

Regarding the new media (besides the dominant video content in all networks in the 2020s) much effort is given to adaptation of capabilities of 5G networks and devices to XR (eXtended Reality) services. In that way, boundless XR in 5G-Advanced defines XR traffic patterns with KPIs and QoS requirements. It also provides application awareness as well as higher energy efficiency (reduced power consumption per bit per second as well as adaptation of mobile devices such as smartphones for convenient use of XR content) and capacity improvements (more bits per second considering that XR services on average require much higher speeds than video content), all aimed at better supporting XR experiences over 5G-Advanced mobile networks and end user devices.

One of the targets of 5G since the start are V2X (vehicular to everything) services, however, their deployment will certainly start widely in 2030s (during 6G era) and continue later in follow-up generations after it (e.g. 7G in 2040s, etc.), due to required harmonization between the mobile industry and the vehicle industry (manufacturers of cars, trucks, buses, etc.). In that way, there are several enhancements for V2X services, such as extended sidelink as an upgrade to the cellular V2X from previous 3GPP releases, and expansion of V2X into new spectrum types and bands such as unlicensed and mmWave, as well as sidelink relays that are targeted to further extend the 5G coverage options.

Furthermore, so-called advanced positioning in 5G-Advanced is aimed to provide improved accuracy regarding the positioning which is needed in many use cases for location-based services [4], in different verticals, such as logistics, smart industry, autonomous vehicles and drones, localized sensing for various XR services (which includes AR and VR), and so on. But, is positioning new in mobile networks?

Well, it was initially introduced in the 1990s, in the 2G era, for emergency services. And, it has been continuously used in follow-up mobile generations from 2G to 5G for different location-based services (e.g. for fleet tracking). However, in the 5G and beyond era, the positioning is not just a localization of a device, but in smart factories (e.g. Industry 4.0) it comes with high requirements on QoS parameters, such as ultra-high reliability and very low latency (e.g. for URLLC services). The central function in 5G positioning is Location Management Function (LMF), which uses the measurement information from user devices and 5G RAN via Access and Mobility Management Function (AMF) by using control plane interfaces. For better positioning in 5G NR compared to 4G LTE there are added two new positioning reference signals, one in the downlink, called NR Positioning Reference Signal (NRS), and one in the uplink, called Sounding Reference Signal (SRS). Considering that 5G uses multi-antenna systems and beamforming for mid- and high-frequency bands, the 3GPP standardized (from 3GPP Release 16 onwards) three main methods for positioning, which are power, angular, and time measurements. For example, each beam can be noted as a resource and measurements are collected from one or multiples such resources. While LTE introduced new location techniques in the radio network (new for 4G) such as Observed Time Difference of Arrival (OTDOA) and Uplink Time Difference of Arrival (UL-TDOA) and also positioning based on signal power measurements, the 5G NR introduced angle-based location and high accuracy regarding the time-location pairing due to low Round-Trip Time (RTT) in the radio part. Future mobile services will require open 5G positioning with the aim to be used for different commercial services.

What are other important novelties and enhancements in 5G-Advanced? Such enhancements include drones, i.e. Unmanned Aerial Vehicles (UAVs) and enhanced satellites, i.e. 5G NTNs. Both can be also related to one another considering that drones can be managed by terrestrial or non-terrestrial 5G networks. In a separate section in this chapter we cover the NTN.

There are several other goals in 5G-Advanced besides the above, including improvements to the use of edge computing, system support for AI/ML-based services, further network automation for 5G, improved support for non-public (private) networks, further improvements to Network Slicing, 5G further improvement in terms of network slicing of IP Multimedia System (IMS) telephony service, personal IoT networks, and others.

5.1.2 Time Synchronization and Time-Sensitive Communication in 5G/5G-Advanced

Different vertical services with strict QoS requirements need time synchronization, which is not new in telecom networks. In fact, Synchronous Digital Hierarchy (SDH) was used to build telecommunications circuit-switching transport networks in the 1980s and 1990s [3], and there are also deployments of them that work in the 21st century to transport IP traffic over them. On the other hand, IP/Internet is asynchronous in nature, given that it was initially standardized as a packet switching technology and best-effort networking approach. However, the openness of the IP protocol stack to various heterogeneous applications at the top and various underlying transport technologies at the bottom of the stack has contributed to unprecedented innovation of digital applications and services. However, for future critical services, synchronization is again coming into focus, so 5G networks and beyond need synchronization, however, this time mainly over their IP-based packet switching networks.

The main change for time synchronization in 5G systems was done in 3GPP Release 17, which extended 5G support for time synchronization and time-sensitive communications for any application. With that Application Function (AF) can provide in its 5G domain the QoS requirements, traffic characteristics, and means for activation/deactivation of the time synchronization. In cases when the AF is in a different network domain than the used 5G system, it can use Network Exposure Function (NEF) application programming interface. On the other side, when AF is in the same trust domain as the 5G system, it uses the so-called Time Sensitive Communication Time Synchronization Function (TSCTSF), as shown in Figure 5.2.

There are two main synchronization methods supported in 5G/5G-Advanced (Releases 17, 18, and beyond):

1) 5G Access Stratum-based Time Distribution: It is used for RAN synchronization and also distributed to mobile devices [5]. In this case, the gNodeB needs to be synchronized with 5G Grand Master (GM) clock.
2) (g)PTP (generalized Precision Time Protocol) Synchronization: It provides a time synchronization based on IEEE Std 802.1AS or IEEE 1588 (IEEE stands for Institute of Electrical and Electronics Engineers) operation standards. It can be implemented as time-aware system based on IEEE Std 802.1AS, or it can be based on IEEE Std 1588 in which case it can be implemented either as boundary clock, as peer-to-peer Transparent Clock, or as end-to-end Transparent Clock.

The 5G System (5GS) needs to be compliant with IEEE Std 802.1AS or IEEE Std 1588. Therefore, in 5GS are defined so-called network side TSN Translator (NW-TT) located at the edge of the 5GS. They include device side DS-TT (located in user device, such as smartphone) and network side NW-TT (located at the User Plane Function (UPF) in 5G core). These translators are responsible for provision of functionalities related to standards IEEE Std 802.1AS or IEEE Std 1588.

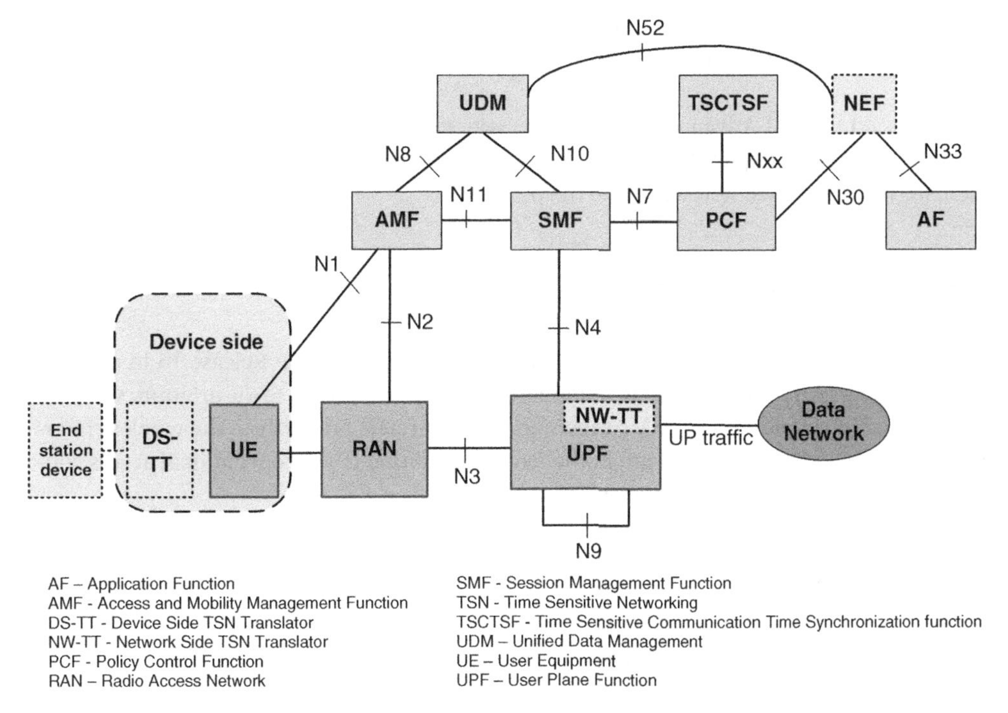

Figure 5.2 5G/5G-Advanced architecture for time-sensitive communications and time synchronization services.

But did this synchronization appear with 5G/5G-Advanced? Well, the IEEE standard 1588 is a widely adopted technique for synchronizing devices even before the appearance of 5G/5G-Advanced standards; however, it also updates over time (last version is IEEE Std. 1588-2019 [6], which is an update of the previous version IEEE Std 1588-2008, which was an update of IEEE Std 1588-2002). So, in this case, the synchronization in IP networks is not re-invented for 5G networks, but it is applied with certain elements that were needed to be standardized by 3GPP for 5G/5G-Advanced time synchronization.

5.1.3 Discussion on 5G-Advanced

5G-Advanced introduces many features that aim to provide more impressive communication, greater accuracy and coverage anywhere on the planet. While existing mobile subscribers expect immersive multimedia communication, its use will strongly depend on the availability of affordable end-user equipment and device design that is suitable for most end users (e.g. VR/AR glasses are more convenient than VR/AR headsets).

The greater precision of 5G moving towards 5G-Advanced, applying time synchronization in the 5G system required for time-sensitive services, is actually primarily aimed at industrial IoT and driverless vehicles and drones. However, industrial IoT is not a novelty introduced by 5G-Advanced; it already exists with Industrial Ethernet standards, but with 5G-Advanced it will also be possible with a 5G mobile network, which can be more convenient because there is no need for cables on the factory floor.

Also, satellite broadband access already exists (without 5G), but with 5G-Advanced, satellite 5G access (i.e. 5G NTN) is expected to emerge, which will be useful as a complement to terrestrial 5G networks for various human- and machine-oriented services.

So, 5G-Advanced provides various improvements in existing networks, but also extends 5G to other areas with the idea of having a single mobile technology for almost all possible use cases in the future.

So, it can be said that 5G-Advanced will pave the way to 2030 and beyond to have one unified mobile technology for all required use cases in all spectrum bands. And, of course, it will drive digital transformation (which never ends) to the next level.

5.2 Integrated Access and Backhaul (IAB)

Integrated Access and Backhaul (IAB) was originally introduced in 3GPP Release 16 in order to extend NR to support wireless backhaul in a standardized way (in addition to the arbitrary way that previously existed in cellular networks). At the same time, it refers primarily to the wireless interconnection of the network nodes that are in the RAN, in addition to the conventional access connection between the base station and the devices.

5.2.1 IAB Architecture

IAB enables wireless backhaul in 5G RAN. The relay node is known as an IAB node, and it is intended to support NR access. On the other hand, the final node of NR backhauling using IAB, which is located on the network side, is called IAB donor (it is actually a gNodeB with additional functionality for IAB). There may be one or more wireless hops within the IAB to connect two entities in the Next Generation Radio Access Network (NG-RAN), i.e. 5G RAN.

The IAB donor includes an IAB-donor-CU (central unit) and one or more IAB-donor-DU(s) (Distributed Unit), as shown in Figure 5.3. In case of separation of gNB-CU-CP (control plane) and gNB-CU-UP (user plane), the IAB donor may consist of IAB-donor-CU-CP (central unit-control plane), multiple IAB-donor-CU-UP (central unit-user plane) and multiple IAB-donor-DU (distribution units) [7].

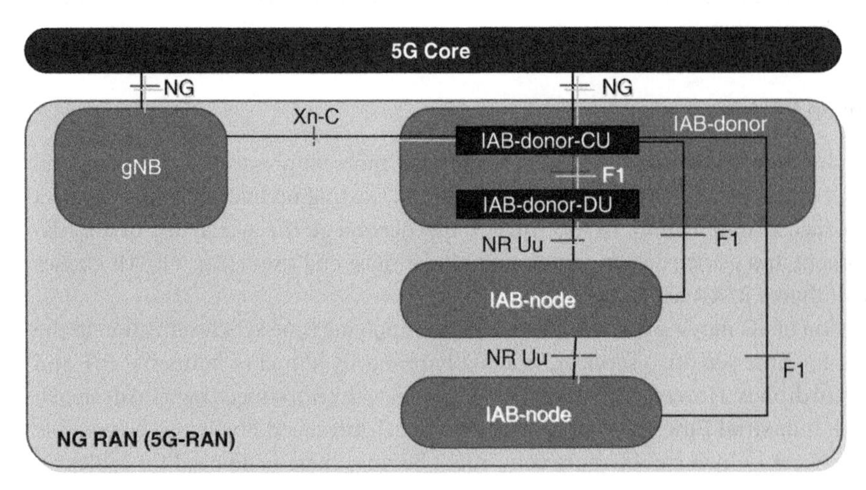

Figure 5.3 Overall architecture of IAB in NG-RAN.

How does the IAB work? Well, the IAB node connects to an upstream IAB node or an IAB-donor-DU via the NR Uu interface of the mobile terminal, i.e. User Equipment – UE. This is named IAB mobile termination, i.e. IAB-MT function of IAB node. The IAB node provides wireless backhaul to the downstream IAB nodes and mobile terminals via the network functionalities of the NR Uu interface. This interface is named the IAB-DU function of IAB node.

The interface between Central Unit of the IAB donor (IAB-donor-/CU) toward its Distribution Unit (IAB-donor-DU) and to IAB nodes in the RAN is called F1 interface. It has split functionalities in CP (control plane) and UP (user plane), which refers generally to NG RAN (i.e. 5G RAN). There, F1-C (control plane) traffic between an IAB node and IAB-donor-CU is backhauled via the IAB-donor-DU and the optional intermediate hop IAB nodes. In the same manner, F1-U (user plane) traffic between an IAB node and IAB-donor-CU is also being backhauled via the IAB-donor-DU and the optional intermediate hop IAB nodes [7].

So, one may say that IAB provides flexibility in deployment of 5G RAN, especially considering that there are needed many small cells and it is not always possible or feasible to have optical connections to the end radio units.

Figure 5.4 shows the protocol stack for carrying the user traffic in IAB for F1-U between IAB-DU and the IAB-donor-CU-UP. In this case, IAB uses tunneling of user data by using GTP-U (GPRS tunneling protocol-user plane), a 3GPP standard (not from the IETF, as the name GPRS), which was also used in 4G core network, but also previously in 3G and 2.5G (i.e. GPRS) systems from 3GPP (of course, newer versions of GTP-U appear over time).

Figure 5.5 shows the protocol stack for F1-C between IAB-DU and the IAB-donor-CU-CP. In this case, the control traffic (F1-C) is also carried over two backhaul hops, through IAB node 1 and IAB node 2.

For both user and control plane traffic, the defined protocol for backhauling in IAB there is Backhaul Adaptation Protocol (BAP) [8]. BAP is positioned below the IP layer (below layer 3) and above layer 2 protocols defined for 5G NR, i.e. above Radio Link Control (RLC) protocol.

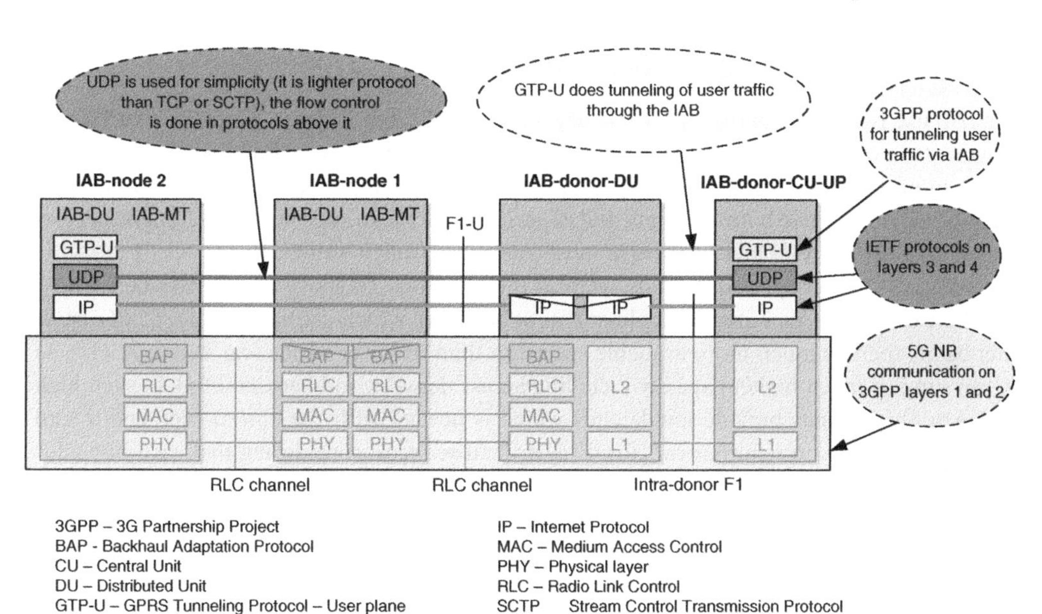

Figure 5.4 Protocol stack for user traffic over IAB.

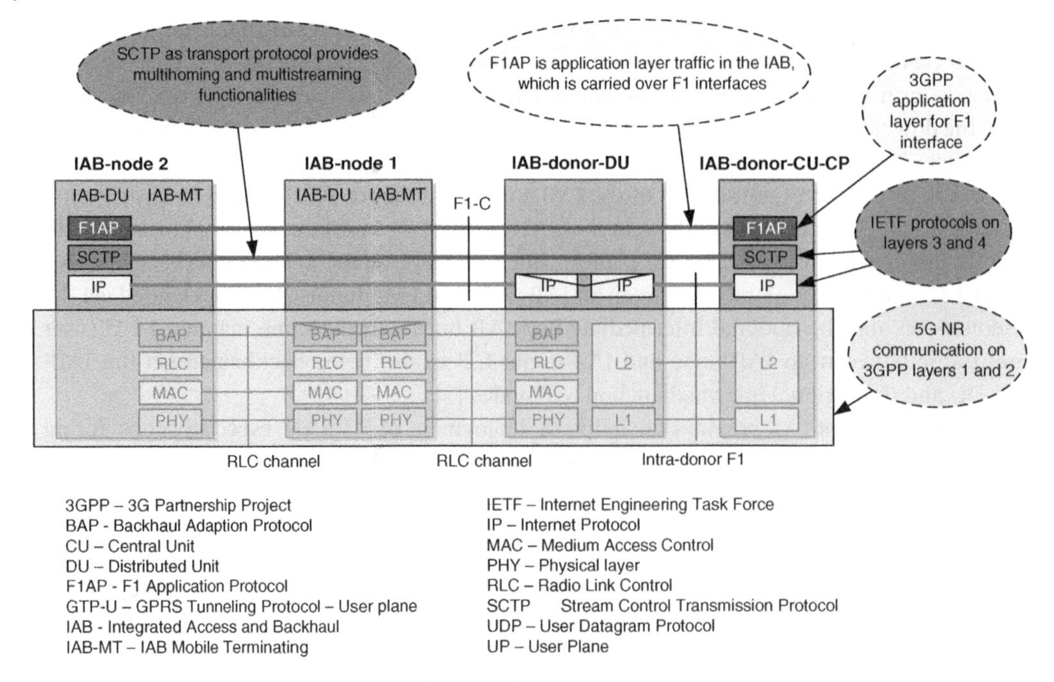

Figure 5.5 Protocol stack for control traffic over IAB.

5.2.2 Spectrum Considerations for Implementation of IAB

There are two different possible implementations of IAB for two main deployment models for 5G, SA (Stand-Alone) and NSA (Non-stand-Alone); they are the following:

1) IAB node using SA mode with 5G Core Network (5GCN);
2) IAB node using EN-DC (E-UTRA-NR Dual Connectivity).

The IAB use is not mandatory in 5G and beyond networks, but it typically applies to backhaul deployments in cases where optical or dedicated wireless backhaul is unavailable or not possible (e.g. due to no possibility for civil engineering works). With IAB based on NR as radio interface some IAB nodes can serve both backhaul and as access network. In such scenarios, mobile devices operate in Line of Sight (LOS) spectrum in most conditions (that is sub-7 GHz).

Regarding the higher spectrum bands, such as FR2-2, which go above 53.6 GHz (where FR2-1 ends), there exist many commercial deployments of backhaul which use bands above 52.6 GHz. In higher bands there is much more available spectrum than in lower bands (e.g. below 7 GHz), so they can support much higher capacity including access network and backhaul link in such high bands. Also, high bands provide low latency which is needed with the aim to keep end-to-end latency within the given requirements for delay critical services, considering that each hop adds additional latency in its radio interface (e.g. if 5G NR latency for a given numerology is, for example, 1 ms, then for two hops in IAB and 5G NR access network as a third hop will result in latency of approximately more than 3 ms, assuming the time needed for processing at IAB nodes, which is certainly not zero, although it should be much smaller than 1 ms).

So, in all cases where optical infrastructure is not feasible, one may use different various wireless point-to-point technologies, such as microwave links, to provide backhaul/relay service. That approach is used since the 2G era in the 1990s until the present time and will continue also in future. However, IAB makes possible to have the maximum possible compliance between the RAN and the backhaul by using the same radio technology (in this case 5G NR) in both access and backhaul. Regarding the 5G standardization in this regard, 3GPP Releases 16 and 17 defined backhaul/relay service with the IAB umbrella specification (concrete implementation can differ from one scenario to another one, even in a single mobile network, depending upon the terrain, type of services provided, and operational costs). However, in Releases 16 and 17 the IAB was limited to carrier frequency below 52.6 GHz (that is FR2-1, the first FR2 band for 5G). However, it is beneficial for IAB to operate also in carrier frequency above 52.6 GHz, which was later extended in 3GPP specifications [9]. So far, there is defined FR2-2 sub-band of FR2 (Frequency Range 2, the reader may refer to Chapter 4 for information about 5G/5G-Advanced spectrum), which extends the 5G NR above 52.6 GHz in the range 52600–71000 MHz. When higher bands, such as FR2-2, are used for IAB, a possible use case is given by a node (e.g. gNB) to provide mobile broadband data services to several mobile devices (e.g. smartphones) and to provide relay over several IAB hops to other mobile devices in parallel. However, the coverage range is typically limited when using such high bands (such as FR2-2), which is in the range of several hundred meters, up to 500 m [9].

Distributed IAB architecture can be used for both private 5G services (e.g. for Industry) and public 5G services (e.g. eMBB for mobile data, voice services, mobile TV, etc.). This distributed architecture generally can be applied to multiple use cases, particularly for deployment of 5G private networks, targeted to enterprise and industrial customers. Also, IAB can be used for intra-RAN local breakout/edge computing services on a private network. However, the main benefits of IAB initially are at mobile network rollout and then eMBB growth, considering the future bandwidth demanding services (e.g. XR services).

The main target of IAB in 5G-Advanced is to enhance reliability and capacity and reduce end-to-end latency (latency is reduced due to shortening the path to certain nodes, such as gNBs).

Additionally, in the focus regarding the IAB (or wireless self-backhauling, as terminology used in 5G-Advanced specifications by 3GPP, in Releases 18 and 19) is on the mobile IAB mounted on vehicles, called Vehicle Mounted Relay (VMR), providing 5G coverage/capacity enhancement supporting single hop in-band, out-of-band backhauling, device handover, and dual connectivity.

5.3 Future WLAN: Wi-Fi 6 (IEEE 802.11ax) and Wi-Fi 7 (IEEE 802.11be)

Wi-Fi (Wireless Fidelity) refers to the IEEE 802.11 family of standards developed for wireless broadband access with high data rates over a limited coverage area, with high fidelity. While 3GPP standardizes all existing mobile networks (including 4G and 5G), IETF standardizes all major Internet technologies (IP, IPv6, TCP, DNS, etc.), IEEE has also standardized several other very important technologies that we use in our daily life and work (as the last part in the Internet access network), such as Ethernet (IEEE 802.3 for fixed Local Area Network – LAN worldwide). The purpose of all IEEE 802.11 standards is to provide wireless connectivity for fixed, portable, and mobile stations within a Wireless Local Area Network (WLAN), also known as Radio LAN (RLAN).

Wi-Fi is a technology that belongs to WLAN, or in other words (used by certain SDOs) that is RLAN. So, it belongs to the LAN segment of IP networks, where the wired part is provided by Ethernet (IEEE 802.3 standards). One should note that the International Telecommunication Union (ITU) uses terminology RLAN to denote WLAN technologies such as Wi-Fi. Also, European

Telecommunication StandardsInstitute (ETSI) uses RLAN terminology in its specifications. Therefore, WLAN and RLAN may be used interchangeably.

Why is Wi-Fi so dominant in the wireless LAN segment? To answer this question, we need to go back in time, to the start of Ethernet standards back in the 1980s, which were later "taken" (for standardization) by IEEE. From the beginning, the Ethernet included standardization (by IEEE) only on the two lowest protocol layers physical layer – PHY (OSI layer 1) and medium access control – MAC layer (OSI layer 2). The same approach continued in Wi-Fi in 1997 when the 802.11 standard appeared (and generally in all IEEE 802 standards later). Why is that important?

Well, because from the beginning, Ethernet and then Wi-Fi were built to have IP (Internet Protocol) stack above it, that is, IP on layer 3 (either IPv4 or IPv6). And that was the approach from the beginning. What happened next?

Well, the Internet technologies overtook the packet-switching world in the telecom/ICT fields by the end of the 1990s, and with that from the beginning of this century both Ethernet and Wi-Fi have become unified LANs. What do we mean by unified? Well, that means there are no other technologies in (W)LAN segments as their competitors.

But what is the primary use of Wi-Fi? Well, it is used as a wireless extension of wired Ethernet. Therefore, many Ethernet and Wi-Fi specifications have similarities, but there are also many differences due to wireless local access in Wi-Fi while there is wired local access with Ethernet.

So, Wi-Fi initially appeared as a standard in 1997, almost simultaneously with the introduction of IP connectivity in 3GPP second-generation mobile networks, with the introduction of General Packet Radio Service (GPRS) in Global System for Mobile communications (GSM) by 3GPP. From that time, in each next mobile generation, including 3G, 4G, and 5G and beyond, Wi-Fi is considered as a technology that may help mobile networks for traffic offload in dense areas.

5.3.1 IEEE 802.11 Standards for Wireless LAN

The scope of those IEEE 802.11 standards (i.e. the Wi-Fi standards) is to define the MAC and PHY specifications, while from protocol layer 3 (where IP is placed in the protocol stack) to the application layer (on top) standardized IETF protocols (e.g. HTTP/TCP/IP) have been used since the beginning of Wi-Fi. Of course, at the application layer you can run any OTT application or any managed IP application as in other broadband Internet access networks.

One should note that after freezing of the initial IEEE 802.11 standard, all further versions of the standard for different functionalities on physical or MAC layers are called amendments and are denoted with a letter or two letters of the alphabet (e.g. IEEE 802.11b, IEEE 802.11n, IEEE 802.11ax, IEEE 802.11be, etc.).

The timeline of Wi-Fi standardization in the past two and half decades (from the initial standard in 1997 until 2022) is shown in Table 5.1.

The IEEE 802.11 standards on PHY define the bitrates by specifying the spectrum bands, modulation schemes, coding rate, number of antennas, etc. The first IEEE 802.11 standard was published in 1997, which provided bitrates up to 2 Mbit/s. It was followed by IEEE 802.11b with up to 11 Mbit/s (in 2.4 GHz), then appeared IEEE 802.11g (at 2.4 GHz) and IEEE 802.11a (at 5 GHz) with up to 54 Mbit/s. That is how the Wi-Fi story has started in the late 1990s and early 2000s.

Thus, 802.11a/g is the 3rd generation, 802.11n is the 4th generation of Wi-Fi, 802.11ac is the 5th generation and the latest generation of Wi-Fi, 802.11ax, is noted as the 6th generation. The future is IEEE 802.11be as Wi-Fi 7th generation, also known as Extremely High Throughput (EHT), which should be completed in 2024 (according to IEEE).

Table 5.1 IEEE 802.11 (i.e. Wi-Fi) standards and their characteristics.

Main WLAN standards	Released	Frequency (GHz)	Maximum data bitrate
IEEE 802.11-1997 (initial standard)	1997	2.4	2 Mbit/s
IEEE 802.11b (Wi-Fi 1st generation)	1999	2.4	11 Mbit/s
IEEE 802.11a (Wi-Fi 2nd generation)	1999	5	54 Mbit/s
IEEE 802.11g (Wi-Fi 3rd generation)	2003	2.4	54 Mbit/s
IEEE 802.11n (Wi-Fi 4th generation)	2009	2.4 and 5	600 Mbit/s
IEEE 802.11ac (Wi-Fi 5th generation)	2013	5	6.9 Gbit/s
IEEE 802.11ad	2012	60	6.9 Gbit/s
IEEE 802.11ah	2016	0.9	8 Gbit/s
IEEE 802.11ax (Wi-Fi 6th generation)	2021	2.4, 5, and 6	9.6 Gbit/s
IEEE 802.11ay	2021	60	20–40 Gbit/s
IEEE 802.11be (Wi-Fi 7th generation)	2024	2.4, 5, and 6	40–50 Gbit/s
Aggregate IEEE 802.11 standards			
802.11-2007	2007	2.4 and 5	Max. 54 Mbit/s
802.11-2012	2012	2.4 and 5	Max. 150 Mbit/s
802.11-2016	2016	2.4, 5, and 60	Max. 6.9 Gbit/s
802.11-2020	2021	2.4, 5, and 60	Max. 10 Gbit/s

5.3.2 Wi-Fi 6 – Next Generation Wi-Fi (IEEE 802.11ax)

The next generation of Wi-Fi that became standardized in the early 2020s is Wi-Fi 6. It refers to the IEEE 802.11ax standard [10]. In the following part, we will focus on the benefits of Wi-Fi 6 compared to previous Wi-Fi generations [1].

One of the benefits of IEEE 802.11ax is the improved use of unlicensed spectrum, which uses both the traditional unlicensed frequency bands of 2.4 and 5 GHz, but can also use the newly opened parts of the 6 GHz band specifically for bandwidth latency-critical applications, such as VoIP or video streaming services, but also for IoT services.

Performance gains are achieved by increasing the average throughput for individual WLAN clients in high-density environments. In doing so, the performance increase is achieved by higher possibility for parallel transfers, not just by higher transfer rates as such. A special role here is played by parallel multi-user MIMO transmissions, including the uplink connection, unlike the previous standard W-iFi 5, IEEE 802.11ac, as well as the introduction of Orthogonal Frequency Division Multiple Access (OFDMA).

Previous WLAN standards (before Wi-Fi 6) used Orthogonal Frequency Division Multiplexing (OFDM), which covers the entire frequency range per unit time when transmitting. For the first time OFDMA is used for WLAN with 802.11ax.

At the same time, the modulation scheme is increased from QAM-256 to QAM-1024, which achieves an increase in the bitrate in the same frequency range and the same number of antennas (by 10/8 times, considering that 256QAM transmits 8 bits per symbol, since 256 = 2^10). With two antennas (based on MIMO) and a channel width of 80 MHz, the potential gross data transfer rate increases to more than 1 Gbit/s, and the theoretical maximum is close to 10 Gbit/s.

Consequently, IEEE 802.11ax is less about increasing absolute speed and more about improved "simultaneity" of transmissions to multiple clients, i.e. increasing efficiency compared to IEEE 802.11ac (Wi-Fi 5) devices. What does that mean? It means that Wi-Fi 6 provides possibility for multiple users to be simultaneously connected to a single AP and to have high performance (without degradation of the signals). That is accomplished by using for the first time in Wi-Fi standards the so-called MU-MIMO (multi user MIMO), which is different than single user MIMO (or just MIMO). What is MU-MIMO? It is directing of radio signals to a device or group of devices based on their location in the coverage area, which is accomplished by using beamforming on the side of Wi-Fi access point – AP (it is usually integrated within a small router, therefore also referred to as wireless router). So, when users are spread on different locations of the coverage area (e.g. in office or at home) then MU-MIMO splits the capacity toward different users by parallel communication via different pairs of antennas to different user or users. With MIMO without MU, all antennas are serving all users, so traffic simply enters one queue on the principle first-come first-served, while with MU-MIMO there can be served multiple users (or multiple groups of users) simultaneously.

In addition to the improvements from parallel transmission, IEEE 802.11ax also offers other benefits to users, such as an improved Target Wake Time (TWT) that increases the battery life of wireless clients (i.e. wireless stations), especially suitable for IoT devices that use Wi-Fi connectivity. They "sleep" better or longer because they need to "wake up" less often to listen to their AP messages.

In general, the key benefits of Wi-Fi 6 technology include higher data rates, increased capacity, higher performance in environments with many connected devices, and improved energy efficiency.

The next generation of broadband Wi-Fi access, IEEE 802.11ax (Wi-Fi 6), is designed for high-density connectivity. It allows many devices to be served simultaneously per AP, unlike previous Wi-Fi standards (where the number of users per AP was limited, and therefore many APs were needed in a given area to serve a large number of Wi-Fi users). It also has optimal performance in dense environments with many APs, and multiple APs can be used on shared channels.

On the throughput side, the gain over 802.11ac (that is, Wi-Fi 5th generation) is evident in both downlink and uplink relative to the number of concurrent clients (wireless stations) connected to the same AP. For example, with only four wireless stations (e.g. laptops, smartphones with Wi-Fi connectivity, IoT devices, etc.), 802.11ax downlink throughput (with large 1500-byte IP packets) is only 10% higher than 802.11ac. On the other hand, the uplink throughput is 2.2 times that of 802.11ac [11]. In short, the "power" of 802.11ax, i.e. Wi-Fi 6 (relative to Wi-Fi 5) is evident when the number of users is higher. In fact, Wi-Fi 6 solves the "weak" point of Wi-Fi in all previous generations, which refers to lower individual bandwidth in a dense environment (with many active Wi-Fi devices connected to the same AP).

Regarding the packet errors in the radio interface, 802.11ax provides extended range by increasing the length of symbols up to four times (from 3.2 to 12.8 μs), thus reducing noise interference (which is always high in unlicensed spectrum bands). Also, considering that Wi-Fi 6 uses Subcarrier Spacing (SCS) four times smaller than the one of Wi-Fi 5, which increases the number of subcarriers on the same frequency carrier. In that manner, 802.11ac (Wi-Fi 5) has 52 subcarriers on 20 MHz carrier, while 802.11ax (Wi-Fi 6) uses 234 subcarriers on the same 20 MHz frequency carrier. However, 802.11ax also uses carriers of 40, 80, and 160 MHz (as 2×80 MHz), with proportionally higher number of subcarriers.

With 802.11ax standard finally approved and published by IEEE in 2021, IEEE 802.11ax (known as Wi-Fi 6) tends to be a "life companion" for 5G mobile networks, which can be used to offload not only mobile Internet traffic, but also certain services that have more stringent quality requirements due to the improved generation of Wi-Fi.

5.3.3 Wi-Fi 7 – Extremely High Throughput Wi-Fi (IEEE 802.11be)

The work on Wi-Fi 7 started in the IEEE 802.11be Task Group which was formally established in 2019, to be finished by 2024 [12]. As with every new Wi-Fi generation, there are several new technologies that come with it.

First, what is the target in Wi-Fi 7 for the throughput? Well, Wi-Fi 7 increases the bitrates up to 30 to 50 Gbit/s, and has lower latencies. For this purpose, there are modifications on both protocol layers which are typically standardized in Wi-Fi standards, that is, changes in the PHY and in data-link layer (MAC). Speeds above 40 or 50 Gbit/s (as theoretical maximum) look extremely high from today's perspective. Therefore, the "alias" for 802.11be, EHT, is completely deserved. Of course, such high throughputs can be achieved only by support of the backhaul network with such capacity (e.g. wired 100 Gbit/s Ethernet as wired LAN and optical connection between the LAN and the telecom transport networks with at least 40 Gbit/s).

What are the other features and novelties in Wi-Fi 7 (802.11be)? Well, as given in Table 5.2, the novelty is the introduction of 320 MHz transmission bandwidth (the previous Wi-Fi generation, Wi-Fi 6, supports up to 160 MHz, the generation before supports up to 80 MHz, and so on). Another improvement is much higher modulation, which is expected to go up to 4096-QAM modulation (which uses 12 bits per symbol, because 2^12 = 4096). Such maximum modulation scheme gives 20% more bandwidth (due to better modulation only) when compared to Wi-Fi 6 (where we have up to 1024 QAM, where we have only 10 bits per symbol). Also, there are enhancements regarding MIMO, by adding more spatial streams (up to 16 spatial streams in Wi-Fi 7, two times higher number that in previous generations, Wi-Fi 6). So, EHT in Wi-Fi 7 is achieved by using multiple radio "tools," such as higher modulation and coding schemes (up to 4096 QAM), larger frequency carriers (up to 320 MHz) and higher order of MIMO. Also, Wi-Fi uses MU-MIMO, similar to Wi-Fi 6, in both downlink and uplink. Then, the achieved theoretical throughputs depend upon several noted parameters, and it can be calculated by using the following equation:

$$\text{Bitrate (bits / s)} = \frac{N_{\text{subcarriers}} \times N_{\text{bits\_per\_OFDM\_symbol}} \times R \times N_{\text{spatial\_stremas}}}{T_{\text{OFDM\_symbol}}} \tag{5.1}$$

where R is error control coding rate, which can be 1/2, 2/3, 3/4, or 5/6 (e.g. for $R = 2/3$ coding rate on each 2 information bits is added 1 redundant bit for error control at the receiving end). $N_{\text{subcarriers}}$ is number of subcarriers (it is 3920 for maximum carrier width of 320 MHz), $N_{\text{bits\_per\_OFDM\_symbol}}$ is

Table 5.2 Comparison of Wi-Fi 7 with previous Wi-Fi generations.

	802.11ac (Wi-Fi 5)	802.11ax (Wi-Fi 6)	802.11be (Wi-Fi 7)
Frequency bands	5 GHz	2.4, 5, and 6 GHz	2.4, 5, and 6 GHz
Multiplexing scheme	OFDM	OFDMA	OFDMA
Channel bandwidth (MHz)	20, 40, 80, 160, 80 + 80	20, 40, 80, 160, 80 + 80	20, 40, 80, 160, 320
Number of spatial streams	8	8	16
Multi-user (MU) MIMO technology	MU-MIMO: downlink only, up to 4 users	MU-MIMO: downlink and uplink, up to 8 users	MU-MIMO: downlink and uplink, up to 8 users
Modulation	Up to 256 QAM	Up to 1024 QAM	Up to 4096 QAM
Maximum theoretical data rate	6.9 Gbps	9.6 Gbps	>40 Gbps

number of bits transmitted over a single OFDM symbol (it depends on the modulation scheme, so for the highest scheme, 4096 QAM, it is equal to 12, because 2^12 = 4096), $N_{spatial\_strema}$ is number of spatial streams and it can be 2, 4, 8, or 16 for Wi-Fi 7, and finally T_{OFDM_symbol} is the duration of the OFDM symbol (it equals 12.8 μs) and guard interval – GI (which can be 0.8, 1.6, or 3.2 μs). If we use all maximum values and minimum GI value, we get theoretical maximum speed of 46 Gbit/s for 802.11be standard (i.e. Wi-Fi 7). However, real speed experienced by end users is typically lower than this. While in Wi-Fi generations up to Wi-Fi 5 was used as a rule of thumb that real speed is 70% of the theoretical maximum; in Wi-Fi 6 and 7 the efficiency is higher due to the use of MU-MIMO, hence, the utilization is somewhere between 70% and 100% of the theoretical maximum speeds.

What about the spectrum that Wi-Fi 7 works in? Well, Wi-Fi 7, similarly to Wi-Fi 6, will continue to use 2.4, 5, and 6 GHz frequency bands. So, regarding the unlicensed spectrum Wi-Fi 7 continues the "path" already traced by Wi-Fi 6. However, the allocations on 6 GHz spectrum (unlicensed or licensed IMT) may be region and country dependent.

Regarding the legacy unlicensed spectrum at 2.4 and 5 GHz these bands are smaller (especially 2.4 GHz) and already congested by many Wi-Fi devices (and many other) operating in them. So, they are limited and often congested, and that results in lower QoS (Quality of Service) when running bandwidth "hungry" applications, such as VR/AR or 4K/8K video. Better throughput definitely requires more spectrum, such as 320 MHz in 802.11be (Wi-Fi 7), but there is no place for it at 2.4 GHz or at 5 GHz spectrum. The only solution is the new 6 GHz spectrum, which enables Wi-Fi 7 to work with 320 MHz transmission bands. Of course, Wi-Fi 7 can use smaller bandwidth allocations as well, which will be probably a default case in "cheaper" devices targeted for home use (as usual) and in ordinary offices.

Are there any QoS improvements with Wi-Fi 7 and Wi-Fi 6? Well, 802.11ax first brings finer granularity for QoS support due to the use of OFDMA in the radio interface and due to the uplink scheduler, which is defined in the standard. Due to that, Wi-Fi 6 networks have the possibility (for the first time in Wi-Fi history) to allocate resources in the uplink (i.e. to schedule traffic from different Wi-Fi clients in the uplink), which brings Wi-Fi networks step closer to QoS in mobile networks.

What is contribution to QoS in Wi-Fi 7? There are two main novelties which contribute to much better QoS in Wi-Fi, and they are Multi-link Operation (MLO) and Multi-AP Cooperation. What are their contributions?

Well, the MLO mechanism in fact provides possibility of multi-link aggregation, which will allow Wi-Fi devices that have support for multiple unlicensed bands (2.4, 5, and/or 6 GHz) to operate on all of them simultaneously. That is aimed to increase the performances in dense environments, including improvements on most critical QoS parameters such as throughput, latency, and reliability (higher number of available links always results in higher reliability and that also means higher QoS than before).

What else contributes to higher QoS in Wi-Fi 7? That is the multi-AP coordination. In the legacy 802.11 protocol framework, there is not much coordination between neighboring APs, but their coverage areas are overlapping and causing interference to one another. However, APs in Wi-Fi 7 use inter-AP coordinated planning in the time and frequency domains. Such coordination between Wi-Fi 7 APs is expected to reduce the interference between them and with that to improve the utilization of radio resources.

So, overall, all the new features coming with Wi-Fi 7 will significantly improve QoS, including throughput, latency, and reliability. It makes it possible to use Wi-Fi 7 for many applications that have stricter QoS requirements, starting from real-time video streaming with 4K and 8K (e.g.

watching Netflix or YouTube over Wi-Fi with such resolutions), online gaming, collaboration in real time, various emerging cloud applications, as well as for some industrial IoT use cases, AR/VR, and so on.

5.4 5G – WLAN Interworking

The mobile and WLAN interworking is being used since the introduction of the Internet in 3GPP mobile networks with the GPRS (i.e. 2.5G). At the beginning of the 2000s, there started to appear Wi-Fi hotspots in many public locations, which were loosely or tightly coupled with the mobile network architectures. They were primarily intended for traffic offload of mobile traffic to Wi-Fi networks in the hotspot area (with dense end users' traffic). But, why is Wi-Fi attractive for traffic offload in mobile networks?

Well, due to several reasons. First, Wi-Fi is a cheaper technology (than mobile technologies, such as 4G and 5G today, and 6G in the future, in the 2030s) and works completely in unlicensed bands (that results in zero costs for spectrum, unlike the licensed bands), so it is a perfect solution for data traffic offload from mobile networks in dense areas. Also, one should note that Wi-Fi offloads the traffic via fixed access networks, thus contributing to less congestion in the mobile access network or, in other words, contributing to higher capacity in hotspots (e.g. in dense urban areas, downtowns, shopping malls, airports, public places, and so on). Further, on average, fixed broadband access can offer higher aggregate and individual speeds than mobile broadband access (on one side, that is due to the huge capacity of the fiber and possibility to deploy as many cables as needed; on the other side, that is due to scarce and expensive radio resources in mobile networks), and Wi-Fi is connected to fixed access infrastructure. Another important fact is that the theoretical peak data rates have historically always been higher in the leading Wi-Fi standards (e.g. Wi-Fi 6 today and Wi-Fi 7 tomorrow) compared to the leading mobile technologies (e.g. 4G and 5G/5G-Advanced).

What is the approach with Wi-Fi access in 3GPP mobile networks? Well, there are several approaches [13]:

- Untrusted non-3GPP access networks – it refers to loose coupling between WLAN (as main non-3GPP access network) and 5G network.
- Trusted non-3GPP access networks – it refers to tight coupling between WLAN (as main non-3GPP access network) and 5G, regarding various 5G functionalities.
- Wireline access networks – this is a similar scenario with trusted non-3GPP access, with Wireline Access Gateway connected to wireline data network on one side and with 5G core elements AMF and UPF on the other side. It is used for Fixed Wireless Access (FWA) solution with 5G network, and we refer to it in Section 5.7 of this chapter.

Based on the above classification, in the following sections, we refer to trusted and non-trusted WLAN access in 5G network.

5.4.1 Untrusted WLAN Access in 5G Network

In untrusted non-3GPP access network scenario, the untrusted network (e.g. WLAN, i.e. Wi-Fi) is connected via dedicated Non-3GPP Inter-Working Function (N3IWF) which is connected to AMF (via N2 interface) for mobility management and to UPF (via N3 interface) for the user traffic. So, N3IWF is the main component for untrusted access in 5G, which uses IPsec for transfer of traffic to/from mobile terminal (i.e. UE), while the traffic between N3IWF and UPF is transferred via GTP-U (GTP for tunneling the User plane traffic). One should note that in untrusted access, UE is connected simultaneously to gNB in 5G RAN and to untrusted access nodes such as Wi-Fi AP. In non-roaming architecture (where gNB, AMF, and N3IWF are all located in the same 5G network), the same AMF serves both access connections, via 5G RAN and via Wi-Fi (as the non-trusted access).

The network architecture for untrusted WLAN access interworking with 5GCN is shown in Figure 5.6. A mobile terminal (UE) that is accessing the 5GCN through an untrusted WLAN shall support Non-access Stratum (NAS) signaling and needs to register and authenticate with the 5GCN via the N3IWF. As a difference from earlier untrusted 3GPP architectures (e.g. 2.5G–4G), the registration and authentication procedures are similar to 3GPP access procedures where the access and mobility management function (AMF) is used to register the UE while the Authentication Server Function (AUSF) is used to authenticate the UE. However, before registration of the UE, the establishment of an IPsec tunnel for CP between the UE and N3IWF is needed for securing the exchange of mobility and session management messages with 5G core functions, AMF and SMF. The UE establishes Protocol Data Unit (PDU) sessions using the IPsec signaling with the SMF via the AMF.

When establishment of a PDU session is initiated, the N3IWF set up IPsec security associations with the mobile terminal (i.e. UE) for various QoS flows of that PDU session. For transfer of user data traffic GTP-U tunnel is used between N3IWF and the UPF in 5GCN. So, we have concatenated tunnels end to end in WLAN access network and 5G core network, through which flows both uplink and downlink traffic between the UE and external data network by using IPsec tunnels between UE and N3IWF and the GTP-U tunnel between N3IWF and UPF. The Network Access Stratum (NAS) messages are transported over the NWu interface which is defined between the N3IWF and UE.

During the session establishment procedure, the N3IWF initiates the establishment of the IPsec Security Associations (SAs) with the UE.

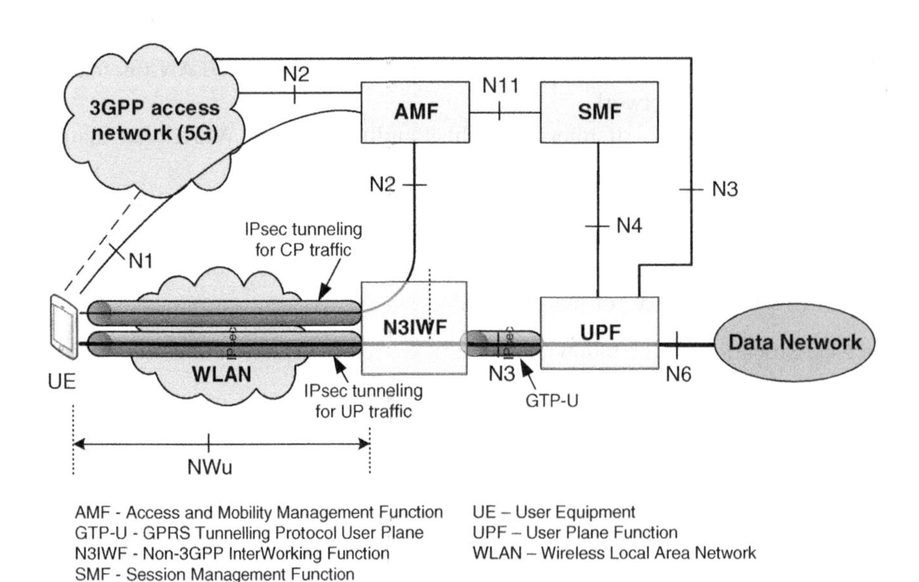

Figure 5.6 Untrusted non-3GPP access network architecture in 5G.

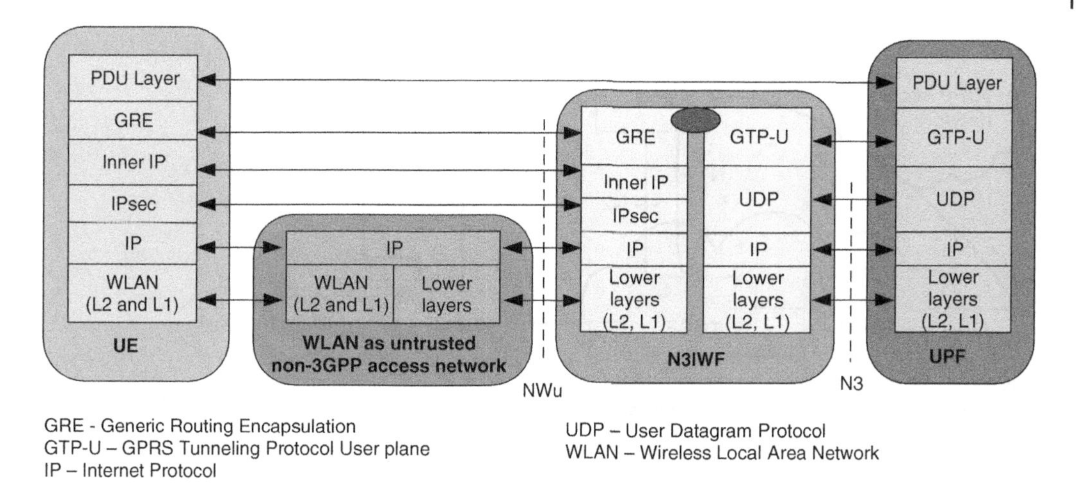

Figure 5.7 User Plane (UP) protocols for untrusted 5G-WLAN interworking.

The UP protocol stack is shown in Figure 5.7. It includes the protocols used in the UE, WLAN, N3IWF, and UPF to carry UP traffic between the UE and the data network through the mobile core network. IPsec tunnel mode is typically used there to protect and encrypt the original IP user data and the port numbers (e.g. TCP port number, UDP port number) that are typically used for IP/Internet communication end to end.

5.4.2 Trusted WLAN Access in 5G and Wireline Access

The other approach is trusted non-3GPP access. In this case, there is standardized Trusted Non-3GPP Gateway Function (TNGF) which is connected with TNAP (Trusted Non-3GPP Access Point) toward the mobile device (i.e. UE), and with 5G core functions, in particular with AMF and UPF (as shown in Figure 5.8).

The connection between the mobile terminal (UE) and TNAP can be by Point-to-Point Protocol (PPP), including wired LAN (e.g. Ethernet, IEEE 802.3) or WLAN (e.g. Wi-Fi). When registering in non-3GPP trusted access, NAS messages are always exchanged between the mobile device (i.e. UE) and the AMF, where the UE can be authenticated using the security context available in the AMF.

The Wireline 5G Access Network (W-5GAN) works by connecting to the 5GCN through a Wireline Access Gateway Function (W-AGF), which connects the CP and UP functions of the 5GCN through the N2 and N3 interfaces, respectively.

The 5G Residential Gateway (5G-RG) is a node in this architecture that is simultaneously connected to the 5GCN via the 5G RAN and to the W-5GAN. Thus, it can be served by the same AMF in a given 5G network.

The 5G-RG maintains the NAS signaling connection which is established with the AMF via the W-5GAN after all PDU sessions (for the 5G-RG) are finished or handed over to continue over the 5G RAN.

What kind of access is wireline 6G access network? Well, it can be provided in both trusted and untrusted manner, that is, with a trusted or untrusted non-3GPP access network. So, mobile terminals (i.e. UEs) which are connected to 5G-RG can access 5GCN via N3IWF (for untrusted access) or Fixed Network Residential Gateway (FN-RG) access via the TNGF (for trusted access). In the case of an untrusted non-3GPP access network, user traffic is connected to the Internet (i.e. to the

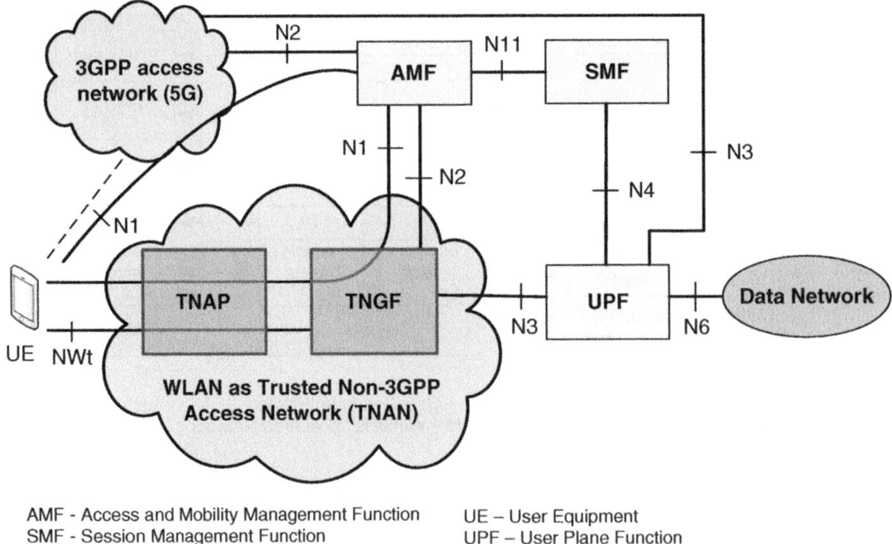

Figure 5.8 Trusted WLAN access to 5G.

packet data network) via W-AGF. In case of trusted wired access, the user traffic goes through UPF in the 5GCN. Overall, wireline 5G access is similar to WLAN access, the difference is the access part which is typically Ethernet (instead of Wi-Fi). Of course, it is possible to have both WLAN and wireline 5G interworking, either in trusted or untrusted mode, depending on the need or customer choices.

5.4.3 Discussion on 5G and WLAN

For devices to access both 5G and WLAN, they must be equipped with 5G and Wi-Fi chips. While smartphones support both types of connectivity, the vast majority of connected devices (e.g. laptops) that enterprises use are Wi-Fi only. Thus, until 5G chips or modules become more widely available and used in a larger number of devices, 5G-WLAN convergence will likely remain limited to specific use cases (e.g. use cases that depend on the mobility that enables the ability to switch from Wi-Fi to 5G seamlessly, or mission-critical use cases in verticals such as manufacturing that can justify an investment in (private) 5G as a backup to Wi-Fi).

On the other hand, one can expect 5G (and beyond) and Wi-Fi interworking to continue as usual when and where needed by mobile operators actually deploying Wi-Fi for additional "cheaper" capacity in their networks in hotspot locations with heavy mobile traffic.

5.5 5G Non-Terrestrial Networks (M2M/IoT Over Satellite)

We have covered satellite broadband in Chapter 3. As a reminder we have Geostationary Earth Orbit (GEO) satellites which are orbiting at 35786 km around the Earth and have a stationary footprint on Earth. However, Low Earth Orbit (LEO) satellites which are orbiting below 2000 km, and sometimes even Medium Earth Orbit (MEO) satellites which are orbiting in the space between

LEO and GEO, could be used for two-way broadband communications. LEO and MEO move relative to the Earth's surface that is why they are called non-GSO.

Non-GSO satellites are an important leap forward in satellite technology innovation for provision of higher capacity and better performances regarding the latency via the satellite broadband network (the GSO provides global coverage with fewer satellites, but incorporates larger latencies and lower capacity due to a larger distance from the Earth's surface). What are the advantages of non-GSO satellite broadband systems regarding their use with 5G/5G-Advanced?

Well, there are multiple advantages. For example, non-GSO satellite systems include near-instant coverage across wide geographies and over any topography, which is not possible with terrestrial mobile networks in many unpopulated areas on the planet, including deserts, high mountains, oceans, and seas. Also, non-GSO has high reliability due to its "virtual" immunity to many risks that other networks may face, such as theft or damages of the equipment, conflict areas in certain countries, as well as natural disasters (e.g. earthquakes, hurricanes, wildfires, etc.).

With the development and completion of the initial 5G standards by 3GPP (with Release 15), there has been (from Release 16 onward) an increase in the interests of the satellite communication industry to combine 5G land mobile networks and satellite networks. In 3GPP's terminology, the satellite networks belong to so-called NTNs. All possible types of satellite platforms for NTN are given in Table 5.3.

In general, 5G NTN refers to a network (or segment of networks) that are using 5G radio communication on board a satellite or Unmanned Aircraft System (UAS) platform. Here, UAS refers to systems operating in altitudes typically between 8 and 50 km including High Altitude Platform Stations (HAPS), as shown in Table 5.3.

5.5.1 5G NTN Architectures

The interworking of Terrestrial Networks (TNs) and NTNs starts with 5G/5G-Advanced standardization, and it is expected to continue toward 6G in the next step. While 5G-Advanced is aimed to provide integration of TN and NTN, 6G is expected to provide unification between these two main types of networks, on the ground and in the sky. Such TNs and NTNs may be operated by the same telecom operators or different telecom operators (in such case roaming will apply, which will further extend in the NTN segment).

What are the initial use cases of 5G NTN? Well, the initial target is to provide eMBB services, especially in cases of unserved or underserved areas. The second target are Massive MTCs (mMTC), that is, massive IoTs. The third main use case are V2X services, considering that vehicles may be

Table 5.3 Possible types of satellite platforms for NTN.

Platform type	Typical altitude	Orbit	Footprint size
LEO satellite	300–1500 km	NTN stations moves around the Earth	100–1000 km
MEO satellite	7000–25000 km		100–1000 km
GEO satellite	35–786 km	NTN station is keeping its position fixed in terms of elevation/azimuth with respect to a given point on the Earth's surface	200–3500 km
UAS platform (including HAPS)	8–50 km (20 km for HAPS)		5–200 km
HEO (high elliptical orbit) satellite	400–50000 km	Elliptical path around the Earth	200–3500 km

moving also on paths not covered by the terrestrial networks, which is always true for ships on seas and planes in the sky. What about URLLC service types?

The URLLC have stricter demands on delay and reliability. While satellites can contribute to reliability, due to multiple paths available via terrestrial and non-terrestrial connections, they cannot provide very low delays which are required for URLLC (i.e. critical IoT) services. The reason lies in the fact that the maximum speed of the radio signal is always below the speed of light which is approximately 300000 km/s. Of course, UAS platforms can provide lower latencies, and also LEO provides comparable latencies in NTNs as those found in TNs in rural areas (where cells are larger). However, MEO and GEO have higher latencies. For example, one-way latency at 7000 km (lower MEO attitude) is 7000 km/300000 km/s = 23 ms, while for GEO satellites one-way latency is 35786 km/300000 km/s = 119.2 ms, which gives an RTT for MEO of at least 46 ms and for GEO of approximately 240 ms. If we compare these to the requirements set for delay critical 5G services (given in Chapter 4 of this book) then it is evident that MEO and GEO cannot be used for delay critical services. Overall, one of the best use cases for Mobile Satellite Services (MSSs) is their use as a complement of terrestrial eMBB services [14].

For completeness, one should note that MSS started in the 1990s with systems such as Iridium, so they are now new to the telecom world. However, such systems were based on different technologies than terrestrial mobile networks. However, MSS interfaces may be integrated in the smartphones, and we have examples provided by different vendors, such as Apple iPhone and Globalstar, Huawei Mate, and BeiDou, as well as adding the noted Iridium to the Qualcomm Snapdragon Satellite [14]. The first standardized NTN solution for 5G was completed with Release 17 and then continued in 5G-Advanced Releases 18 and 19.

What are the possible 5G NTN architectures? Two typical scenarios of a satellite-based 5G NTN providing access to UE can be defined [15]:

1) Transparent Payload (shown in Figure 5.9): In this case, standardized with Release 17, the gNodeB is located on the ground (not in the satellite), between the satellite gateway (on the side of the satellite connection) and 5GCN (on the other end). The satellite acts as a repeater, without processing of packets at the satellite. Hence, the waveform signal that repeats the carrier is unchanged.
2) Regenerative Payload (shown in Figure 5.10): In this case, satellite payload implements regeneration of the signals received from Earth. The satellite (or UAS platform) effectively has gNodeB functions and hence can perform decoding, processing, and forwarding of packets (e.g. to next hop satellite via inter-satellite link).

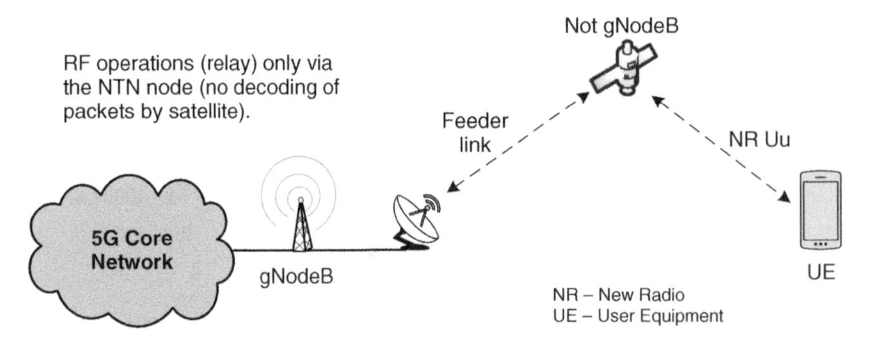

Figure 5.9 5G NTN with transparent payload.

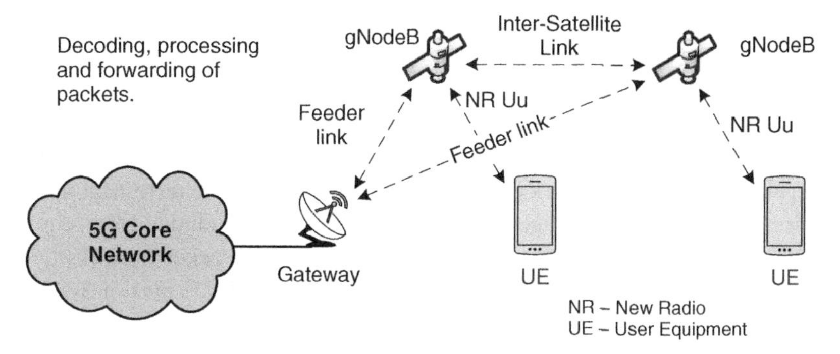

Figure 5.10 5G NTN with regenerative payload.

So, in transparent payload scenario, the satellites (i.e. NTN nodes) appear only as a bitpipe in the sky, while all 5G communication, including 5G RAN and 5GCN are deployed on the ground. For 5G NTN communication this is the initial implementation. The regenerative use case is a long-term scenario which, in practice, implements a 5G access network in the sky. The protocol stack for the regenerative use case is shown in Figure 5.11. In such a case, 5G QoS flows are established between the 5G UE (e.g. smartphone) and the UPF in the 5GCN for the given PDU session. In this way, the regenerative NTN mode is a fully integrated TN-NTN scenario, which can be expected to be widely deployed in the 6G era (in the 2030s).

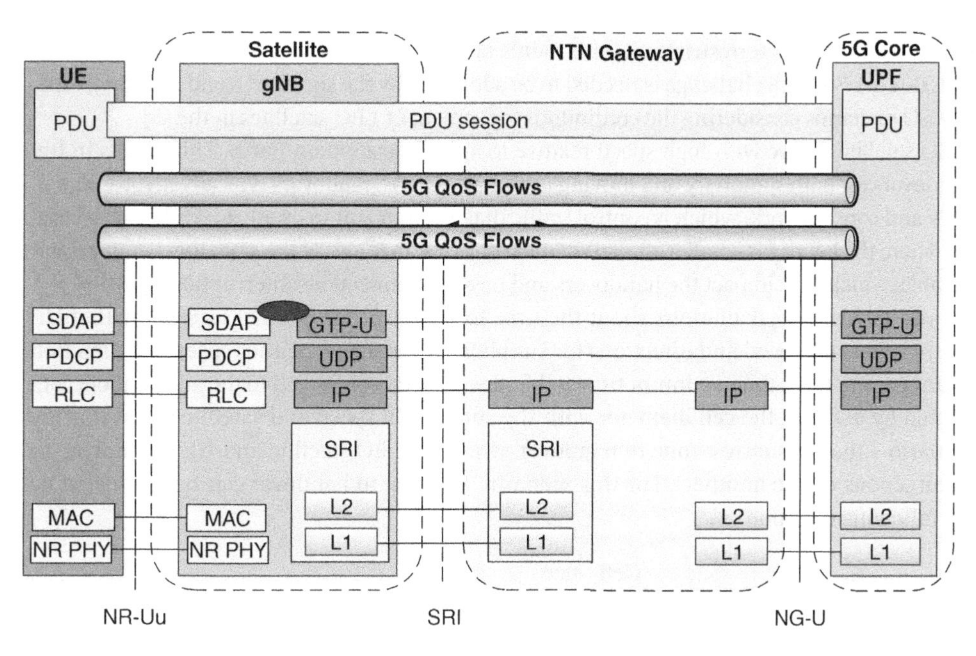

Figure 5.11 Protocol stack for NTN regenerative mode.

The 5G and further 6G system is expected to support service continuity between the 5G TN and the 5G NTN, whether owned by the same operator or by different operators that have a TN-NTN roaming agreement.

However, one of the most interesting use cases (especially in agriculture, farming, disaster recovery, Earth exploration, etc.) is IoT services over NTN. This way, 3GPP includes support for NB-IoT and eMTC over NTN. In general, satellite networks can be used to supplement the terrestrial network to provide IoT services for low or no coverage areas. 3GPP in Release 17 included NTN support for NB-IoT and eMTC. However, it should be noted that only Evolved Packet Core (EPC) connectivity was in scope in Release 17 for NTN IoT, while 5GCN connectivity is part of 3GPP Release 18 and beyond, i.e. part of 5G-Advanced framework.

5.5.2 Mobility and Handovers in NTN

In land mobile network the terminal (UE) determines that it is near a cell edge due to a clear difference in signal strength (Reference Signal Received Power – RSRP) when compared to signal strength of that same cell when UE is close to the center. That also influences the algorithms for handovers (of a connection from one cell to another) in terrestrial mobile networks. However, it is not applicable in the same manner in NTNs, considering that there may be a small difference in signal strength between two adjacent satellite beams which normally overlap in some region on ground. Hence, the UE may have issues in distinguishing which cell is the better cell. Also, there can be a ping-pong effect, even with defined hysteresis (difference of signal thresholds for handover in the opposite direction for avoiding ping-pong effect), because the cells are larger in size (than terrestrial cells), sometime even more than 1000 times (e.g. terrestrial cell with radius of 1 km in comparison with satellite cell with radius of 1000 km). So, such challenge is needed to be addressed by the standards (and vendors), especially for LEO scenarios considering the continuous movement of LEO satellites in the sky.

Non-GSO satellites move with high speed relative to a fixed position on Earth. This results in frequent handovers for both stationary and moving terminals (UEs). Handovers typically involve signaling in RAN and core network, which is control traffic that is redundant to user load. Unlike terrestrial networks where the latency is smaller, due to smaller cell sizes, in 5G NTN the signaling latency is not unneglectable, which may impact the handovers and possible connections interruptions.

In this way, let's make calculations about the time to handover for UE. Let us assume that UE travels with a constant speed and direction (for simplicity, in practice, neither speed nor direction are constant parameters), the maximum time to handover from a given cell to the adjacent one can be calculated by dividing the cell diameter with the sum of UE speed and satellite speed (for the worst scenario – that is, shortest time to handover, when the LEO satellite and UE are moving in opposite directions one to another). For this scenario, the time to handover can be calculated by using the following equation:

$$\text{Time\_to\_Handover} = \frac{\text{Cell\_size}}{\text{Sattelite\_speed} + \text{UE\_speed}} \tag{5.2}$$

On the other side, for NTN LEO satellites, in cases where the UE is moving in the same direction as the satellite, the cell size is divided by the relative speed of the satellite with respect to UE (e.g. absolute value of their subtraction). In this case, the time to handover can be calculated by the following equation:

$$\text{Time\_to\_Handover} = \frac{\text{Cell\_size}}{\left|\text{Sattelite\_speed} - \text{UE\_speed}\right|} \tag{5.3}$$

Based on Eqs. (5.2) and (5.3), it is evident that bigger cell diameter results in lower handover intensity, and vice versa. If we assume that UE movement is neglectable in respect to the speed of the satellite, which is around 8 km/s at height above the Earth's surface of 1000 km [14], then for the NTN LEO cells ranging from diameter of 50–1000 km the time to handover ranges from (50 km/8 km/s) = 6.25 seconds to approximately (1000 km/8 km/s) = 125 seconds. One should note that 8 km/s is in fact 28800 km/h, so the impact of the UE movement is up to ± several percent (e.g. for a speed of 200 km/s on ground, with fast trains, the impact is less than 1%). For UE speed equal to 500 km/h, such handover intensity is comparable to terrestrial UE handover intensity when it is served by cells with a diameter ranging approximately 1–20 km. Of course, the mobility of UEs on the ground is typically with much lower speeds, especially considering that the majority of connections in terrestrial mobile networks comes from nomadic users such as pedestrians or people at home and in offices who have speeds from 0 up to 5 km/h.

On the other hand, it is unlikely that there will be high handover intensity in the case of GEO satellites due to their large footprints and thus larger cell sizes. As we discussed, in the case of LEO satellites, the main factor for the handover intensity is the cell size and the speed of the satellite orbiting the Earth, while the speed of the UE is negligible.

For the equilibrium case (which is usually assumed in mobile networks, that is, the rate of UEs leaving a cell is equal to the rate of UEs entering a cell), the maximum number of UEs will remain connected to the cell at all times (neglecting the movement of users in respect to the movement of the satellite).

5.5.3 Spectrum for NTN in 5G-Advanced

The NTN and TN is possible to operate in different frequency ranges (e.g. FR1 and FR2), or in the same frequency range (e.g. FR1 or FR2).

The spectrum for satellite communications is divided into spectrum for providing MSSs and FSSs. The reader should refer to Chapter 3 and section 3.8.1 for FSS bands. The L-bands (1–2 GHz) and S-bands (2–4 GHz) are examples belonging to the MSS domain, while the Ku-bands (11–14 GHz) and Ka-bands (20–30 GHz) are used for the FSS.

Regarding the 5G satellite services 3GPP Release 17 specified L-band and S-band support for NR bands n255 and n256, as given in Table 5.4, where each of these Frequency Division Duplex (FDD) bands provides approximately 30 MHz of spectrum in each direction, uplink and downlink.

In 5G-Advanced, Release 18 and beyond, other MSS FDD bands are added with UL in the L-band (1610–1626.5 MHz) and DL in the S-band (2483.5–2500 MHz), which provides about 80 MHz of downlink and 80 MHz of uplink bands that are appropriate for providing direct communication between the mobile terminal and the satellite by using 5G NR.

Generally, throughput of all satellite systems, including LEO, MEO, and GEO, is increased by using High-Throughput Satellites (HTS) approach [16]. HTS systems utilize much concentrated spot beams, so coverage area by the footprint of the satellite is divided into many smaller cells (e.g. with cell diameter as small as 50 km), by utilizing frequency reuse to significantly increase capacity

Table 5.4 NR bands for support of mobile satellite services.

NR band	General name	Uplink (MHz)	Downlink (MHz)	Duplexing
n255	L-band	1626.5–1660.5	1525–1559	FDD
n256	S-band	1980–2010	2170–2200	FDD

and speeds. This way, each beam (which defines a satellite cell) uses the same bandwidth (e.g. 30, 80, 400 MHz, etc.), depending on the bands used and their availability for the given NTN services. In that manner, considering bandwidth over one satellite beam [17], the peak data rates are given in Table 5.5. As expected, the peak data rates are much higher with larger channel bandwidth per beam, such as 400 MHz, which is typically available at a higher spectrum. With the estimated user density of 10 UEs per beam [17], and using channel bandwidth of 400 MHz per satellite beam, the provided individual bitrates can approach approximately 100 Mbit/s in downlink and uplink, which is expected broadband capacity per user in the 5G era, the 2020s.

In 5G-Advanced, starting with Release 18, there are initially specified three bands (n510–n512) in the Ka frequency range 17.7–20.2 GHz (downlink) and 27.5–30 GHz (uplink). While the noted L/S-band targets handheld mobile devices such as smartphones, in the Ka-band are higher gain antennas (e.g. very small-array terminals) that are usually mounted on buildings (for FSS) or on top of vehicles (for MSS) [14]. Going toward 6G the number of NR bands to be used by NTN will increase considering that in the 6G era is an expected unification of TN and NTN [16].

Regarding the HAPS, WRC-19 agreed that the frequency bands 31–31.3 GHz and 38–39.5 GHz also be identified as being allowed for use by HAPS worldwide. It was also confirmed that the bands 47.2–47.5 and 47.9–48.2 GHz, which were already identified before WRC-19, are available for use worldwide by administrations wishing to implement HAPS. The WRC-19 agreed that the frequency bands 21.4–22 and 24.25–27.5 GHz could be used by HAPS in the fixed service in Region 2. Concerning future HAPS developments, ITU-R in WRC 2023 agenda considered also bands below 2.7 GHz for HAPS to be used as IMT base stations (e.g. gNodeB in 5G era, 2020s).

5.5.4 M2M/IoT Over Satellite

M2M/IoT (Machine-to-Machine/Internet of Things) services can be provided in various configurations utilizing both TN and NTN networks. Regarding the NTN, they can utilize both GSO and non-GSO satellite networks. Depending on the availability of terrestrial access networks, such M2M/IoT services can be provided either with standalone satellite connectivity or using hybrid TN-NTN connections.

M2M/IoT services are provided by using MSS spectrum which is below 3 GHz (including L-Band, S-Band, and VHF) [18].

Satellite terminals already provide M2M/IoT connectivity to many verticals in all areas where TN is not present or cannot be used continuously. As such, NTN networks can be used for M2M/IoT services in the following fields [18] such as transportation (including land, maritime, and aeronautical), asset tracking, logistics, security applications, government services, smart meters in the oil, gas, and energy industries, agriculture (e.g. on distance or automated watering and fertilization), environmental data monitoring and reporting (e.g. about pollution), emergency management, fishing monitoring, animal tracking, various sensors and meters in mining and construction,

Table 5.5 Peak data rates for satellite radio interfaces of IMT-2020/5G.

Channel bandwidth (MHz)	Peak data rate downlink (per satellite beam) (Mbit/s)	Peak data rate uplink (per satellite beam) (Mbit/s)
30	70	2
400	900	900

as well as use cases in telecom operators' networks for accessing data networks and solutions for provision of M2M/IoT applications/services via satellite.

However, for rollouts of all these possible M2M/IoT services, harmonization and availability of MSS spectrum is required. In that manner, WRC-23 considered possible new allocations for mobile-satellite service for future satellite-based M2M/IoT applications. According to ECC [18], there are several possible frequency bands in FR1 between 1685 and 3400 MHz on a regional basis, including also 2010–2025 MHz within ITU Region 1, to be used for M2M/IoT services.

Particularly interesting are hybrid M2M/IoT applications, aimed to meet service cost, performance, and coverage targets. In that way, TN can offer urban coverage whereas NTN is more suitable for coverage of rural and less populated areas. Various solutions also exist for dual-mode hybrid terminals which typically use MSS bands below 3 GHz. The integration of TN and NTN in 5G-Advanced and their foreseen unification in 6G era is expected to make use of both FR1 and FR2 bands by hybrid terminals, which can have both TN and NTN connections simultaneously (e.g. using handheld terminals such as smartphones).

5.6 Fixed-Wireless Access (FWA)

As stated earlier, access to broadband is the key for socio-economic development in all countries. However, the deployment of wired/fixed broadband networks requires high capital costs and medium- to long-term investments. The costliest in network deployment is the last mile (or last kilometer) due to high cost of civil engineering works, which may be also time-consuming in many countries due to the required administration procedures. Then, the solution to such time-consuming and costly last mile deployment is FWA. What is FWA?

FWA is a service that provides primary broadband access through the mobile network for fixed location of the Customer Premises Equipment (CPE), such as home or office. Depending on the location and building, there are various form factors of CPEs, such as indoor (e.g. desktop, window) and outdoor (e.g. rooftop, wall mounted).

The FWA is not new in the mobile world. In that manner, ITU published its first book on FWA in 2001 [19], which at that time referred to 2.5G and 3G mobile networks. Even 2G mobile networks, such as GSM, were used for FWA access for telephony. Later, with higher speeds in mobile networks (e.g. in 3G and especially in 4G networks) the FWA was used for provision of Internet access service besides the telephony. At the end of 2020, there were around 500 mobile operators (approximately 50% of the sector) with launched 4G FWA services. The FWA is used as replacement of fixed access in rural areas and in the cases when there is no fixed access infrastructure, but also in cases when FWA is the competitive alternative to other broadband solutions. Additionally, FWA is an approach for different IoT services in which devices are not required to be mobile.

FWA was one of the early drivers for 5G, which can be explained by the much higher bitrates possible with 5G mobile networks, which are more comparable to fixed optical access (when compared to previous mobile generations).

There are several parameters that can determine the decision on use of 5G FWA, which include the revenue per Hz (primarily driven by spectral efficiency and mobile data transferred per Hz), customer Average Revenue per User (ARPU), and mobile data usage. One should note that the revenue rises in parallel with ARPU and spectral efficiency, while on the other side they are inversely related to the data usage. But, the crucial parameter for provision of 5G FWA is the increased spectral efficiency in 5G compared to 4G and before. However, it is also typical that each new mobile generation will come with increased spectral efficiency than previous mobile generations due to new modulation and coding schemes and higher order of MIMO. In that way, FWA has the possibility to maximize the bandwidth usage considering the relatively constant signal-to-noise ratio (which is not the case with mobile terminals). Of course, disadvantages of FWA are normally lack of mobility and need for higher usage caps (than in mobile networks) comparable to fixed broadband networks. But, in many countries (e.g. developing economies) which lack fixed access infrastructure due to different reasons, the 5G FWA can be used as replacement of fixed broadband access in homes and offices, of course by appropriate design of the 5G access and core network, as well as interconnection point, regarding the QoS.

5.6.1 5G FWA Architecture and Spectrum

The 5G continues the provision of FWA services to customers, but with bitrates comparable with fixed broadband access. The FWA also can use types of 5G FWA implementations, which include the following (Figure 5.12):

- 5G UE (user equipment) is replaced 5G RG (residential gateway) and
- 5G UE (at fixed location) is used as residential gateway.

5G RG is in fact an upgraded version of the legacy RG for broadband access via fixed networks (e.g. optical access network). It is connected as a UE to the 5GCN through a fixed or mobile access network. When 5G RG is connected to 5GCN via 5G NR link then we refer to such connection as FWA. However, 5G RG can be connected to the core network also via hybrid links. On the other side, NG UE is an upgraded version of the original UE which is connected to the 5GCN through 5G mobile network (although it is possible to be connected also via wireline connection, considering that 5G also supports wireline access). In all cases, the RG is a 3GPP trusted node which can provide local access (e.g. in home, office, or public place) via Wi-Fi or Ethernet as typical (or more concrete, globally unified) WLAN and wired LAN technologies.

5G-RG can be connected via gNodeB, using 5G NR, which is in fact the FWA. However, it is possible for 5G-RG to be connected via wireline access to 5G core, and in such a case it goes through W-5GAN. W-5GAN is a wireline access network that connects to a 5GC via N2 and N3 reference points. The W-5GAN can be either a W-5GBAN or W-5GCAN, where the Wireline 5G cable access

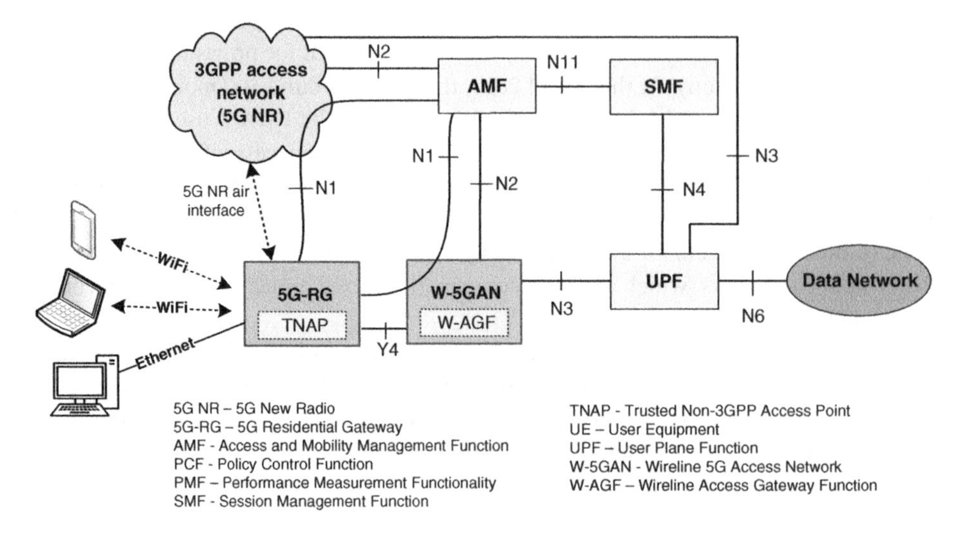

Figure 5.12 5G FWA architecture.

network (W-5GCAN) is defined in CableLabs while the Wireline 5G BBF Access Network (W-5GBAN) is defined in BBF [13]. There, the function that provides connectivity for 5G-RG to 5GCN is noted as W-AGF. It interfaces the 5GCN CP and UP functions via N2 (toward AMF) and N3 (toward UPF) interfaces, respectively.

Regarding the connection between the 5G-RG and the base station (e.g. gNodeB), the FWA can be deployed in any available 5G spectrum, depending upon the business decision of the mobile operator.

Typically, FWA is provided with medium and high spectrum bands (not in lower bands, such as sub-1 GHz spectrum); however, when the demand for FWA services is low then it is possible to also use lower frequency ranges. Overall, FWA deployments are mainly in the mid frequency bands such as 2.3, 2.6, or 3.5 GHz.

Regarding the 5G FWA initial deployments in the early 2020s, the most used is 3.5 GHz spectrum, while in later deployments (when mid spectrum will be needed for capacity in dense urban areas), the 5G FWA is deployed in mmWave 5G spectrum. According to case studies [20], deployment in mmWave bands together with mid bands (such as 3.5 GHz) can result in three times higher capacity for FWA.

In the case of mmWave 5G spectrum, the primary challenges include deployments, the main concern is greatly reduced signal quality due to higher attenuation of signals in mmWave spectrum and no possibility for LOS visibility between the antenna systems of gNodeB and receiver antennas (e.g. at user's home), which influence expected throughput and quality of experience. To combat such issues in FWA deployments, there should be used antennas with beamforming, that is, to accurately target each individual device on the end user side. In general, the beamforming provides the possibility to limit the radio signals energy into a smaller area, thus contributing to extension of penetration and range of signals in mmWave bands.

5.6.2 5G FWA Services

What are the possible services that are offered via FWA? Well, they are very similar as FWA services offered via 4G mobile networks in 2010s. Overall, there are several use cases for 5G FWA:

- FWA as replacement (or backup) of fixed broadband access: This is typically an option when a telecom operator does not possess fixed broadband infrastructure at a given location (e.g. rural area), or it is a mobile-only operator that wants to enter the fixed broadband market. Also, FWA can be used for seasonal broadband provisioning in tourist places as well as for certain IoT services. Considering the high capacity of 5G networks and beyond, when using mmWave spectrum bands, 5G FWA may be even a competition to new FTTx deployments in regions with low fiber penetration.
- FWA for 5G RAN backhaul: This use case is deployment of 5G FWA solution as a backhaul for gNodeBs. It is complementary with IAB approach in 5G-Advanced, for provision of backhaul for 5G macro and small cells.

Overall, 5G FWA is aimed at mobile operators to provide alternative for fixed broadband services, for provision of broadband Internet access with sustainable bitrates as in fixed broadband networks, IPTV and VoD as carrier grade service (not over the public Internet), as well as advanced cloud and IoT services with guaranteed QoS. FWA services are attractive for both developed and developing markets. However, in developing telecom markets with the absence of terrestrial optical infrastructure and in underserved rural and large areas, FWA may be the preferable approach for delivery of different services.

5.7 5G-Advanced Non-Public (Private) Networks

Network slicing in 5G provides native capability for provision of NPNs for various use cases by private entities such as companies (e.g. industry, logistics, etc.). 3GPP defines two major categories of NPNs [21]:

1) Standalone Non-public Network (SNPN) and
2) Public Network Integrated NPN (PNI-NPN).

5.7.1 Standalone Non-Public Network (SNPN)

SNPN is a private 5G network that is deployed as a standalone network, which is in fact completely isolated from public mobile networks (physically and logically). Regarding the spectrum, SNPN usually uses unlicensed 5G spectrum rather than licensed spectrum used by national mobile operators. However, in certain countries, certain 5G licensed bands can be allocated for NPN use, which depends upon the national regulations.

In the case of SNPN, both CP and UP functions including gNodeB as well as 5G core are deployed in the premise of the private entity (e.g. enterprise, institution, etc.), as shown in Figure 5.13. The advantage of SNPN deployment by the private entity is that independent 5G network allows application of innovative 5G technologies which are relevant to that company (e.g. SNPN in a smart factory), without limitations of legacy Ethernet or Wi-Fi.

There are several advantages of a standalone private network which include the following:

- Higher security, because SNPN is completely separated from the public mobile networks, thus ensuring security and privacy of internal data and personal information of users.
- Ultra-low latency, which is a feature of 5G mobile networks in general, but this is only really possible in limited deployments in a given geographic area with all network functions located in the room. Due to shorter distances between UE and gNodeB as well as between gNodeB and 5G core functions, the use of 5G (especially in higher bands) results in real ultra-low latency that is required for certain critical processes such as motion control of machines in the industry.
- QoS as needed, considering that private entity is not offering a wide range of services as mobile operators, so it can focus on QoS parameters (data rate, reliability, latency, etc.) that are of higher importance for traffic over the SNPN.
- Independent network from public mobile networks, so any failures or regulatory interventions in public mobile networks will not directly affect the SNPN.

In addition to the advantages of deploying and operating a standalone NPN, there are also disadvantages. The biggest drawbacks are CAPEX, capital investment in all network elements including access and core parts, as well as lack of expertise in mobile network operation and maintenance. However, for such cases, vendors of telecommunication equipment (used for SNPN), as well as available expertise from mobile operators, can be used to overcome the lack of expertise or at least until it is built by the owner of the standalone non-private 5G network who is also called a vertical operator. Also, a mobile operator can implement and operate SNPN for a given customer, independently from its own public networks. Hence, 3GPP defines the following models for SNPN management [22]:

- Mobile Network Operator (MNO) managed mode: In this case the NPN operator is in fact the mobile network operator who also plays the network operator role for public mobile services. In this case, MNO can use 5G unlicensed spectrums or, in some cases, part of its licensed spectrum for implementation of the SNPN.

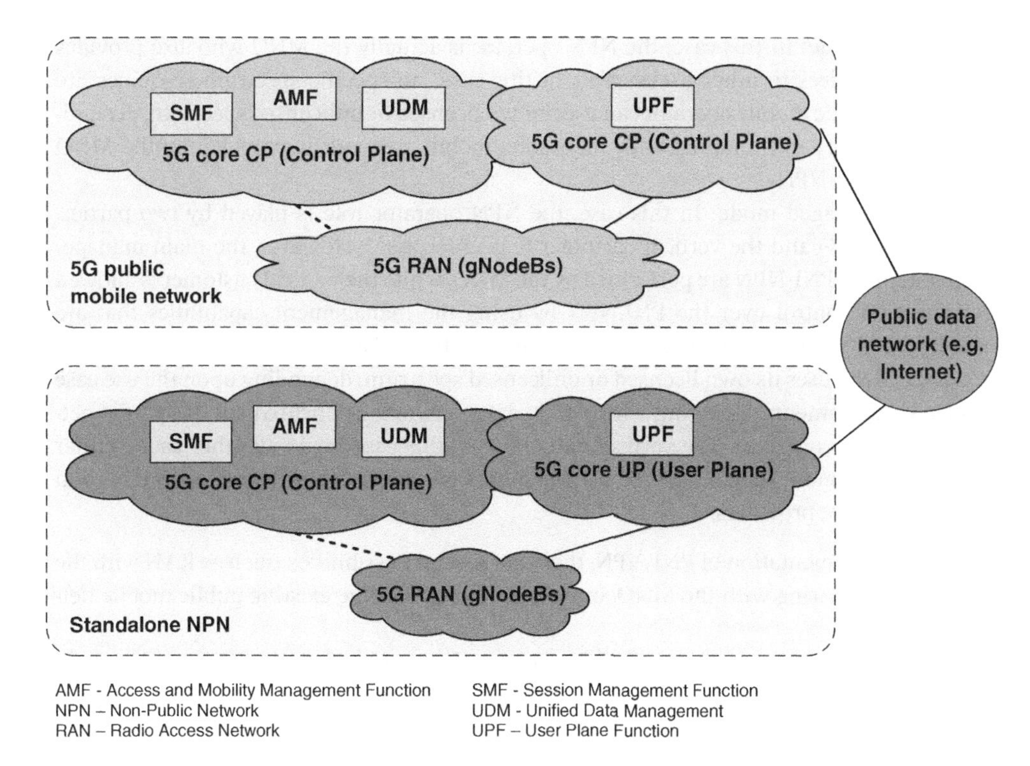

Figure 5.13 Standalone Non-Public Network (SNPN) isolated from public 5G network.

- MNO-vertical managed mode: In this case the NPN operator role is played by both MOP and a vertical customer (e.g. enterprise). Due to their own expertise, the MNO does the main management tasks for the SNPN, while allowing certain control over the SNPN by the vertical (e.g. customization of SNPN services, such as configuration, start/stop, etc.). This is based on business agreement (SLA) between the two parties, the mobile network operator and the vertical entity. The vertical entity can also do outsourcing of its SNPN management tasks to other Operation and Maintenance (OAM) service provider, if that is allowed by the SLA with the MNO (the other party in management of the SNPN).
- Vertical Managed Mode: In this case, the NPN operator's role is entirely done by a vertical entity. This is in fact the "true" standalone NPN. However, in this case, there are also required 5G spectrum resources (e.g. use of unlicensed spectrums at 6 GHz) as well as connectivity to external data network, i.e. to public services, such as Internet or telephony. In this case, when needed, the vertical entity can autonomously outsource the SNPN management tasks to a third party OAM service provider.

5.7.2 Public Network Integrated Non-Public Network (PNI-NPN)

Public Network Integrated Non-Public Network (PNI-NPN) is NPN deployed with the support of public mobile network, i.e. MNO. However, depending upon the level of integration (only RAN sharing by MNO and NPN, or NPN as a service of MNO), there are two main types of PNI-NPN deployment [22]:

1) MNO managed mode: In this case, the NPN operator is actually the MNO who also provides public mobile services to other users. Also, in this case, no specific spectrum resources are required (because the mobile operator can use its own licensed or unlicensed spectrum, depending on the use case) and roaming with the public mobile network is provided by the MNO operating with PNI-NPN.
2) MNO-vertical managed mode: In this case, the NPN operator role is played by two parties, which include MNO and the vertical customer (e.g. enterprise). However, the main management operations in PNI-NPN are performed by the MNO, while the vertical customer is allowed to retain certain control over the PNI-NPN by using the management capabilities that are exposed by the MNO (based on SLA between the two parties, MNO and vertical entity). Also, in this case, the MNO uses its own licensed or unlicensed spectrum, depending upon the use case and its QoS requirements. Roaming with public data networks is ensured via the public network of the mobile operator. The vertical entity also in this case (as in all other cases, either SNPN or SPI-NPN mode) can outsource management tasks to other third party operations and maintenance service providers.

Regarding the implementation of PNI-NPN, there are several possibilities, such as RAN with the MNO, RAN and CP sharing with the MNO, or NPN deployed into the existing public mobile network of the MNO.

In the NPN RAN sharing scenario, a 5G NPN network shares part of the 5G RAN, such as base stations and gNodeB, with the public network, while other network functions are deployed on the enterprise premise and remain separated from the public network. This deployment scenario can be divided into two sub-scenarios, which depend upon the spectrum management approach, that is, whether NPN uses unlicensed spectrum or if it uses the MNO's licensed spectrum (in this case, it is shared with the public mobile network). Data traffic and signaling traffic (i.e. UP and CP) of the NPN is delivered to the 5G core UP in the NPN. Here, the private 5G core CP is also deployed in the vertical entity, so NPN users and operation information do not leak outside; thus, security and privacy of the NPN can still be guaranteed. This scenario is suitable for URLLC services, considering that all 5G network functions are located in the premises of the vertical entity.

The other deployment case is when CP traffic goes to 5G core of the MNO, besides the RAN sharing between the MNO and NPN. All control functions, such as mobility or authentication of users, are performed in the CP part of 5G core of MNO. In this case, the devices of the NPN appear as customers of the MNO's public network. The user traffic between the NPN and MNO can be isolated by using 5G network slicing. In this scenario, URLLC services can be provided by using edge computing in the premises of the vertical entity.

Finally, the NPN can be hosted in MNO's public network, as shown in Figure 5.14. In this case, all data traffic of the NPN flows through the public network of the MNO which is located outside the premises of the vertical entity; however, the traffic of each network is logically separated of one another by using 5G network slicing approach. In this case, the MNO as a service provider has full control of all functions, so any failure in the public mobile network results in the failure of the NPN. In this scenario, it may not be feasible to implement URLLC services. On the other hand, it is easier to implement roaming between the private and public networks, since the mobile operator has full control over both.

We have noted that in many of the deployment scenarios, the NPN and the public 5G network share parts of 5G infrastructure, network functions, and resources in radio access, backhaul, and core networks. Due to this sharing, there is the possibility of traffic from one segment (public

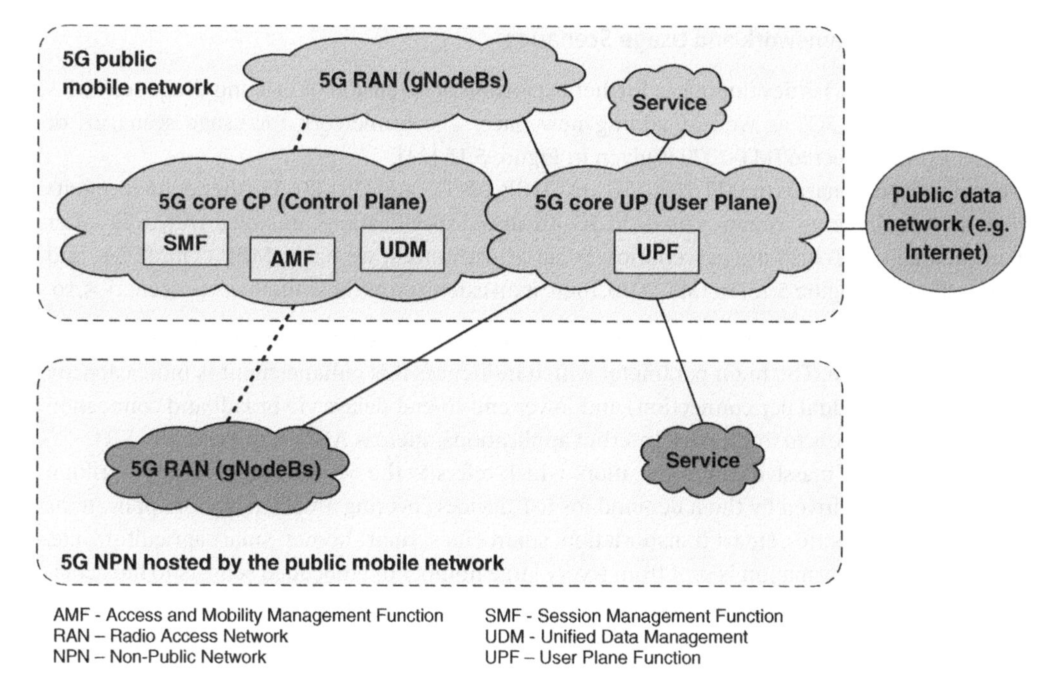

Figure 5.14 NPN hosted by 5G mobile operators.

mobile network) to impact the traffic from the other segment (NPN), and vice versa. To avoid such a situation, there is required logical isolation of the traffic via the network slicing end to end, in 5G RAN and 5GCN. Network slicing provides logical isolation between NPN and public mobile network; however, in cases where QoS has the highest priority in the design of the NPN (e.g. for URLLC services), it may be necessary to consider physical network resource isolation in which case the network resources for the NPN and for the public mobile network are physically segregated from each other (e.g. separate base stations, separate frequency bands, etc.). Also, QoS requirements and KPI targets [23] vary from one 5G NPN deployment to another. For example, for massive IoT services, there are modest requirements for latency and reliability, while for URLLC services (e.g. critical IoT services for industry), we have stringent QoS requirements such as latency in range of 1 ms and very high reliability (e.g. many nines, such as 99.999999% of the time service to be available to the end devices).

5.8 Future Mobile Broadband: IMT-2030 and Beyond

The next mobile generation after 5G/5G-Advanced is 6G (6th Generation of mobile systems). Similar to previous mobile generations (from 3G onwards) ITU-R with its Wording Party 5G (WP 5D) is responsible to set the requirements on the global mobile stage for 6G. Similar to 5G requirements that were set in umbrella specification IMT-2020, 6G requirements will be set by ITU-R in umbrella specification IMT-2030 (International Mobile Systems 2030), which is denoted by the year (2030) when it is expected to be available, while the technology trends denoting the IMT-2030 are already set [24].

5.8.1 IMT-2030 Framework and Usage Scenarios

The IMT-2030 (i.e. 6G) is developing as further expansion of extension of existing usage scenarios from IMT-2020 (i.e. 5G) as well as adding new ones. The framework for usage scenarios of IMT-2030/6G in respect to IMT-2020 is given in Figure 5.15 [24].

The main usage scenarios in IMT-2020/5G are eMBB, eMTC, and URLLC. Further enhancements to these three main usage scenarios made in 5G-Advanced specifications are noted with a "+" after each of the main IMT-2020 usage scenarios. So, in 5G-Advanced, we have eMBB+, mMTC+, and URLLC+ shown in Figure 5.15. In IMT-2030, there is an extension of these main usage scenarios, so:

- eMBB+ results in immersive communication which refers to encompassing intense human and machine interaction. The main parameter which influences this enhancement is more capacity (overall and individual per connection) and lower end-to-end delays via broadband connection (e.g. for mobile access to immersive Internet applications, such as AR/VR or generally XR).
- mMTC+ results in massive communication, which refers to the 6G support of nearly a trillion devices, primarily driven by the n demand for IoT devices covering a wide range of applications such as smart industries, smart transportation, smart cities, smart homes, smart agriculture, etc. This means that information is sent from a very large number of embedded sensors to the actuators through reliable connections that are available all the time. Such massiveness requires low-cost devices and the power supply of those devices, where one possibility is to use energy harvesting (from ambient energy from vibrations, light, temperature gradients, or even from the radio frequency waves themselves) to support a huge number of embedded devices in 6G to work without batteries, the so-called zero energy devices (no need for battery replacement or charging).

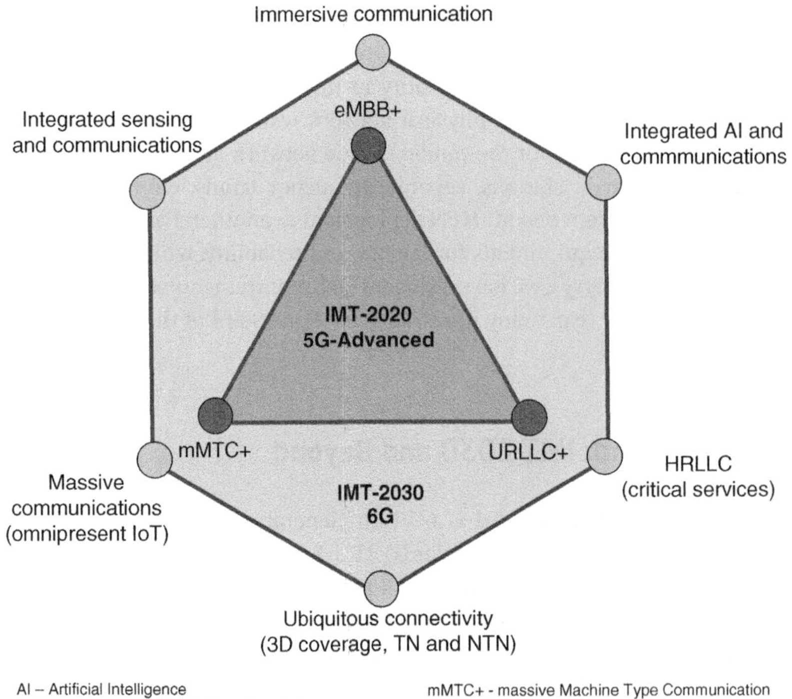

Figure 5.15 Usage scenarios in IMT-2030/6G.

- URLLC+ results in hyper reliable low latency communication (HRLLC), which is targeted for provision of extremely critical services which require hyper reliability. The HRLLC use cases include services in industry with full automation in terms of control and operation, as well as robotic interactions, telemedicine, emergency services, monitoring for transmission and distribution of electricity, and others that require hyper reliability and very low latency.

Besides the further enhancements of three main usage scenarios in 5G/5G-Advanced, IMT-2030/6G defined three new which include the following (Figure 5.15):

1) Ubiquitous connectivity – this usage scenario targets wide range of coverage and mobility by combining legacy TNs with satellite NTNs. The coverage extends into the 3D space (of course, in kilometers above the surface of Earth), and integration of TN and NTN should extend the broadband access to Internet and IoT services to all unserved and underserved locations worldwide.
2) Integrated AI and communication – this refers to the design of IMT-2030 and beyond network architectures, protocols, and algorithms to learn from mobile/cellular big data collected from the mobile network and services that run through it. One of the main differences of the use of AI in IMT-2030 compared to IMT-2020 is that it will use mobile technologies to enable the spread of AI and use radio networks to augment ubiquitous, distributed ML as the current main form of application of AI. So, there will be AI-native radio interface, AI-native radio network which will be able to support AI services [24].
3) Integrated sensing and communication – it refers to object detection, ranging, tracking, positioning, imaging, etc. Unlike IMT-2020/5G systems, which provide only positioning services, IMT-2030 will use very high frequency bands (from mmWave up to THz) which will require larger antenna arrays, denser deployments of radio units, as well as AI-based collaborations between mobile network nodes. In such fully integrated and dense systems with unified beamforming techniques, sensing appears as new function for provision of higher accuracy (than before) for new innovative services such as high accuracy tracking, imaging for security and biomedical applications, monitoring of nature (e.g. pollution, disasters, etc.), gesture and activity recognition, flaw detection, as well as provision of sensing data from the surroundings to various AI models and services, creation of XR content/services and digital twin applications.

To support new usage scenarios, IMT-2030/6G has enhanced capabilities that are also present in IMT-2020/5G [25] as well as 6 new capabilities as given in Table 5.6.

IMT-2030/6G will enable the cyber-physical continuum by creating a digital representation of the physical world (consisting of senses, actions, and experiences) using the huge number of sensors and a high level of accuracy when positioning objects in 3D space. Such a 6G cyber-physical continuum also includes the metaverse as a digital environment in which avatars (representatives of people in the digital world) interact in a XR environment that is a close replica of the real physical world. Generally, in the cyber-physical continuum, it will be possible to project digital twins of real objects from the physical world, thus creating the cyber-physical continuum.

5.8.2 IMT-2030/6G Radio Interface and Spectrum

The IMT systems of 2030 are expected to provide ubiquitous ultra-speed mobile Internet access that should emerge as an extension of optical fiber, then extraordinary communication such as holographic services or XR that are at 6 degrees of freedom, etc. All this requires enormous bitrates that can only be provided with a very high bandwidth, heading toward the THz spectrum. Such bitrates also increase the challenges for network backhaul infrastructures, which will require hundreds of Gbit/s to connect to or between cell towers and remote radio heads. As we have already discussed in

Table 5.6 Capabilities of IMT-2030/6G.

Capability of IMT-2030/6G	Enhanced or new (in relation to IMT-2020/5G)
Peak data rate	Enhanced
User experienced data rate	Enhanced
Spectrum efficiency	Enhanced
Area traffic capacity	Enhanced
Connection density	Enhanced (up to 10^8 devices/km^2)
Mobility	Enhanced (up to 1000 km/h)
Latency	Enhanced (down to 0.1 ms)
Reliability	Enhanced (10^{-5} to 10^{-7})
Security, privacy and resilience	Enhanced
Coverage	New (Global 3D coverage, TN and NTN)
Sensing-related capabilities	New
AI-related capabilities	New
Sustainability	New
Interoperability	New
Positioning	New (1–10 cm)

this book, speed can be increased by three main strategies, that is, additional frequency bandwidth, improving spectrum efficiency with new modulation and coding schemes and massive (or ultra-massive) MIMO, and deploying small cells (which increases the density of the nodes of the access network, since for smaller cells they are usually at short distances from each other). As usual, all three strategies are used. So, as far as spectrum is concerned, 5G has used a large part of the spectrum in the mmWave range, so new spectrum can be allocated around and above 100 GHz in the so-called sub THz spectrum (sub-THz because it is above 100 GHz but below 1 THz = 1000 GHz). Long-term predictions for future mobile systems, as given in the first ITU report for IMT-2030, consider both sub-THz and THz spectrum, ranging from about 100 GHz to 3 THz [24].

However, according to the industry [25], it can be expected that the IMT-2030/6G will focus on the sub-THz spectrum, which includes frequency ranges roughly between 90 and 300 GHz, as shown in Figure 5.16.

Communication in the sub-THz range is different from the 5G New Radio (NR) frequency bands due to the challenging conditions for propagating radio waves at such high frequencies (high propagation loss – the propagation loss increases with the frequency due to very small wavelengths, in nanometers),

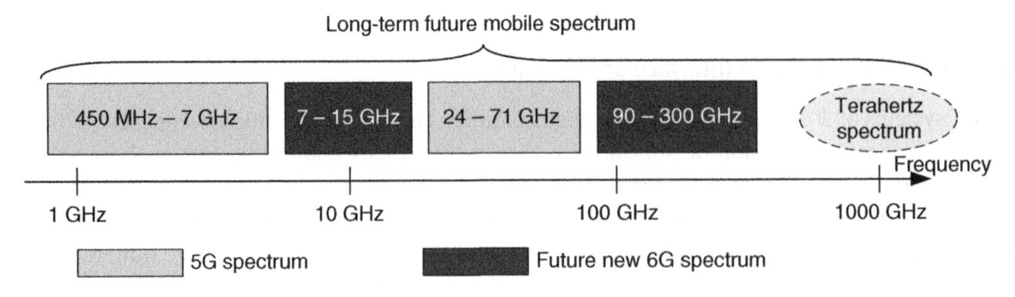

Figure 5.16 Spectrum for IMT-2030/6G and beyond.

the extreme data rates (when compared to 5G and before), the limited availability of commercial RF components, as well as small coverage areas. All this affects the design of systems that will operate in the sub-THz spectrum. Therefore, the design of radio components (transceivers) for sub THz communication is based on two important approaches:

1) Extensive use of beamforming (to combat propagation loss at such high bands) and
2) Efficient processing which is required for very high data rates and/or super low latencies.

Also, there is required new RAN, with changes in multiple access techniques. While OFDM was successfully used in 4G and then in 5G (with different numerologies, which we covered in Chapter 4, in 5G New Radio section), One such solution is DFTS-OFDM (Discrete Fourier Transform-Spread-Orthogonal Frequency Division Multiplexing) which works with a single frequency carrier that is suitable for low peak-to-average signals and hence supports good power efficiency. The DFTS-OFDM does not require spacing between users but combines all of them orthogonally (by using different Fourier coefficients assigned to each of them) so the peak of one user coincides with the null of all other users. Of course, this refers to sub-THz bands (although it has similarities to SC-FDMA, i.e. Single Carrier Frequency Division Multiple Access), while OFDM can continue to be used in bands below 100 GHz, the lack of frequency-domain multiplexing is not so important in the sub-THz range. With this technique, the transmissions can start at any DFTS-OFDM symbol, which reduces latency to a time needed to transmit a symbol, which is extremely low latency. For short duration of DFTS-OFDM symbols it can go below 1 ms [26].

Further, it is needed to perform parallel processing (e.g. via carrier aggregation in 6G) in the radio access part to achieve the extreme data rates in the order of 100 Gbit/s and above.

For sub-THz and further in THz spectrum, beamforming is required on both ends of the radio connection, at 6G base station and at UE, with the aim to achieve the needed antenna array gain. In sub-THz bands, to achieve the needed data rates, there are required many beams in parallel between the base station and the end user, which requires narrower beam formations (when compared to 5G NR in mmWave bands). Such need increases the importance of the beam alignment between the 6G base station and UE when operating in sub-THz spectrum. Also, LOS channels will need to be available for work in sub-THz bands.

Overall, the sub-THz spectrum is expected to be one of many technical novelties in the 6G mobile system, which is expected to offer extreme data demands in some use cases, as well as to provide ultra-low latencies and hyper-reliability in other deployment scenarios in the future.

IMT-2030 is expected to have 6G RAN which will be user-centric access network, which is using different frequency bands. Also, each RAN resource in 6G should be flexibly logically partitioned (sliced) to support packet data flows with the same or similar QoS requirement. While network slicing was introduced in 5G networks (in IMT-2020), which mainly referred to the core network functions, user-centric RAN slicing is expected to be the main novel approach (regarding the slicing) in IMT-2030/6G.

In 6G, the QoS requirements may vary from one user to another to a larger extent than in 5G networks, the future 6G RAN should be flexible, resilient, and soft, i.e. user-centric and service oriented, with the possibility to provide QoS guarantees and consistent end user experience.

For the purposes of a user-centric and service-oriented RAN, it should be divided into several basic radio network functions or services and elements that can be flexibly combined to support the generation of radio network functions on demand. In addition, AI/ML techniques can be applied for different purposes in RAN, including but not limited to real-time traffic demand forecasting and classification (not offline traffic planning as in previous mobile networks), dynamic resource availability prediction over different radio interfaces and different bands, application of call admission/session control for managed services and flows over the RAN, and adaptive scheduling approaches for efficient QoS/QoE management in the future 6G RAN.

Considering the large deployment of 6G networks with a large number of base stations (much more than in previous generations), the construction costs of telecom operators can be reduced by sharing the RAN infrastructure. On the other hand, such RAN sharing adds new challenges in terms of reliability, transparency, QoS support, reliable data storage, etc., among multiple mobile operators. When RAN sharing is used, distributed AI technologies such as federated learning (FL) may also be required, which is appropriate to use when we have datasets from different mobile operators.

Overall, future 6G mobile networks have multiple technical as well as business and regulatory challenges, which however is not new in the telecom world in this century.

5.9 Business and Regulatory Aspects of Future Mobile and Wireless Broadband

Every new development in mobile networks gives rise to business and regulatory activities. What are the new business options with 5G-Advanced?

Well, 5G-Advanced improves the 5G network in terms of enhancing the experience for human end users as well as machines (for IoT services) [27]. Also, 5G-Advanced improves uplink directionality, further increasing the available capacity, which can be used for outstanding communication. It also begins the gradual introduction of AI/ML techniques into network functions and services (in terms of its configuration, healing, optimization, etc.), which make the network more automated than before (similar to a telecom operator as a plug and play concept of an end-user application on a computer or smartphone). What are the business aspects of it? Well, the relational improvements are reduced OPEX because automation will do many things that were done manually in the past. However, the biggest benefit is automating the design of network components, their creation and their decommissioning (when they complete their life cycle, e.g. due to the termination of the provision of 5G services to a given party, such as an enterprise).

Then in this chapter we discussed Wi-Fi 6 and Wi-Fi 7, what is their "business" relationship with 5G? As usual, Wi-Fi can provide "cheaper" capacity in hotspot zones of the mobile operator's 5G network and thus save money for the operator. Deploying Wi-Fi over wired broadband also contributes to load balancing between fixed and mobile access networks, which ultimately (in simple words) results in higher QoS and a higher user experience in the mobile network (because the load on mobile network is reduced). Typically, all smartphones have Wi-Fi interfaces (while they do not have a wired Ethernet interface), so Wi-Fi (when connected to fixed broadband access such as fiber) indirectly contributes to better performance in mobile networks, be it 4G, 5G, or future 6G.

Then, we have 5G non-terrestrial networks, which are expected to appear in the 5G-Advanced era, 2025–2030. It will provide new opportunities to cover the uncovered areas, either for human users or for IoT services. Also, it has multiple benefits, including mobile operators and various verticals that will use 5G IoT services over satellite connectivity, either in transparent mode (initially) or in regenerative mode (later).

There are also 5G FWA networks, which are actually one of the first use cases for mobile broadband networks, with many deployments around the world even before the initial 5G deployments. In fact, both 4G LTE (later releases in the 2010s, which have higher total capacity) and 5G NR have much higher capacity than the mobile networks before them (e.g. 3G, 2G), so they can indeed serve as replacements for fixed broadband access networks in all areas without optical infrastructure (e.g. rural areas). Also, with 5G FWA, mobile operators can become TV providers (if they want to), XR providers (with XR content as a managed service, similar to IPTV as a managed service), and providers to novel immersive services.

Finally, we also covered private 5G networks (i.e. NPNs). What are the business opportunities with them? Well, there are as many opportunities as there are verticals (other sectors) looking to deploy private 5G networks, based on their business needs and plans. This could be a business option for mobile operators which can offer NPN services to other parties (using network slicing) via NPI-NPN modes, but also possible new business players, for example, 5G NTN providers, to appear in certain countries or globally. So, many business options are on the table with 5G/5G-Advanced and then toward 6G. Many of the business opportunities of 5G-Advanced will be implemented in the 6G era, the 2030s.

What about regulatory aspects for 5G-Advanced and further for 6G? One of the most important aspects of mobile networks is the radio spectrum. Each new generation adds new frequency bands to the global International Mobile Telecommunications (IMT) spectrum. This is most evident in 5G and 5G-Advanced systems, which first entered the mmWave spectrum (including FR2-1 and FR2-2 bands), while 6G is expected to enter sub-THz bands (between 90 and 300 GHz) and later even in the THz bands (up to 3 THz), which still depends on the development of radio technology in the 2030s (6G era, i.e. IMT-2030) and beyond. The basic regulatory activity regarding the spectrum is their licensing or provision as unlicensed bands. The mobile spectrum allocations require international harmonization which is usually done at ITU World Radiocommunication Conferences (WRCs) held every 3–4 years, such as WRC 2019, WRC 2023, WRC 2027, etc. The results of the WRCs are then implemented by National Regulatory Agencies (NRAs) in every region of the world, because without standards and harmonization in all regions, none of the mobile generations will work as we know them.

Thus, 5G/5G-Advanced requires more spectrum (than 4G) to provide very high bitrates and very low latencies (in the range of 1 ms), so spectrum should be available in shorter periods of time than before and at lower prices, especially in the high bands that have large parts of the spectrum. So, the price per MHz or GHz cannot be calculated with the same equations as the prices in the low and mid spectrum bands (e.g. below 7 GHz). Typically, the cost of spectrum decreases per Hertz as the amount of spectrum increases.

But spectrum is not the only issue for regulators (i.e. the NRAs). The other issue is infrastructure, where optical infrastructure is most needed for 5G and beyond. Optical is the main type of infrastructure (where feasible) that can carry the large amount of traffic that will be transferred from/to 5G and future 6G access networks.

Also, the new spectrum in the mmWave bands requires much smaller cells. So fronthaul and midhaul will need regulation for easy access, e.g. on street furniture and other public (open) areas and facilities for mounting different radio units [28].

Overall, the new spectrum in all kinds of bands, fiber-optic infrastructure (which can be shared) and small cells are coming hand in hand with 5G-Advanced and future 6G (IMT-2030). Also, they require faster regulatory and policy actions.

The mobile future, such as 5G-Advanced and 6G, is expected to be more aggressively focused on verticals (i.e. other sectors, including non-ICT sectors). Such a regulatory framework (or at least a guide) is needed for NPNs for their easy deployment (and removal when necessary).

Finally, consistency is also needed in all regulatory actions in order to build a sustainable digital market and digital society, which is expected to depend heavily on 5G/5G-Advanced mobile technologies in many countries, and even more toward the next mobile generation, the 6G.

References

1 Janevski, T. (April 2019). *QoS for Fixed and Mobile Ultra-Broadband*. USA: John Wiley & Sons.
2 Janevski, T. (November 2015). *Internet Technologies for Fixed and Mobile Networks*. USA: Artech House.
3 Janevski, T. (April 2014). *NGN Architectures, Protocols and Services*. UK: John Wiley & Sons.

4 3GPP TS 23.273 V18.2.0, "5G System (5GS) Location Services (LCS); Stage 2 (Release 18)", June 2023.
5 3GPP TS 38.331 V17.5.0, "Radio Resource Control (RRC) protocol specification (Release 17)", June 2023.
6 IEEE 1588-2019, "IEEE Standard for a Precision Clock Synchronization Protocol for Networked Measurement and Control Systems", June 2020.
7 3GPP TS 38.401 V17.5.0, "NG-RAN; Architecture Description (Release 17)", June 2023.
8 3GPP TS 38.340 V17.5.0, "Backhaul Adaptation Protocol (BAP) Specification (Release 17)", June 2023.
9 3GPP TR 38.807 V16.1.0, "Study on Requirements for NR Beyond 52.6 GHz (Release 16)", March 2021.
10 IEEE 802.11ax-2021, "Wireless LAN Medium Access Control (MAC) and Physical Layer (PHY) Specifications Amendment 1: Enhancements for High-Efficiency WLAN", May 2021.
11 Lancom, "The New WiFi Standard IEEE 802.11ax—Its Features and the Challenge for Europe", June 2018.
12 Khorov, E., Levitsky, I., and Akyildiz, I.F. (May 2020). *Current Status and Directions of IEEE 802.11be, the Future Wi-Fi 7*. IEEE Access.
13 3GPP TS 23.501 V18.2.2, "System Architecture for the 5G System (5GS); Stage 2 (Release 18)", July 2023.
14 Ericsson Technology Review, "Using 3GPP Technology for Satellite Communication", June 2023.
15 3GPP TR 38.821 V16.2.0, "Solutions for NR to Support Non-Terrestrial Networks (NTN) (Release 16)", March 2023.
16 ITU-R M.2460-0, "Key Elements for Integration of Satellite Systems into Next Generation Access Technologies", July 2019.
17 ITU-R Report M.2514-0, "Vision, Requirements and Evaluation Guidelines for Satellite Radio Interface(s) of IMT-2020", September 2022.
18 ECC Report 305, "M2M/IoT Operation via Satellite", February 2020.
19 ITU, "Fixed Wireless Access", 2001.
20 Ericsson Technology Review, "Fixed Wireless Access—Using mmWave Extended Range", November 2022.
21 3GPP, "Non-Public Networks (NPN)", https://www.3gpp.org/technologies/npn, Last accessed in July 2023.
22 3GPP TS 28.557 V17.1.0, "Management of Non-Public Networks (NPN); Stage 1 and stage 2 (Release 17)", March 2023.
23 3GPP TS 28.554 V18.2.0, "5G End to End Key Performance Indicators (KPI) (Release 18)", June 2023.
24 ITU-R M.2516-0, "Future Technology Trends of Terrestrial International Mobile Telecommunications Systems Towards 2030 and Beyond", November 2022.
25 ITU, "IMT Towards 2030 and Beyond", https://www.itu.int/en/ITU-R/study-groups/rsg5/rwp5d/imt-2030, Last accessed in August 2023.
26 Stefan Parkvall et al., "A Concept for Evaluating Sub-THz Communication for Future 6G", Ericsson and Intel Research Paper, 2023.
27 GSMA, "Advancing the 5G Era—Benefits and Opportunity of 5G-Advanced", September 2022.
28 European Commission, "Accelerating the 5G Transition in Europe—How to Boost Investments in Transformative 5G Solutions", February 2021.

6

Internet of Things (IoT), Big Data, and Artificial Intelligence

The development of telecom/ICT world in this century was initially based on transition from circuit-switching telecommunications from the 20th century to packet-switching telecommunication in the 21st century by using established telecom practices for provision of services with guaranteed Quality of Service (QoS) [1], based on Internet technologies [2], and using standardized framework for telecom sector by the International Telecommunication Union (ITU), the well-known Next Generation Network (NGN) framework [3]. The NGN included transition to all-IP networks of legacy telecom services such as carrier grade voice and TV services, but also set the basis for global standardization effort of Internet of Things (IoT).

6.1 Internet of Things (IoT) Framework

IoT is a large framework which refers to the interconnection of physical and even virtual things via telecommunication infrastructure with an aim to provide different services, going from no-real-time services based on sensors and actuators to real-time services such as surveillance cameras and critical services such as intelligent/smart transportation or smart factories (e.g. Industry 4.0 and beyond). Of course, the IoT framework is too broad to be captured with a single sentence. If we look into the standardization work, e.g. from ITU [4], the definition says that IoT is a global communication infrastructure enabling advanced services by interconnecting (physical and virtual) things based on existing and evolving interoperable Information and Communication Technology (ICT).

The IoT approach in fact adds a new dimension to the telecom/ICTs, which is referred to as "any-thing communication" besides the existing (before the IoT) "any-time communication" and "any-where communication" (Figure 6.1).

In the general IoT definition, various things are objects of either the physical world (physical things) or of the information world (virtual world). Each such thing needs to be capable of being identified and integrated into telecommunication networks, either that is public Internet network or specialized/managed IP (or non-IP) networks.

Physical things exist in the physical world while virtual things exist in the information world. In that way, physical IoT devices are capable of sensing, actuating, and connecting by using electrical equipment with sensors and actuators. In the broad definition, virtual things include various multimedia content and application software. However, in practice, the IoT term mainly refers to physical things. However, a physical thing may be represented via one or multiple virtual things (in the information world). On the other side, a virtual thing can exist without any association to any physical thing.

Future Fixed and Mobile Broadband Internet, Clouds, and IoT/AI, First Edition. Toni Janevski.

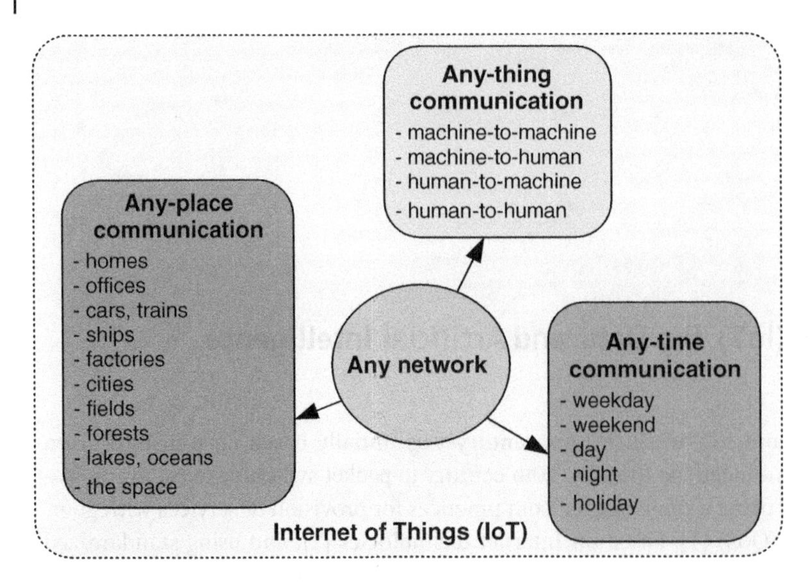

Figure 6.1 Internet of things dimensions.

Unlike legacy telecom services, such as voice and TV services, which have certain QoS requirements based on human senses to hear (for audio) and watch (for video), the IoT services may have different requirements on QoS, ranging from very low (e.g. low power devices that use low bitrates in range of kbit/s with infrequent sending of messages via radio communication) to very high performance requirements such as high reliability and very low latency.

6.1.1 Massive IoT and Critical IoT Technologies

The IoT services can differ in the QoS they require. Thus, certain IoT services have lower QoS requirements, exchanging only a small fraction of data over longer periods of time, while other services have strict performance requirements (e.g. latency, jitter, losses, bitrate, etc.). Regarding the QoS, different IoT services may have different requirements on different Key Performance Indicators (KPIs):

- Power efficiency: It is important for massive deployment of low power devices, such as sensors and actuators in Smart City services or personal wearables which need to have longer battery life or perform energy harvesting.
- Complexity: IoT devices for massive deployments are simpler in design and have lower prices (e.g. in smart homes, remote sensors, etc.) while critical IoT services use devices which have higher complexity.
- Mobility and range: Some IoT devices are static (e.g. smart home) while other are mobile (e.g. devices on vehicles for fleet tracking are one example). Also, coverage range is important for many IoT services in both urban and rural areas such as utility metering for public services (e.g. water supply, electricity, heating, etc.) as well as for fleet tracking services, unmanned vehicles and drones, etc.
- High reliability and very low latency: These parameters are important for critical IoT services, such as URLLC services in 5G and beyond mobile networks. However, critical IoT services are initially implemented by using Industrial Ethernet via wired connections to IoT devices. Typically, the requirement in this case includes also low end-to-end packet losses (e.g. 99.9999%

packet success rate, which may vary per service and its requirements) and very low latency (e.g. as low as 1 ms or even lower with 6G in 2030s). Such services include autonomous vehicles, connected robots, industrial automation, drones, etc.
- High availability: This is needed for critical IoT services that require services availability all the time. Such IoT services are energy smart grids, medical IoT services, critical V2X (Vehicular to everything) services, and so on. In the telecommunications world, higher availability is usually provided by enabling multiple links/paths of comparable performance, given that the probability of two or more paths being unavailable is much lower than the probability of a single path failure.

Based on the above KPIs, we may conclude that different IoT services have different contexts and, with that, different QoS requirements. In general, it is possible to distinguish between two main types of IoT services regarding their QoS requirements, which include the following (Figure 6.2):

- Massive IoT: These are IoT services that are based on low-cost, low-power devices that generate and transmit small volumes of data and are characterized by massive deployments. These include all possible sensor networks, smart agriculture, smart metering, smart transportation, smart homes, smart buildings, etc. Massive IoT does not have strict QoS requirements, meaning they are more tolerant of packet losses and longer delays as well as low bitrates in range of kbit/s. Massive IoT can be provided over both open Internet networks as well as managed telecom networks (not part of Internet). For example, in 5G mobile networks, massive IoT services can be provided over Extended Mobile Broadband (eMBB) or Massive Machine Type Communication (mMTC) network slices. There are specific mobile technologies for IoT that are standardized in both 4G and 5G mobile networks, such as enhanced MTC (eMTC) and Narrow-Band IoT (NB-IoT).
- Critical IoT: These services require ultra-reliability, very low latency, and very high availability. Such services include motion control in smart factories, remote manufacturing, remote real-time medical operations (e.g. surgery), energy smart grids, traffic safety and control in V2X services, and so on. These services require very strict QoS, therefore they are not likely to be provided via the public Internet network which is based on best-effort principles (the network does its

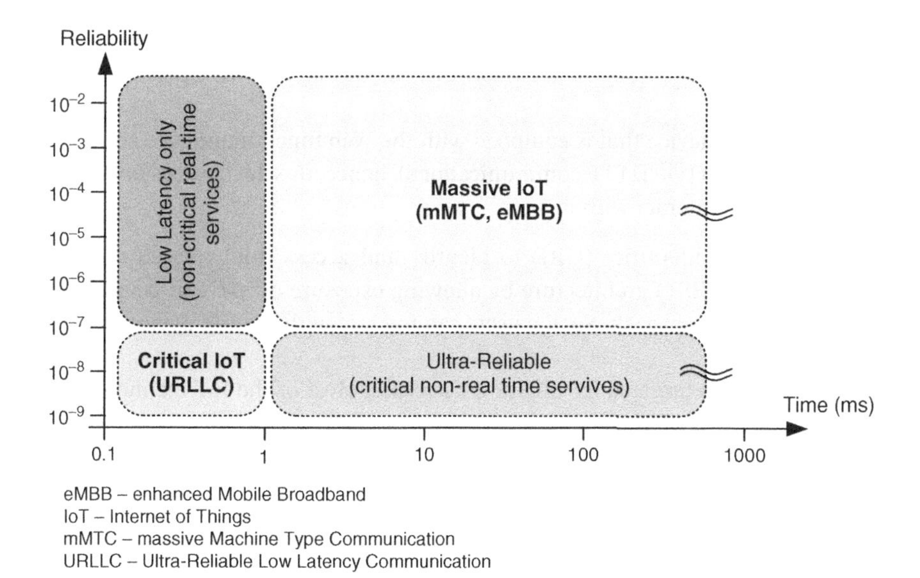

Figure 6.2 QoS comparison of massive IoT and critical IoT.

best effort to deliver the packet, but without specific guarantees on it) and network neutrality policy. Therefore, critical IoT services are typically provided via privately managed IP or non-IP networks. For example, dedicated type of service in 5G and beyond mobile networks for critical IoT services is Ultra-Reliable Low Latency Communication (URLLC). Hence, such services are provided via URLLC network slices, which can be implemented by the telecom operator's network or using a private network (e.g. private 5G network, standalone or operated by the telecom operator).

Regarding the KPIs for IoT services, similar to the legacy telecom networks, they include coverage, interference (on the physical layer), packet losses (on the network layer), latency (on the connection i.e. end-to-end layer), and bitrate [1]. However, depending upon type, the IoT services may extend the set of KPIs to new ones such as meter report success rate (from IoT devices to network nodes), software/firmware upgrade (from network nodes to IoT devices), tracking report delay, surveillance video quality (e.g. are objects and people on the video identifiable), etc. However, the applicable important KPIs and their target values strongly depend upon the IoT service type (e.g. massive or critical IoT service) and type of data being exchanged by the end IoT devices (e.g. messages or data transfer uplink and downlink, video stream uplink, etc.).

In all cases, for efficient management of the IoT devices and services, there is required a digital identity, e.g. a globally unique identifier which is used to represent IoT entities, e.g. over TCP/IP network. In general, the IoT devices can use different protocols and be accessible over private and public networks or through the open Internet.

Although the IoT is targeted to solutions for interconnecting things with Internet technologies as the name IoT indicates, creation of applications that run on the top of heterogeneous devices is a problem due to many heterogeneous networks and technologies that are used for provision of services. The common graphical user interface for many services is Web technology, which is based on HTTP protocol (the latest version is HTTP 3.0, which runs on QUIC/UDP/IP in the protocol stack). The Web technology provides the possibility for users to interact with devices (i.e. the things) in the IoT by using the standardized Web interfaces. Such approach is referred to as Web-of-Things (WoT). In the WoT approach we have two general types of devices:

- Constrained device, which has no Internet protocol stack so it cannot use Web technology, therefore, it requires so-called WoT broker placed between the device and the open Internet network.
- Fully-fledged device, which is a device that is equipped with the Web functionalities to connect with any other service on Internet (via HTTP communications), hence, this device does not need the WoT broker although it can interact with it.

WoT uses a Uniform Resource Identifier (URI) to identify and access things based on the Representational State Transfer (REST) architecture by allowing exposure of IoT end devices as resources on the web using a WoT approach. So, the users can interact with the IoT devices using Web interfaces.

There are different IoT standards targeted for different use cases. Most of the IoT technologies are wireless (with wired backbone, as shown in Figure 6.3):

- RFID, Near-Field Communication (NFC) and Bluetooth for personal area networks (e.g. via smartphone);
- Low Power Wide Area Networking (LoRaWAN), Sigfox, and other proprietary solutions for longer range indoor and outdoor massive IoT services;
- Wi-Fi as the main WLAN solution for various IoT devices in home, office, and public places;

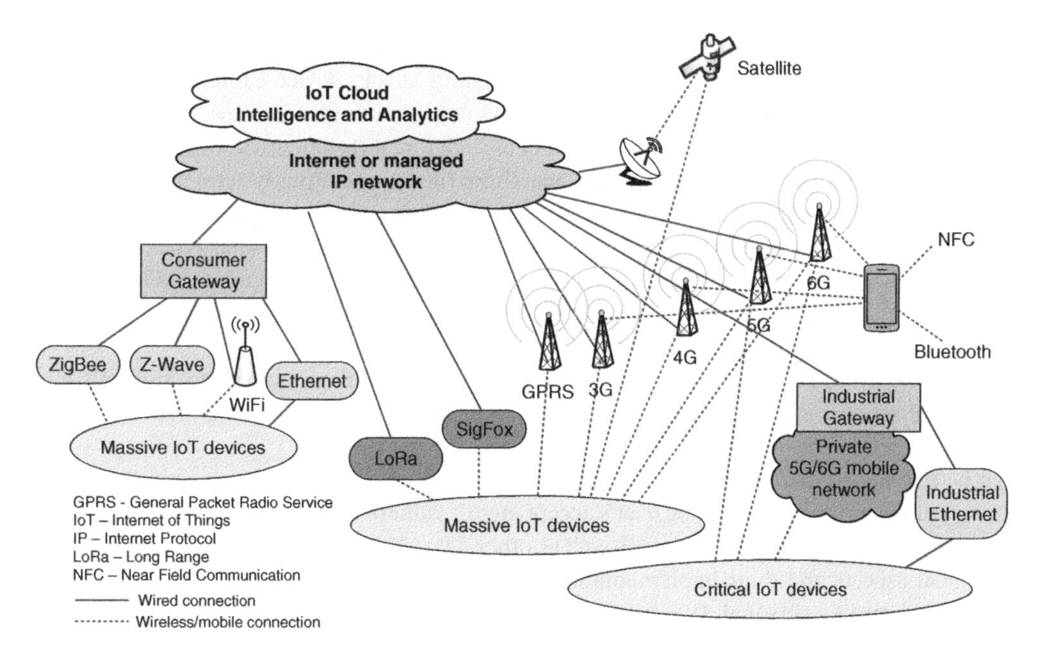

Figure 6.3 Wireless and mobile access technologies for IoT.

- Industrial IoT services which can be provided through fixed access (e.g. industrial Ethernet) and mobile access (e.g. 5G private network, URLLC type of service in a network slice for critical IoT services for industry);
- Cellular/mobile IoT, based on mobile technologies from 2.5G (e.g. GPRS) to 5G and beyond (6G), for provision of wide area IoT coverage for both massive and critical IoT. Mobile networks can be used for massive IoT service (via specific standardization such as eMTC, NB-IoT, Light NR, RedCap – Reduced Capability NR) as well as critical IoT services (via URLLC network slices).

Due to various heterogeneous noncompatible IoT technologies that were deployed since the beginning of the 21st century, IoT convergence and interoperability (e.g. via standardization) is essential for achieving the full potential of IoT and future widescale deployments in existing and new vertical sectors. In that way, Common (machine-to-machine (M2M) service layer, as middleware between application on top and network layer (e.g. IP works on the network layer, that is, layer 3) [2], can be used by different industry sectors, across IoT vertical markets in a multi-vendor environment (instead of using customized solution where customers are locked in by IoT equipment vendors or IoT service providers).

6.1.2 Security and Trust in IoT

Security and trust in IoT services are critical because they can harm both security and privacy, because we have various entities engaged in IoT services, including IoT devices, hardware and software, IoT gateways, and IoT service platforms.

If we look at the IoT chain, data acquisition is done by the IoT devices that are at one end of the "chain". The collected data is collected through the IoT gateways and finally stored in the cloud (for example, in a data center). All of these elements in the IoT data collection and transmission chain are potential targets for attacks, and they all have specific security threats. Which elements of the IoT network are most vulnerable to various threats?

Well, naturally, they are IoT devices, which can have very limited capabilities and this makes them vulnerable to various attacks (for example, flooding attack, device capture, and so on). However, there are also security threats to IoT gateways (e.g. unauthorized access, rogue gateway, denial of service attacks, and so on). But the number of such threats to gateways is less than threats to IoT devices, because gateways have greater capabilities (better operating system, more processing power, power supply, etc.). Threats to IoT traffic (i.e. IoT data transmission) on the network are similar for different types of traffic, because IoT traffic is multiplexed with other types of traffic, whether logically isolated or not, or transmitted over the public Internet or via managed IP address networks. There are also threats to the IoT platform and services (e.g. denial of service, malicious code execution, network eavesdropping, etc.).

To reduce security risks, security solutions are needed, such as data encryption, firewalls for platforms and servers, etc. Applied security measures increase trust in data (e.g. data collected from IoT devices).

What is trust in this case, what does it refer to? Well, trust is the user's confidence in the trustworthiness of the entity, including the user's acceptance of vulnerability in a potentially risky situation. It can also be applied to IoT data (as well as other data). While security mainly refers to the technological aspects and their implementation, and privacy mainly refers to the user (human user) aspects regarding personal data, trust is a broader concept. In fact, trust includes both security and privacy and their interrelationships (e.g. whether there are security and policy measures applied in telecom/ICT networks and services to protect users' privacy). However, trust applies to all domains, the social domain, the cyber domain, and the physical domain.

Trust has a "memory"; it is built on actions in the past, and past experience builds expectations for the future. Trust is also part of many services, especially for OTT services (since for managed IP services trust for all services comes from trust in the telecom operator). Trust in a particular service is built from the experience of many users who have used a particular service in the past. In the same way, we can now talk about trust in data, such as IoT data. We need to trust data in order to use it, and that is part of our common human reasoning. Finally, reliability is not expected to be the same for all IoT services. End-to-end security should be much higher for critical IoT services than massive IoT services. Why?

Well, because they are critical and any vulnerability can be dangerous, resulting in machine hazard in a smart factory, or a smart grid hazard, or a driverless vehicle hazard. So, not all IoT services are the same.

6.2 Mobile Internet of Things (e.g., NB-IoT)

IoT as a concept refers to any communication between machines and generally physical (and virtual) objects over the telecom networks, including open Internet (IoT includes in its name the term "Internet") and managed IP networks (and non-IP networks which are connected via gateways to global telecom IP networks). The main challenge for all IoT services is standardization and harmonization, based on global standards and avoiding being locked in by IoT vendors. The development of 4G and 5G mobile networks has dedicated a specific focus to the IoT segment in standardization. In general, mobile networks are globally harmonized, use the same spectrum in most regions, and have wide terrestrial coverage. Also, with 5G and beyond they expand to NTNs in addition to TNs, and are expected to provide 3D coverage.

When did the provision of IoT services in mobile networks begin? Well, that happened with the introduction of IP connectivity in 2G, which in terms of 3GPP standardization is the introduction of General Packet Radio Service (GPRS) in the late 1990s. IoT services provided over GPRS may only use tens of kbit/s, but in fact these are the required bitrates even decades later for massive IoT deployments (e.g. sensors, remote control of public utilities or measurement units, mobile credit/debit card payment, fleet tracking and other massive IoT services). This is because most of the IoT devices (e.g. sensors, meters) require narrowband bitrates to control the IoT devices and to transmit data to/from the IoT devices. So, it can be said that broadband Internet access (e.g. multiple Mbit/s or higher) is not required for most of the massive IoT services. However, IoT services were not in the focus in 2G and 3G mobile networks, which were targeted for the provision of mobile Internet access via mobile broadband, beside the legacy mobile telephony.

The standardization of specific solutions for mobile IoT (or cellular IoT, which is interchangeably used with the term mobile IoT) started with 4G standardization by 3GPP. The last 3GPP releases that were focused on 4G (releases 13 and 14), labeled with LTE Advanced Pro, focused largely on the IoT segment. Mobile networks are natively first choice for IoT services because most of the IoT devices (e.g. sensors) are spread over a wide area, while mobile networks have nationwide (WAN) coverage in each country.

6.2.1 Cellular IoT in 4G

LTE is a mobile broadband technology that does not fit well with services that rarely use low data rates. For such services and devices, the so-called cellular IoT approach in LTE targets low-power devices and long-range real-time communication. Why? Because legacy data transmission in LTE/LTE Advanced networks consumes too much energy and too much bandwidth; it is therefore not suitable for massive IoT deployments.

The LTE Advanced Pro has focused on mobile IoT with introduction of new narrowband technologies. In particular, LTE Advanced Pro defines the following three technologies targeted at IoT devices in Low Power Wide Area Network (LPWAN) segment (Table 6.1):

- Long Term Evolution of Machines (LTE-M) provides further LTE enhancements for machine type communications. It uses only 1.4 MHz (the smallest LTE carrier width) to provide throughput up to around 1 Mbit/s. The device categories defined for this technology are low-cost Cat-M1 and Cat-M2.
- Narrow Band Internet of Things (NB-IoT) is developed as a new radio technology (uses different radio interface than LTE in 4G or NR in 5G) which is aimed to be used for the low end of the cellular IoT market. It started in 4G networks and continues to exist in 5G networks, considering that it is "standalone" radio access technology which is adapted to legacy 4G or 5G frequency carriers. It supports bit rates from a few tens of kbps up to several hundreds of kbit/s using a 180 kHz bandwidth, which is similar to the GMS/GPRS/EDGE frequency carrier width (2.5G mobile systems from 3GPP). Due to narrowband spectrum and simpler design, NB-IoT has lower costs than LTE-M. NB-IoT can support up to 50000 devices per cell, which is convenient for massive IoT deployments.
- Extended Coverage GSM for Internet of Things (EC-GSM-IoT) is based on enhanced GPRS (eGPRS) intended for use of low cost massive IoT deployments. The EC-GSM-IoT deployments started around 2017 (with technology which is almost two decades old, but suitable for IoT that need only narrowband bitrates and hence need narrowband spectrum). The bitrates vary from 70 to 350 kbit/s when using Gaussian Minimum Shift Keying (GMSK), or up to 240 kbit/s for 8 Phase Shift Keying (PSK) modulation in radio interface.

Table 6.1 Comparison of 4G cellular IoT standards.

	LTE-M	NB-IoT	EC-GSM-IoT
Bitrates	~10 kbit/s to 1 Mbit/s	Reduced data rates (in kbit/s), with mobility support	70–350 kbit/s (GMSK) Up to 240 kbit/s (8PSK)
Cost	Low	Lowest	Low
Extended coverage	>155.7 dB maximum coupling loss (MCL)	164 dB maximum coupling loss (for standalone)	164 dB MCL for 33 dBm UE 154 dB MCL for 23 dBm UE
Spectrum	FDD, TDD and half duplex (HD) in any LTE spectrum	Stand-alone carrier, LTE carrier Guard band, In-band	200 kHz GSM frequency carrier
Radio technology	Reuses existing LTE base stations	NB-IoT	GSM (with GMSK or 8PSK)

If we compare 4G IoT technologies, EC-GSM-IoT and LTE-M are based on existing radio access technologies, and NB-IoT introduces new radio access technology to some extent. In doing so, NB-IoT can operate over a bandwidth of up to 200 kHz in standalone mode or in an LTE carrier or within the protection range of LTE carriers. It also supports a minimum channel bandwidth of only 3.75 kHz Single Carrier Frequency Division Multiple Access (SC-FDMA). This gives spectrum flexibility and system capacity (in terms of number of connected NB-IoT devices). Considering both the very low complexity of the NB-IoT device and the ubiquitous coverage makes NB-IoT the global cellular IoT standard for the LPWAN segment of the IoT.

6.2.2 Cellular IoT in 5G

For massive IoT services there are standardized NB-IoT and LTE-M for LPWAN IoT networks, which are also used in the 5G era (2020s). On the other side, 5G and beyond networks also provide possibility for URLLC service type, which are the most demanding ones regarding the performances. But there exists a need to address a broad range of so-called mid-tier applications, between legacy massive IoT (such as NB-IoT and LTE-M) and critical IoT (such as URLLC). Such mid-tier IoT part is standardized with RedCap (also known as NR Light) [5].

5G RedCap in fact provides range of devices between the low end IoT devices, such as NB-IoT and LTE-M, and legacy NR devices that are used via eMBB network slices, providing a mix of capabilities regarding the throughput, battery life, complexity, and device density, for diverse use cases. For instance, RedCap can be used for medical, personal, and business monitoring use cases, while RedCap sensors can be deployed in factories and campuses as well as other private locations, video surveillance (which does not require the full high data rate capability of 5G NR), etc., which are aimed to facilitate development of existing and new verticals with 5G and beyond mobile networks.

In general, cellular IoT services in 5G networks can be distinguished into several types, which include the following (Figure 6.4):

- Low Power Wide Area (LPWA) cellular IoT technologies, such as LTE-M and NB-IoT support in 5G which can be provided also in 5G mobile networks. 3GPP performed a 5G self-evaluation of

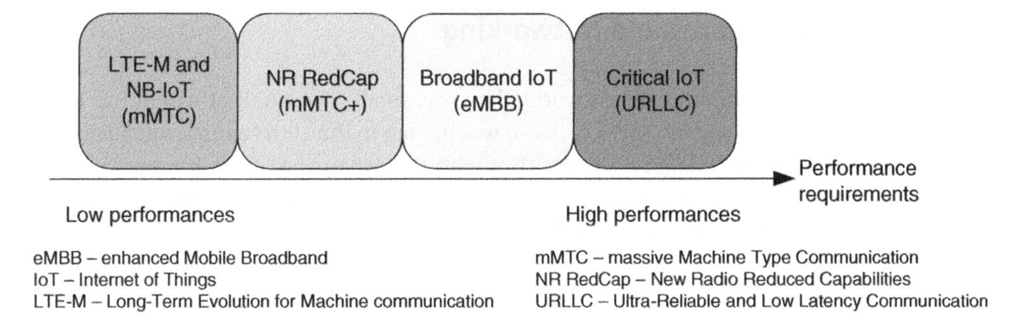

Figure 6.4 Comparison of cellular IoT in 5G.

LTE-M and NB-IoT performance in 5G NR radio access, which proved that LTE-M and NB-IoT both support the IMT-2020 requirement for massive IoT (i.e. MTC) density of 1000000 connected devices per km^2. Both LPWA technologies can be deployed also in-band of 5G NR carriers as shown in Figure 6.5. This IoT type is used for massive MTC in 5G.

- NR RedCap, which provides mid-range IoT devices with capabilities higher than LPWAN, but lower that broadband access in 5G such as eMBB. It started in 3GPP Release 17 and continued with the development in Release 18 and beyond. NR RedCap can be deployed in both FR1 and FR2 bands for 5G, and provides reduced NR modem complexity for 65% in FR1 (i.e. low and mid 5G bands) and about 50% for FR2 bands (i.e. high 5G bands) [6]. The latency for RedCap is not too low, but it is below 100 ms (much higher than URLLC and even eMBB), the bitrates are 150 Mbit/s in downlink and 50 Mbit/s in uplink (less than eMBB, more than eMTC), while battery life is in range of several years for industrial sensors to 1–2 weeks for wearables (unlike mMTC which use batteries that last more than 10 years).
- Broadband IoT, which refers to provision of IoT services through mobile broadband access, such as eMBB network slices in 5G mobile networks which have peak data rate target of 20 Gbit/s in downlink according to the IMT-2020.
- Critical IoT, which includes critical services provided through URLLC networks slices, as well as critical V2X services. These IoT services require highest performances which include ultra-high reliability and very low latency. The target for latency for URLLC is in the range of ms. IoT services to vertical industries provided via URLLC are also referred to as industrial IoT (one may note that industrial IoT are also provided with wired Ethernet).

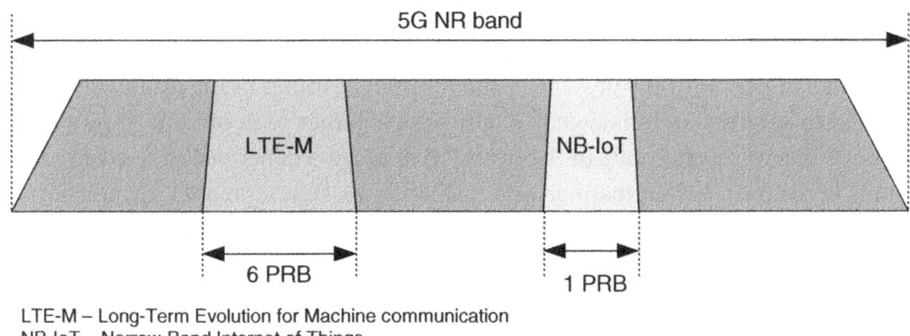

Figure 6.5 Use of NB-IoT and LTE-M in-band 5G NR frequency carrier.

6.3 Big Data Architectures and Networking

At the beginning of the telecommunications world, data was small. The smallest was in the 19th century when telegraphy was the main service. It also was not big in the 20th century when telephony and television were the main legacy services. But with the expansion of the Internet and IP networks to new limits in this 21st century, especially now in its third decade, data is growing extremely fast and will not slow down at all in the future, that's for sure. Why?

Well, because the number of users connected to the network is growing rapidly, in order to connect everyone with a broadband connection. In addition to people, tens of billions of "things" are also connected to the network. And there are many clouds that store and process data for various services. So, the amount of data generated and used by all people (via their devices) and all things and machines (without direct human interaction) is constantly increasing, as connected devices (either user devices or IoT devices) are constantly increasing over time. In addition, bitrates are also constantly increasing, so that in the 2020s the targets are hundreds of Mbit/s individual speeds and more Gbit/s in hotspots in the 2030s, and so on into the future. With ever-increasing bitrates (as well as data caps heading toward infinity, in both fixed and mobile networks) data volumes are increasing extremely fast. The question is what to do with it?

Well, some data is electronic waste. Other data is useful, and some data is very useful. Some data may seem unimportant at the present time, but may appear to be important in the future. So there is a lot of data, but also a lot of variables around that data – how it is collected, whether it is confidential, whether we should keep it longer or not, whether we should allow the data to be forgotten or not, privacy aspects, etc. In addition to these questions, there are answers that come with technological advances as well as data policies and regulations. But it should be recognized that currently, and in the future, data is an asset and has value, which still depends on the context of the data used.

So, when we have a lot of data, we talk about Big Data. With the development of cloud computing services, data is stored and accessed through the cloud infrastructure. The data centers of cloud service providers and their resource providers create vast amounts of data stored (and many times replicated through cache) in various storage locations around the world. Billions of pieces of information (data) are generated every day about individuals and businesses, including supplier data, delivery notes, company employment records, customer complaints about various services (whether public or private), as well as user-generated content such as messages, published photos and videos on social networks or content sharing websites, logging into different web portals, and so on.

The telecom/ICT world, based on Internet technologies, is evolving, from being Internet of contents in the past, to Internet of people of the present, to the Internet of things of the future. Billions of IoT devices connected to Internet and managed IP networks together with billions of humans will generate much more data in several years in the future than all data generated in telecom and Internet infrastructures in the past. In that manner, several Standards Development Organizations (SDOs) such as ITU, ISO/IEC, and NIST, have defined Big Data as a new term around the middle of the 2010s. To what does Big Data refer?

Well, Big Data refers to a dataset that is so large or complex that traditional computational analysis and processing cannot be used. So, Big Data is too complex to be processed with legacy computing approaches; therefore, it is directly benefiting from the development of AI and, particularly its sub-part, ML which are gaining momentum in the telecom/ICT world.

6.3.1 Big Data Ecosystem

By definition, Big Data consists of extensive datasets with certain characteristics in terms of volume, velocity, variety, variability, veracity, etc., as architecture for efficient storage, manipulation, and analysis of data [7].

So, in Big Data we have volume, velocity, variety, variability, and also veracity, at least five Vs. These characteristics can be explained as follows:

- Volume refers to the amount of data that is collected, stored, analyzed, and visualized;
- Variety refers to heterogeneity of data types and data formats that are processed by Big Data technologies;
- Velocity refers to data collection and processing approaches with aim Big Data applications to deliver expected results;
- Veracity refers to the uncertainty of the data;
- Value refers to the business results from using Big Data technologies.

Data is very heterogeneous in any respect when it comes from many different sources and from different time periods. But then, it is structured and analyzed and ultimately used, for example, to improve the same services from which it originates.

Let's take a look at the Big Data ecosystem. Its reference architecture consists of a data provider, a Big Data service provider, and a Big Data service client, which is a logical Big Data ecosystem. The supply of data can be direct or indirect (through a data intermediary).

The Big Data ecosystem, shown in Figure 6.6, depicts the flow of Big Data from its collection to the use of the data by customers, which involves several data transformations. It includes the following roles:

- Data Provider (DP) – its role consists of two main sub-roles which include:
  - Data provider, which provides data from various sources to the data broker and further to the Big Data service provider. Generates data, creates metadata information and publishes it.
  - A data broker, used to connect the data supplier and the Big Data service provider, acts as a clearing house. It provides a meta-information registry to the data provider for publishing data sources, as well as providing tools to the Big Data service provider to search for usable data.

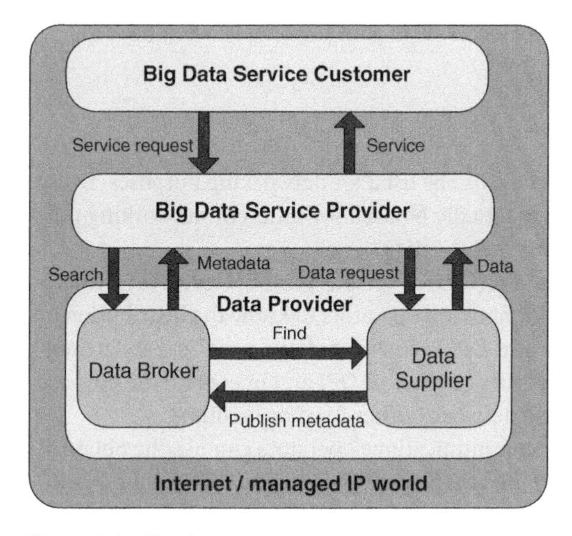

Figure 6.6 Big data ecosystem.

- Big Data Service Provider (BDSP) – it has capabilities for Big Data analytics and builds and operates the infrastructure, using metadata from the Data Broker for getting the required datasets from the Data Supplier. Typically BCSP provides a form of Big Data platform.
- Big Data Service Client (BDSC) – it is the end user or system that uses the services provided by the BDSP. BDSC requests services from the Big Data service provider and uses the results of such service for its own purposes or to third parties outside the Big Data ecosystem.

In general, data can be collected from different sources, in different forms and types (e.g. public data, private data, social data, sensor data, etc.). Typically, similar data sources are bound by similar policies. For example, data from cookies in web browsers is bound by formal policies set individually by the end user, while customer billing records of the telecommunications operator follow stricter policies (internal as well as legal regulation) regarding their processing as data.

After data collection, initial metadata is created to facilitate subsequent collection or retrieval. What is metadata? By ITU's definition [8], metadata is structured, encoded data that describes characteristics of information-bearing entities to aid in the identification, discovery, assessment, and management of the described entities. Listing of all metadata which is made available by a given data broker is called data catalog.

Furthermore, smaller sets of easily correlated data are aggregated into a larger dataset (in which case the datasets have similar security and privacy considerations). On the other hand, with matching, datasets with different metadata (i.e. keys) are also aggregated into larger collections. A typical example of matching is done by the Internet advertising industry (e.g. various commercial advertisements on websites targeted at the individual end user). Thus, matching services can associate HTTP cookie values (obtained from browsing activities) with a person's real name (e.g. real name used for online credit card payments).

Finally, there is also data mining to transform data and provide results from collected data from various sources. Data mining can be defined as the process of extracting data and then analyzing it from different perspectives to produce summarized information in some useful form that identifies existing relationships in the analyzed data. There are two main types of data mining, one is descriptive (provides information about the analyzed data) and the other is predictive (based on the collected data).

What infrastructure is used for Big Data? Well, typically, it consists of servers, databases and storage, various software (e.g. for databases and storage, etc.) as well as networking capabilities that are required for data transfer and storage on remote locations (e.g. in the cloud) when needed.

6.3.2 Big Data Driven Networking (bDDN)

In terms of telecom/ICT infrastructures, Big Data can also be used for networking purposes. Every network carries a lot of data, so the same data can actually be used for different networking purposes. But how do networks (e.g. telecom operators) get the data?

Well, it is mainly done with Deep Packet Inspection (DPI) at various protocol levels (by inspecting headers and payloads). Also, DPI is essential for network operators to know the distribution of service/application traffic in the network. In this way, DPI is considered as a generic core technology (or building block) for future networks. What is the purpose of DPI and in general the purpose of resulting Big Data from traffic carried across the network (either fixed or mobile)?

There are multiple uses for it. For example, telecommunications operators can use the obtained Big Data to further optimize their services (transmitted to their networks), and this is done by efficient bandwidth and traffic management, aimed at improvements in QoS as well as QoE (Quality of experience). In addition, Big Data can result in lower service costs due to higher network efficiency (i.e. fewer average resources to be used for the same services) and this can also impact capital investment by avoiding unnecessary capacity upgrades on the network.

So, the Big Data obtained from the packet inspection (i.e. from DPI) gives important information for network operation and maintenance, network management, control, optimization, and so on. Efficient usage of such valuable information obtained from packet inspection of the network traffic is achieved by bDDN [9]. What is Big Data driven networking?

Well, Big Data driven networking is in fact a group of technologies and methods which facilitate network operation, administration, maintenance, control, and optimization, by using the Big Data generated by the network (via certain algorithms and tools). A large amount of data is generated by the network. Then, such Big Data is used to serve for the improvements of that network. Also, if we want to exploit such bDDN (with the aim to make the telecom networks better, to automatically improve themselves, on the basis of certain intelligent algorithms) then the solution toward the future (to have better networks) is to introduce and apply the Big Data technology in the framework of future networks.

Is bDDN dependent on the type of the network (fixed, mobile, any particular type)? Well, bDDN is defined to be used in any kind of network in any architecture. But, does the bDDN change the network architecture when it is applied? Well, the bDDN does not change the network architecture to which it is applied, it just makes it better. Also, Big Data processing nowadays is also referred toward the use of AI, so we are transiting from the Big Data field to AI/ML use cases in the telecom/ICT world [10].

6.3.3 Big Data Use in the Telecom Sector

Massive use of mobile broadband services and large number of used smartphones and other devices (e.g. IoT devices) enables telecom operators to have access to huge volumes of data sources which are related but not limited to customers' information (e.g. from their profiles), their usage patterns, used devices, and networks or network slices (e.g. in 5G and future networks), location, etc.

The Big Data generated by networks can be for network management, operations, administration, marketing, etc. However, such valuable but tremendous amount of information is not possible to be efficiently used by traditional network architecture. There, bDDN solves this problem by separation of complex data computing and processing functionalities from the network control plane and management plane (Figure 6.7) and defines, in fact, a Big Data plane, which includes

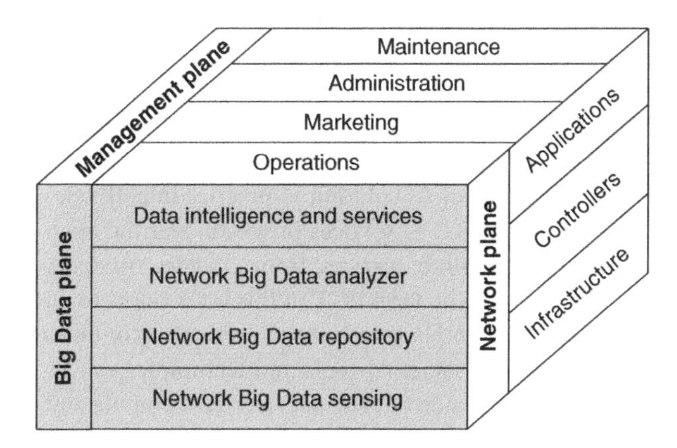

Figure 6.7 Big Data driven networking (bDDN) plane.

network Big Data sensing, storing in repository, analyzing the data, and using for data intelligence and services.

Telecom operators can use such big volumes of data to improve customer experience, to optimize their fixed or mobile networks as well as to create new revenue streams (e.g. by targeted marketing based on the data). In fact, telecom operators with Big Data analytics can have 360-degree view of their customers, including their usage patterns, movement, history of purchases, etc. With that telecom operators may do a kind of micro-segmentation of their customer base and provide adequate offers and products to them, which will be of interest with higher probability. That is already successfully done by online service providers over the open Internet, which uses multiple data sources generated by the user activity, its location, and shown preferences and interests.

Regarding networks, big data analytics have the potential to assist telecom operators in dimensioning their networks, managing and monitoring their networks' capacities, and effectively planning their investments in network extension and evolution which will strongly assist developing countries in 5G rollouts.

Analytics could also help telecom operators for improving the QoS by preventing, identifying, and fixing problems in networks or with services in real time. On the other side, Big Data analytics may help the telecom operators with prevention and detection of frauds as well as cybersecurity threats.

Furthermore, the analytical processing of Big Data can contribute to the reduction of Capital Expenditure (CAPEX) as well as Operational Expenditure (OPEX). In addition to their own needs, telecommunication operators can monetize data for their customers through a better offer of services for other or future businesses and users in different verticals, through appropriate retailing and advertising.

6.4 ITU's Framework for Machine Learning (ML)

AI and ML have been appearing and disappearing for a longer time since the 1950s (with ups and downs regarding their impact). In fact, the first paper on AI was published in 1950 [11]. AI and ML has come again into focus at the end of 2010s and in 2020s with the development and maturation of several telecom/ICT technologies such as cloud computing and IoT and particularly 5G (i.e. IMT-2020) and beyond mobile networks. With a huge number of devices being connected and huge amount of traffic being transferred (with very high speeds in fixed networks as well as 5G and beyond mobile networks), very large amounts of data is being generated, stored, processed, and used for different purposes. As already noted in Big Data section, such large amounts of data cannot be processed by traditional approaches, and there is the place where AI techniques come into the focus of interest for use cases in the telecom world.

In general, the AI is a broad terminology, which is not new (it has existed since the middle of the 20th century), but the novelty today is the application of AI solution in practice in both telecom operators' networks and services and Over The Top (OTT) services provided over the open Internet by the global online service providers (e.g. online services from Google, Amazon, Facebook, Microsoft, and many more smaller providers). The term OTT in this book refers to all services provided over the open Internet including, for example, a website of a university or news site, not only services from big global online service providers. AI tools and techniques can be applied in all OTT services as well as in equipment on the side of providers as well as on the end users' side.

Why AI? Because machines can execute repetitive tasks with complete precision, and machines are now gaining the ability to learn (of course, when they are designed by humans to do it), to improve (e.g. by using the collected data), and to make calculated decisions in ways that will enable them to perform tasks previously thought to rely on human experience, creativity, and ingenuity.

Also, since the end of the 2010s, AI is largely being used across the United Nations (UN) system to help solve some of the world's biggest challenges, from fighting hunger to tackling the urgent climate crisis to building smart sustainable cities [12].

Overall, AI advances in the 2020s at an exponential pace given that the development of technology in the telecommunications/ICT sector allows its practical use which, however, depends on the benefits of its use. Artificially intelligent machines (for example, computers, network nodes, user devices, etc.) are capable of processing huge amounts of data from different sources for different tasks in networks, services, and devices (which is possible due to the availability of more processing power in devices, which approximately doubles every two years, at least according to Moore's Law [13]). For example, the ability to analyze high-resolution images from satellites, drones, or medical scans with the help of AI tools can provide improvements in agricultural productivity, identification of various diseases in medicine, etc. Thus, it can be said that AI as a broad term encompasses various fields and techniques in various sectors, whereas we focus here on its application in telecom/ICT networks and services.

On the other hand, in addition to its use for good purposes, the power of AI also comes with challenges, ranging from issues related to transparency, trust, and security, to various concerns about job displacement and worsening inequalities.

What is, in fact, the AI we are now talking about? Well, the current expansion of AI is the result of advances in ML, which is only one form of AI that is being standardized (at least in umbrella specification) in the telecom sector.

6.4.1 Definition and Classification of Machine Learning in Internet and Telecoms

The main "form" of AI at the present time is ML. ITU has defined architecture for ML in future networks including 5G/IMT-2020 and beyond.

What is ML? According to the ITU's definition (based on ETSI definition) [14]: Machine learning refers to processes that enable computational systems to understand data and gain knowledge from it without necessarily being explicitly programmed. However, ML is based on mathematics and it is commonly implemented via software tools which are programmed (either by humans or by software tools developed by humans).

The ML model is created by applying machine learning techniques to data with the aim to learn from them.

What is the purpose of the ML model? It is used to generate predictions (e.g. regression, classification, clustering) on new (untrained) data.

How is the ML model implemented? Typically, it is deployed in the form of software (e.g. virtual machine) or hardware component (e.g. IoT device).

What do ML techniques include? They typically include certain learning algorithms that map the input data (which is used for feeding the ML) to output data. Normally, the type of data depends upon the use case (e.g. it is traffic data in a case of QoS management or network optimization).

An ML model is typically used to generate some predictions (e.g. regression, classification, clustering) on new untrained data based on previous so-called training of the model with similar trusted datasets (collected in the past, which can be hours and days ago, but also milliseconds ago).

There are two main types of ML algorithms used in the telecom world:

- Supervised ML – in this case, ML algorithms make predictions on a given set of samples. Examples of this ML type include:
  - Linear regression – used for regression problems;
  - Random forest – used for regression and classification problems.
- Unsupervised ML – in this case, ML algorithms group data by organizing it into a group of clusters in order to describe its structure and make initially complex data appear simple and organized for analysis.

There is also a third type of ML algorithms, called reinforcement ML, which uses rewards and punishments, so that the model is rewarded if it gets the job done and, vice versa, punished when it fails.

Which ML algorithm is best depends on the technical or business problem it is supposed to solve, as well as the nature of the set and available resources (e.g. time, computing power, etc.).

6.4.1.1 Naive Bayes ML Algorithm in Internet and Telecoms

Naive Bayes Classifier is a popular supervised ML algorithm in which the classification is done using the well-known Bayes Theorem of Probability to build ML models. If there are two outcomes X and Y, Bayes Theorem provides calculation of conditional probability of an event Y to happen when event X has already happened, which is expressed with the following equation:

$$P(Y / X) = \frac{P(X / Y) * P(Y)}{P(X)} \tag{6.1}$$

Naive Bayes ML algorithm is the best in cases with a moderate or large training dataset. There are different use cases, including the following:

- Use case 1: used by Facebook to analyze status updates expressing positive or negative emotions.
- Use case 2: with this algorithm, Google uses document classification to index documents and finds relevancy scores, i.e. the PageRank.
- Use case 3: email spam filtering (e.g. used by Google mail – gmail).
- Use case 4: classifying news articles on the Web per subject (science, sports, fashion, etc.).

The main advantage of the Naive Bayes algorithm is the fact that it needs relatively small training data to perform well when there are two mutually exclusive variables (for example, “like” and “dislike”, “1” and “0”).

6.4.1.2 *K*-Means Clustering ML Algorithm

The K-means ML clustering algorithm is a popularly used unsupervised ML algorithm. The output of the K means algorithm is K clusters (each with its own mean, hence, the K means) with input data divided among the clusters. The number of clusters is generally K, the value of which depends on the use case. For example, the word “jaguar” can be classified into two groups in search engines; for example, “Ibiza” as an island and “Ibiza” as a car model.

In this algorithm K data points are randomly assigned to clusters. Cluster centroids are calculated sequentially, and each data point is assigned to a cluster by calculating the least squared distance between the centroid and the given data point.

There are various applications of K-means clustering. Examples include search engines such as Yahoo and Google that use the K-means clustering algorithm to group web pages by similarity and identify the "relevance rate" of search results. In this way, this algorithm helps search engines to reduce the computing time for users.

6.4.1.3 Apriori Algorithm

An Apriori algorithm is an unsupervised ML algorithm that generates association rules from a given dataset. In most cases, the rules are based on an "IF THEN" format. For example, IF people buy an iPhone THEN they may also buy iPhone wearables.

The principle for Apriori ML algorithm is generally set as: IF an item dataset frequently occurs, THEN all the subsets of the item set also happen frequently (and vice versa, for infrequently).

The main advantage of this ML algorithm is easy implementation, which results in many use cases in online services provided by OTT providers. Such use cases (applications) of Apriori algorithm:

- Market basket analysis – many online retail companies (e.g. Amazon) use Apriori to extract data about which products are most likely to be purchased together and which are most responsive to promotion.
- Autocomplete apps – when end user types a word (in a search engine field on a web browser), the search engine looks for other related words that people typically type after a particular word.
- Healthcare data analysis – refers to analysis of medications taken by patients, initial diagnosis, etc.

6.4.1.4 Regression ML Algorithms

Regression ML algorithms are also used in the telecom/ICT world and wider. They predict the output variables' values based on different input variables. For example, the output may be price for a given service, while input variables maybe the age of the customer, gender, location, etc.

One type of regression ML algorithms is the linear regression model which belongs to supervised ML algorithms. It shows the changes between two variables – how one (or multiple variables) impacts the other. The goal is to find a simple line (linear equation) that can predict the values of the output variable based on the values of input independent variables, which can be expressed via an equation in the following for:

$$Y = A_0 + A_1X_1 + A_2X_2 + \ldots + A_NX_N \tag{6.2}$$

where Y is the output variable from the linear regression ML algorithm, A_i $(i = 0, \ldots N)$ are average effect on the output Y by each X_i, and X_i $(i = 1, \ldots, N)$ are independent input variables.

Typical use cases of linear regression ML model include estimating sales (e.g. forecasting based on trends), risk assessment (e.g. used for online/offline insurance calculations), market analysis, etc.

Other regression ML algorithm is logistic regression which belongs to the supervised ML algorithms. It is used to estimate discrete values of variable in classification tasks by giving as an output a value between 0 and 1. Typically a threshold value is used to determine which value tends to 0 and which tends to 1, so the output is either "0" or "1" (discrete values). For the calculation of the output logistic regression ML model uses the output of the linear regression function, given in equation (6.2), as input to a Sigmoid function to estimate the probability for the given class, given via the following equation:

$$Y_{\log istic} = \frac{1}{1+e^{-Y_{linear}}} = \frac{e^{Y_{linear}}}{1+e^{Y_{linear}}} = \frac{e^{A_0+A_1*X_1+A_2*X_2+\ldots+A_N*X_N}}{1+e^{A_0+A_1*X_1+A_2*X_2+\ldots+A_N*X_N}} \tag{6.3}$$

The main advantage in regression algorithms is easy implementation and lower complexity. The main disadvantages are that the algorithm requires more training data, may overfit the training dataset if it is sparse, and depends on the ML human designer in predicting the possible variables, i.e. outcomes (e.g. via setting the threshold in logistic regression ML algorithm).

Possible use cases of logistic regression ML include weather forecasting (e.g. rain forecasting), predicting risk factors for diseases (in healthcare), online hotel booking, etc.

6.4.1.5 Random Forest ML Algorithm

Random forest belongs to supervised ML algorithms that are used in classification and regression problems. It uses multiple decision trees at the training process of the algorithm with the given datasets.

Unlike other ML algorithm random forest algorithm can use the datasets containing continuous variables (as in regression algorithms) and categorical variables (as in classification algorithms), so overall it is suitable for classification and regression problems.

The use cases of this type of algorithms include generating demand forecasts for irregular customer demands, such as purchase of equipment of telecom operators from their vendors [15, 16].

There are other ML algorithms, but here we have covered some of the most used and well-known ones in the telecommunications/ICT world. However, ML algorithms cannot be standardized because there is simply no best one, just like there is no smartest human. Therefore, the use of the ML algorithm depends and will depend on the equipment vendor or application/service developer in the open Internet space. But, on the other hand, there is a need for an umbrella specification that will frame ML so that different contributors can develop different parts of the ML solution in a standardized way, which is needed especially for telecom operators who have to adhere to stricter national regulations (when compared to global OTT service providers).

6.4.2 Framework for Machine Learning (ML) by ITU

The main "form" of AI right now is ML. Also, the ITU has defined the architecture for ML in future networks, including 5G/IMT-2020 [17]. The high level ML architecture is shown in Figure 6.8.

One of the main building blocks in that architecture is the so-called ML pipeline. What is ML pipeline? An ML pipeline is a set of logical nodes, each with specific functionalities that can be combined to form an ML application in a telecommunications network, thus providing an abstraction for handling ML.

Then, in the ITU's ML architecture is defined Machine Learning Function Orchestrator (MLFO) as a logical node with functionalities that manage and orchestrate the nodes of ML pipelines based on ML Intent and/or dynamic network conditions.

The third building block in the ML architecture is the ML sandbox, which is an isolated domain that allows hosting individual ML pipelines to train, test, and evaluate them before deploying them to a live network. For training or testing, the ML sandbox can use data generated from simulated ML sandbox networks and/or live networks.

In Figure 6.8, the source node SRC is in fact the source of data that can be used as input to ML pipeline while SINK is the target on ML output on which it takes action. Further, Distributor (D) is a node that identifies the sinking nodes and distributes the output of the ML model node (denoted with M) to the corresponding sink nodes.

The given ML architecture is an umbrella one that can be applied in different networks, including 5G/IMT-2020 mobile networks or others [18], for support or provision of future services toward 2030 and beyond.

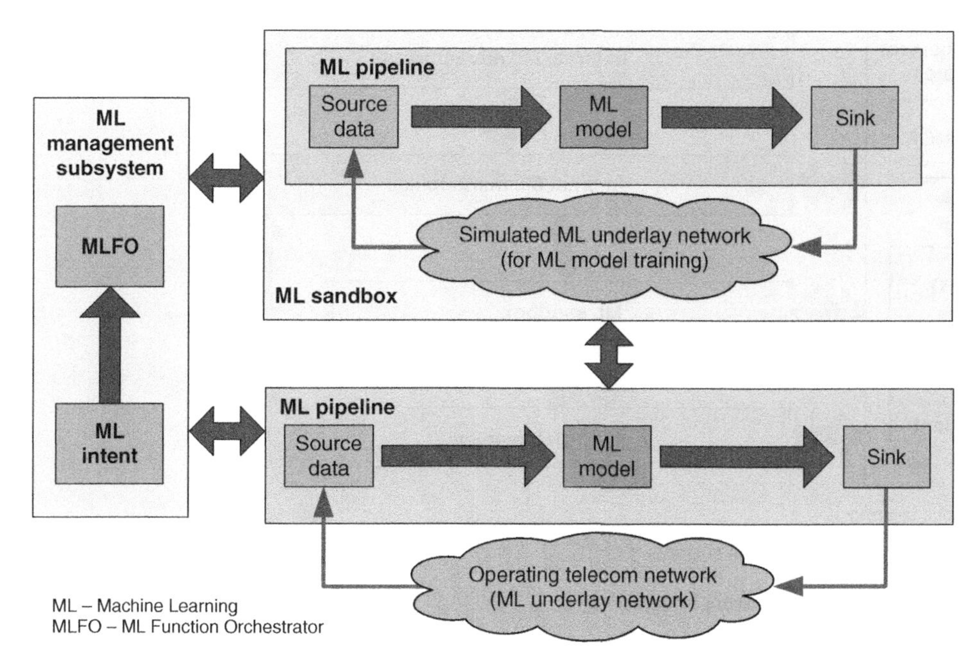

Figure 6.8 Machine Learning (ML) architecture.

The end user devices (e.g. smartphone, IoT devices, etc.) are resource-constrained devices, therefore, only the data source is instantiated in it. Such constraint is specified in so-called ML Intent.

What is ML Intent? Well, ML Intent is a declarative description which is used to specify an ML application. ML Intent does not specify some specific network functions that need to be used in the ML application, but it provides a basis for mapping ML applications to technology-specific network functions which may be diverse (e.g. they have differences in fixed and mobile access networks). ML Intent can use a ML-specific meta-language for defining the ML applications.

6.4.3 Machine Learning Marketplace

There is diversity in the data used in ML which results in increased flexibility and agility which in turn results in more complicated configuration of network parameters and policies.

Then, for exchange and delivery of ML models, there is defined so-called ML marketplace. What is ML marketplace? ML marketplace refers to a repository of ML, with ML models that is possible to be exchanged between multiple parties (e.g. providers and users of such ML models).

The architecture for ML marketplace network integration [19], is given in Figure 6.9. A telecom operator that needs to use ML models from a given ML marketplace may obtain them either from ML marketplace deployed internally and/or externally (in respect to its administrative domains). If one ML marketplace is internal to some telecom operator (i.e. network operator), it appears as external to other operators and vice versa.

The exchange of ML models is done by using querying, selecting, pushing, discovering, and deploying the ML models by the MLFO in the ML sandbox or ML pipeline subsystem.

The ML architecture defines so-called reference points, which are not a definite set considering that they are being extended with the further development of ML framework. For example, there are defined reference points between ML marketplace and the ML sandbox subsystem, then between the ML pipeline and the management subsystem.

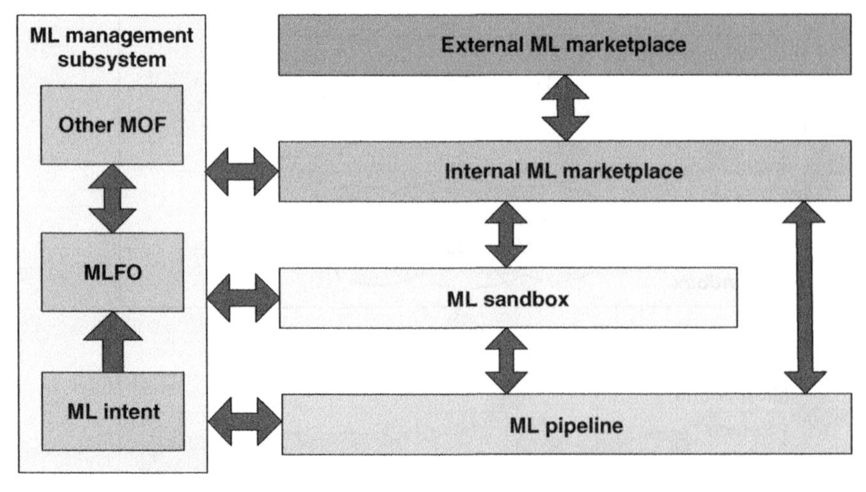

Figure 6.9 ML marketplace network integration.

Finally, one should note that the ML architecture defined in the ITU framework is an umbrella specification, so it covers various existing and future ML implementations that will be targeted to different use cases in future networks (e.g. use case of network and service QoS/QoE, use case of resource and fault management in future networks including IMT-2020 [20], and so on).

6.4.4 ITU's Network Intelligence Levels

Overall, AI and, in particular, its current frontrunner ML, is considered to be a technology to help with the increasing complexity of networks and services and their building components, and to improve the performance of future networks including IMT-2020/5G. In such a case, when AI/ML is used in network operations and management, it can be observed that networks are becoming more intelligent, so it is important to have a methodology to assess their intelligence levels.

Table 6.2 describes the network intelligence level of a whole network Level 0 to Level 5 based on how many (of the total five) dimensions are covered by human and/or AI. Although we have different network intelligence levels shown in Table 6.2, decisions and execution instructions provided by a human being have the highest authority regardless of the level (Level 0 to 5) and dimension.

So, with AI we apply an intelligence in the network. But, in such case it becomes important to adopt methodology for evaluating network intelligence levels. For such evaluation ITU considers five dimensions.

The first dimension is demand mapping, which refers to converting the requirements from operations and maintenance to certain instructions that network elements can execute. Then, we have the data collection dimension, which refers to collecting the data from the networks (e.g. with deep packet inspection). After that we have the analysis dimension, which refers to pre-processing of data to get the contextual information of the network data. Next comes the decision dimension, which is tailored to a decision regarding a specific network or service configuration as a result of

Table 6.2 Network intelligence levels.

Network intelligence level		Dimensions				
		Action	Data collection	Analysis	Decision	Demand mapping
Level 0	Manual network operation	Human only	Human only	Human only	Human only	Human only
Level 1	Assisted network operation	Human and AI	Human and AI	Human only	Human only	Human only
Level 2	Preliminary intelligence	AI	Human and AI	Human and AI	Human only	Human only
Level 3	Intermediate intelligence	AI	AI	Human and AI	Human and AI	Human only
Level 4	Advanced intelligence	AI	AI	AI	AI	Human and AI
Level 5	Full intelligence	AI	AI	AI	AI	AI

data analysis. And finally, the fifth dimension is action implementation, and that is execution of the network or service configuration that resulted as a decision from the analysis of the collected data.

But what kind of intelligence can there be in telecom/ICT networks and services? Well, one kind is the human intelligence, the other is combined human and AI, and the third one is only AI (without direct human interaction).

Each of the five mentioned dimensions can be mapped to one of the three intelligence orders (human, human and AI, AI). Based on such mapping we can have the intelligence level, which goes from Level 0 (in all five dimensions there is only human intelligence) to Level 5 (in such case the network has full intelligence, and all five dimensions are based fully on AI).

What will happen when the telecom/ICT network will be Level 5, fully AI-run? Well, in such case they should run (operate) automatically and update themselves automatically (e.g., update, reconfigure, recover from failures, and so on). For example, that will make the networks similar to operating system in our computers and smartphones, that is, once installed it works without manual configuration of end users, because the operating system is too complex for ordinary users to be able to apply critical settings to it (e.g. it may become not functional in a case of wrong configuration). Another example from the Internet technologies (about the automatic configuration for plug and play access to Internet network), the DHCP does automatic configuration of network interfaces on the devices we use (manual configuration is not feasible for non-technical end users). If we want to apply similar approaches toward the automation of telecom/ICT networks and services, then we need to use AI approaches.

It can be noted that in the 2020s and even in the 2030s, AI can be considered to be at an early stage in the world of telecommunications/ICT. But it can be expected to continuously evolve to improve telecom and Internet networks and services (for example, higher QoE), more resilient, highly accessible and affordable, as well as highly innovative services and support in the future fully digital era. So, the "rise" of AI in the world of telecommunications/ICTs has just begun (in the 2020s), and it is certain that it will last a long time in the future.

6.5 AI (Artificial Intelligence)/ML (Machine Learning) for 5G

High level ML architecture for future networks can be applied to any network, including fixed and mobile networks. Somehow the development and deployment of 5G mobile networks in the 2020s coexists with the rise of AI/ML in the telecom world. Hence, the ML architecture consisted of ML pipeline, ML sandbox, MLFO, and ML underlay networks, is applicable to IMT-2020/5G mobile networks. In such case the underlay network in the ML architecture is IMT-2020/5G network [17].

When mapping the ML architecture on IMT-2020/5G system architecture the UE (e.g. a smartphone, an IoT device, etc.) is a resource-constrained device, hence only the Source (for data used in the ML) is instantiated in the UE. And this constraint is specified in the ML Intent.

The collectors of data in the 5G RAN (Access Network – AN) and 5G Core (Core Networks – CN) are placed by the MLFO which is done based on the specifications of the ML applications in the ML Intent. For different applications/services in 5G network different ML pipelines can be used. For example, for latency sensitive applications in the access network can be used one ML pipeline can be used, while another ML pipeline can be used for latency tolerant applications. The chaining from Source to the Sink in ML pipelines is done according to the requirements that are specified in the ML Intent which is a declarative description used to specify an ML application (e.g. using meta-language to define ML applications).

6.5.1 AI/ML Model Transfer in 5G System

The application of AI/ML in 5G system can be in RAN, in core network, as well as for various intelligent mobile services. Regarding the 5G RAN intelligence the following elements can be distinguished [21], as shown in Figure 6.10:

- Data collection is a function in 5G RAN that provides input data to ML model training and model inference functions. Data preparation for the ML includes data pre-processing and cleaning, formatting, and transformation. Input data can include various measurements from mobile terminal (i.e. UE), 5G base station of other network entities. There are two main types of data

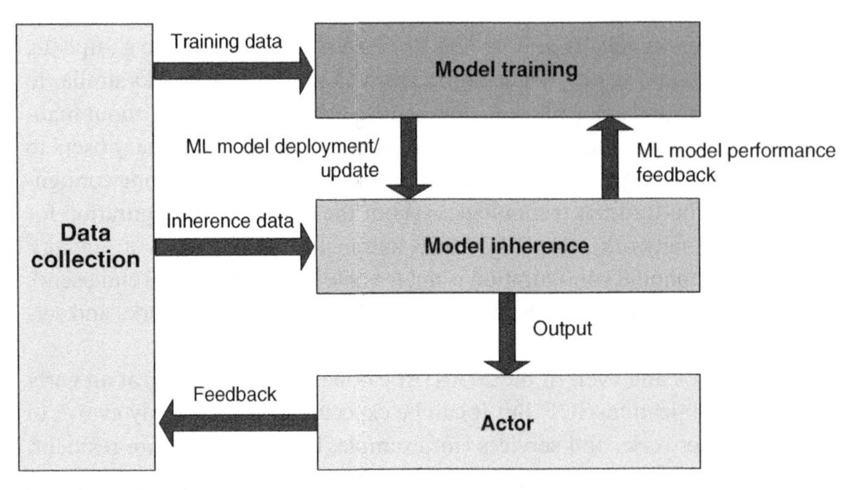

Figure 6.10 5G RAN intelligence framework.

which include the following: (1) training data, which is data used as input to the AI/ML model training function; and (2) inference data, which is data used as input to the AI/ML model inference function.

- Model training is a function that executes the AI/ML model training, validation, and testing. As a result of testing procedure, model training functions may generate model performance metrics. This function also performs data preparation such as data pre-processing and cleaning as well as formatting and transformation.
- Model inference is a function that provides predictions or decisions as an output. It provides model performance feedback to ML model training (may be used for monitoring the performance of the AI/ML model), while in the opposite way receives model deployment/update. The output of model inference, which is use case–specific, is provided to the Actor function.
- Actor is a function that receives the output from the Model Inference function and performs resultant actions, including triggering actions toward other network entities. The output of the Actor function is feedback which is information that may be needed to derive training an inference data, or for monitoring the ML model performance and its influence on the network. That is typically realized via update of a given set of KPIs depending upon the use case.

In mobile systems, mobile devices (e.g. smartphones, cars, robots) are increasingly replacing conventional algorithms (e.g. speech or image recognition, video processing, etc.) with AI/ML models to enable applications. Overall, 5G and future 6G mobile systems can at least support three types of AI/ML operations:

- AI/ML operation splitting between AI/ML endpoints (Figure 6.11) – This is done according to the given operating environment with the main goal of offloading tasks from the end user device to network AI/ML endpoints in cases that require more computing power and are also energy intensive, while leaving it to the User Endpoints (i.e. UE) the parts that are privacy sensitive or delay sensitive.
- AI/ML model/data distribution and sharing over 5G/6G system (Figure 6.12) – In this case is performed online AI/ML model distribution, i.e. new model downloading. Such new AI/ML model can be distributed from the mobile network endpoint to the mobile devices when they need it to adapt to the changed AI/ML tasks and environments. For such scenario of AI/ML operations, there is needed continuous monitoring of the performance of the AI/ML model at the UE.

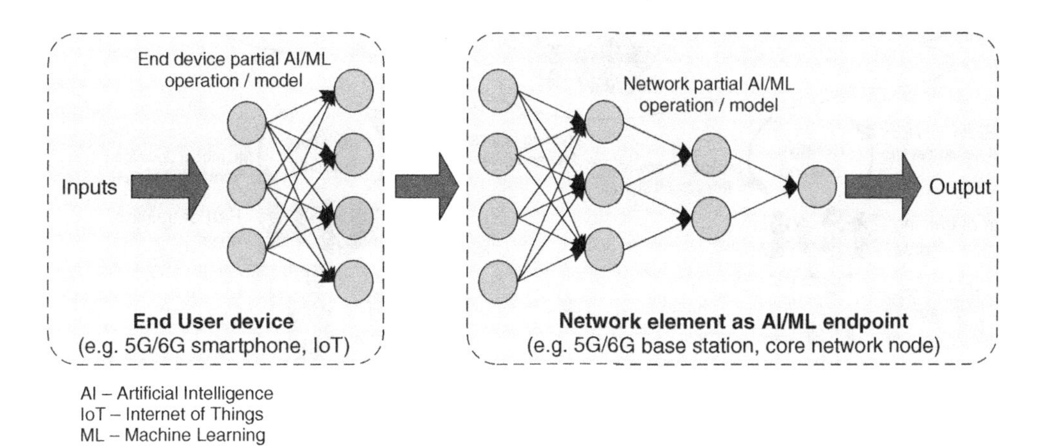

Figure 6.11 Split AI/ML inference in 5G/6G system.

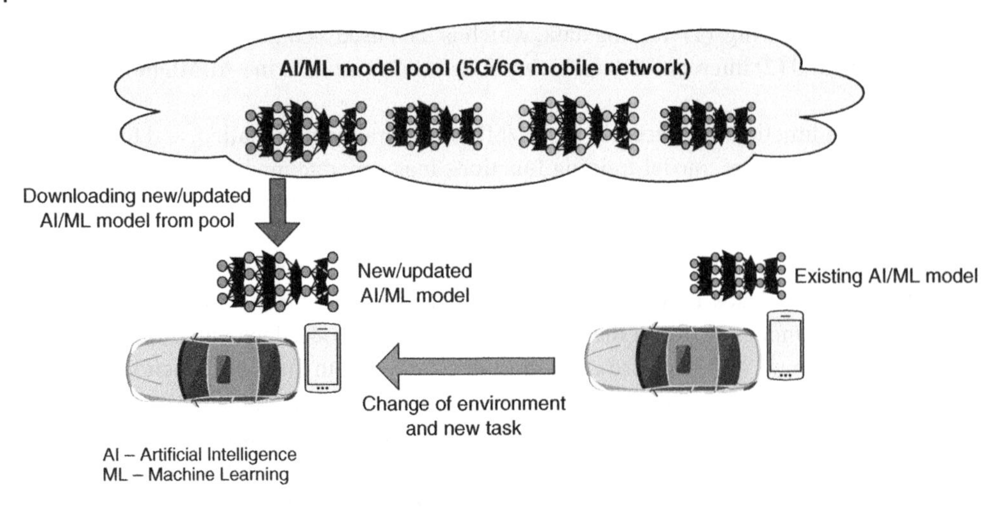

Figure 6.12 AI/ML model distribution in 5G/6G system.

- Distributed/Federated learning over 5G/6G system (Figure 6.13) – In this case, the 5G/6G cloud server contributes to the training of a global ML model based on the aggregation of local models partially trained by each of the mobile end devices. The mobile device receives an AI/ML model from the 5G/6G cloud server and performs local training based on such model. Then, the local training results and the ML model in the UE are uploaded to the cloud. Cloud services collect many such training reports received from UEs and update the global model by federating many local training results. Further, such global ML model is distributed back to the mobile devices, which perform another training cycle and return. Thus there is a continuous optimization of the global ML model using much trainings performed by the mobile devices in 5G/6G networks.

The AI/ML techniques in 5G and beyond mobile networks can be used for different goals. Over the time services are becoming more personalized than before. For example, each OTT mobile application or Web page accessed via the mobile handset is personalized. It can be expected that in

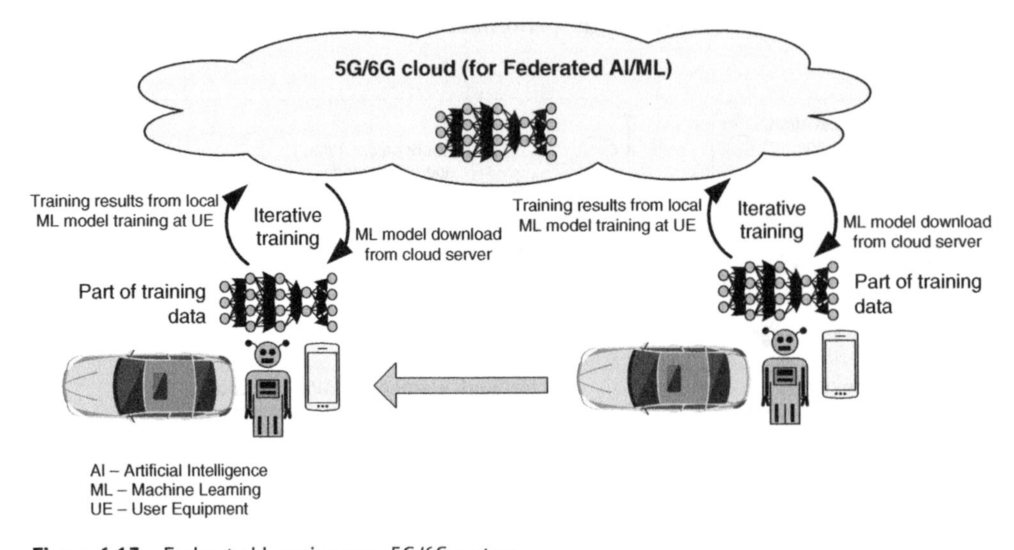

Figure 6.13 Federated learning over 5G/6G system.

the future there will be a special service package for each customer or group of customers of the mobile operator, and even a special network slice (assuming massive network slicing in future mobile systems). Such personalization of services is not possible to be manually handled by humans, and requires digital tools based on AI/ML techniques.

Further, Internet users were content consumers only in the past, but now they have become content producers which increase Internet traffic and make storage and transmission (in UL and DL) a great challenge for providers and operators, especially in case of mobile access. In that way, AI/ML techniques should help operators to efficiently handle the Internet/IP traffic.

Current and future mobile networks (4G and 6G) are/will be based on Network Function Virtualization (NFV) and Software-Defined Networking (SDN). With such approach network management is becoming more precise and complex than before (e.g., before 5G). Network virtualization is not only at the level of network elements, but with the development of mobile systems it goes at the level of components such as processors, memory units, network interfaces, available bandwidth, etc. AI/ML technologies are aimed to provide possibility telecom operators to set up on-demand networks, such as private 5G/6G mobile networks for different vertical sectors.

In the following part we provide certain possible use cases in 5G/5G-Advanced mobile networks.

6.5.2 AI/ML Use Cases in 5G/5G-Advanced

There are different use cases of AI/ML in 5G mobile networks and beyond. In this part we refer to a subset of them which are into the focus, however the application of AI/ML in mobile networks is not limited only to them.

6.5.2.1 Use Cases of AI/ML for QoS, QoE and Energy Saving in 5G and Beyond

The IMT-2020 i.e. 5G mobile network and future 6G network are expected to be able to provide optimized support for many heterogeneous services, with different traffic requirements and different QoS needs. The typical KPIs associated to services include bit rates, user density and mobility (i.e., velocity) and burstiness of the traffic (i.e., variable data rates).

The ML mechanism can learn from network environment data and react to dynamic network situations, especially in mobile networks. Such an ML mechanism is possible to learn from previous QoS data against a given set of target KPIs and reconstruct relationships between recorded past QoS-related data and performance anomalies. After training with such data, ML algorithms can learn the relationships between network settings, target KPI values and traffic (e.g. load, type of services and their traffic distribution, etc.) and can automatically prevent future QoS anomalies from occurring by using mitigation and actions in appropriate network elements (e.g., traffic load balancing, configuration settings adjustment, KPI target values change, etc.).

In the field of QoS and QoE, QoE is more difficult to measure than QoS (easily measured using a set of KPIs, such as service availability, downlink and uplink bitrate, latency, packet losses, and others), because QoE is subjective by definition (it depends on human users and their perception and enjoyment of consuming a given service). However, AI/ML algorithms can be used to estimate QoE for a given service by using complex network QoS analysis for different services and interrelating such QoS data with available non-technical data such as tariff plan, customer age, environment, end-user equipment and other related contextual information that may affect the quality of service experienced by end-users.

For example, who cares the most about latencies in 5G and beyond? The machines care the most about latency, especially ultra low latencies, such as less than 10 ms (in range of milliseconds). The humans, on the other side, need the most bitrates and reasonably low latency, which may be several tens of ms (e.g., for voice and video services) and less for emerging mobile services (e.g., AR/VR/MR).

So, one may say that machines, not humans, really can benefit from the ultra-low latencies which in mobile world are possible only with 5G and beyond networks. Normally humans benefit from all such services indirectly. Management and orchestration of delay-critical services in 5G networks needs AI/ML techniques.

To meet the 5G requirements on QoS and set KPIs on reliability, capacity and latency, much more 5G base stations are needed (with that the cell size decreases and number of cells increases, i.e., higher mobile network density, particularly in urban hotspot areas). But, more equipment results also in high energy consumption, CO_2 emissions and potentially higher operational costs. From all different viewpoint (either telecom operators or environment and society) the energy saving in mobile networks is becoming important use case which may involve different layers of the network, with mechanisms operating at different time scales.

For example, cell activation/deactivation is an energy saving scheme, which may be realized by real time traffic prediction, so when the expected traffic volume is lower than a given threshold (which may be set based on different parameters and updated over time), the cells may be switched off and UEs to be offloaded to another cell.

Efficient energy saving depends on many factors such as traffic load, cell utilization, users mobility, coverage areas, configuration parameters of radio units, etc., and their real-time monitoring and adaptation can be done by AI/ML algorithms, which may use the existing data for training and update the ML model for energy saving in mobile network.

6.5.2.2 AI/ML for Network Slicing in 5G/IMT-2020 and Beyond

Network slicing introduced in 5G mobile networks aims to provide dedicated logically separated networks over the same mobile infrastructure (i.e., network slices) with customer specific functionalities. A network slice, including radio access network, transport network and core network, can be dedicated to certain specific types of service.

Mobile devices (i.e. UEs) can access one Network Slice Instance (NSI) or multiple NSIs simultaneously, for example, vehicles can access V2X and eMBB. Thus, a service of a telecommunications operator to a car vertical for provision of autonomous driving car will use the V2X communication service. However, passengers on board can use information and entertainment services through an eMBB network slice. This leads to different QoS requirements per network segment and service, such as:

- A V2X service (e.g. provided via a URLLC service type) requires very low latency, but not necessarily a high bitrate, and
- Video streaming service (provided via eMBB) requires high bitrates but is latency tolerant (e.g. can accommodate hundreds of ms latency).

Many network slices and much more NSIs with different requirements of QoS and security increase the complexity of network management. In that manner, AI/ML-assisted analysis is beneficial for the whole lifecycle management of each NSI, from its deployment to its termination. Thus, AI/ML-assisted analysis emerges as a medium to bridge the network of the telecom operator and the customers connected to the network slice. Figure 6.14 shows ML-based (i.e. cognitive) end-to-end network slice management and orchestration.

The architecture for cognitive slice management in Figure 6.14 is based on closed-control loop approach which refers to both the control plane and the data (i.e. user) plane. With the aim for closed-control loop to function well several subsystems need to coexist and interact to ensure the high availability and security of network slices and services provided over them. Such subsystems

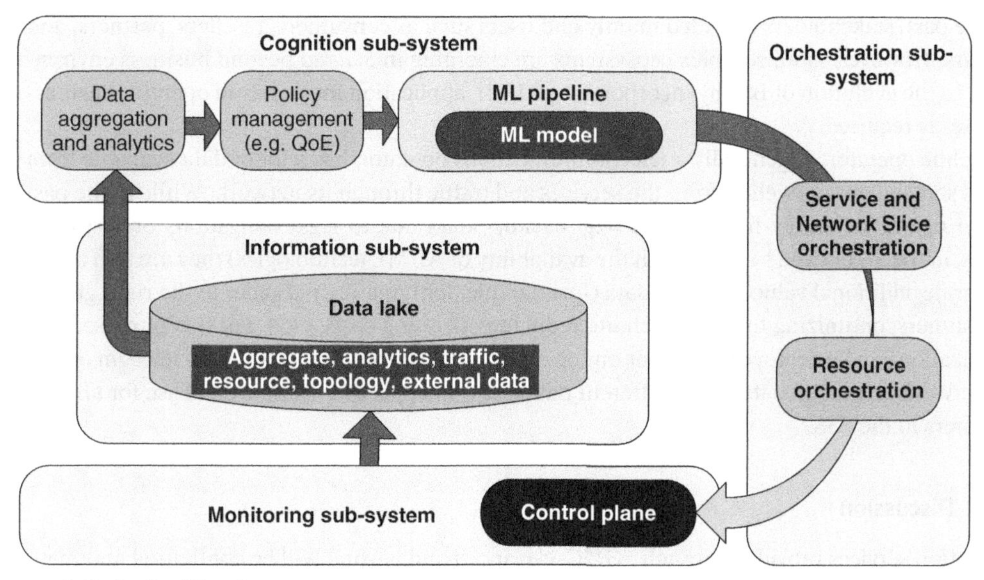

Figure 6.14 Architecture for cognitive end-to-end network slice management and orchestration.

include monitoring subsystem, information subsystem (with data lake), cognitive subsystem (with data aggregation and ML pipeline), and orchestration subsystem (for service, network slice, and resource orchestration). The monitoring subsystem monitors different metrics and counters regarding the traffic, resource utilization, network topologies, etc., from physical and virtual network elements [22]. The information subsystem is a data platform, with a main element referred to as a "data lake" that is used to store metrics and counters related to network infrastructure and network components, services, and thus acts as a tracking repository that stores all available catalogs and supplies. The orchestration sub-plane provides a set of coordination functions required to enable, secure, and maintain the network elements of the network infrastructure. There end-to-end orchestration of multi-domain network slices is achieved through hierarchical approach wherein the orchestration subsystem in the domain of each network slice provider is responsible for orchestrating the slice, while the orchestration subsystem at the service provider (e.g. the telecom operator) orchestrates end-to-end connectivity across multiple network slice domains. Finally, the cognitive subsystem uses AI/ML techniques and tools, so it has ML pipeline which is trained with the data from the data lake of information subsystem in this architecture. The main task of the cognitive subsystem is provision of operational optimization for services provided, network slices and their instances, as well as underlying resources in the network.

This approach contributes to the minimization of human engagement in maintenance and operation tasks, which can significantly reduce operational costs (OPEX) in complex network infrastructures of 5G and further mobile networks, with a large number of network slices with different services for different users and verticals.

6.5.2.3 AI/ML for Business Support Systems (BSS) in 5G and Beyond

The Business Support System (BSS) is present in every telecommunications operator and is used to manage relationships with stakeholders by handling SLAs, managing orders, generating reports, invoicing, and so on.

In the past, stakeholders included mainly end users such as consumers, resellers, partners, and suppliers. However, more complex ecosystems are emerging in 5G and beyond business environments, so the evolution of BSS by incorporating AI/ML application into telecom operator business processes is required.

A mobile operator, or generally a telecommunications operator, has a lot of data available from various stakeholders as well as from the services and traffic through its network. While in the past telecom operators simply had to store (e.g. backup) data due to legal obligations or historical records, in the 5G era and beyond (with the availability of AI/ML technologies) they are also trying to generate additional value from the data (for example, for targeting packages to the right groups of consumers, optimizing the supply chain, reducing operating costs, etc.). For this purpose, data virtualization can be deployed as a layer on top of data repositories that exist in the telecom operator's network so that all data (from different databases) to appear as a single database for all data consumers in the BSS.

6.5.3 Discussion

Application services provided through 5G/5G-Advanced and beyond will be configured and managed using a variety of infrastructure capabilities based on cloud computing, edge computing, and various types of network technologies. Each application service (e.g. provided in various vertical industries, such as manufacturing, transportation, medicine, agriculture, etc.) requires certain capabilities for its business operation.

Then, in 5G and beyond mobile networks with such heterogeneous services, the mapping between services and individual data flows to network parts and QoS profiles can be a very complex task, which needs to be done in an automated way. So, one use case in 5G and beyond mobile systems is the provision of AI-based services with the required quality and security [23].

Also, the control of 5G networks includes design and deployment (based on customer and service data) and operation and management (e.g. resource management, fault management, etc.) [20]. In both cases, the amount of data to be processed is huge, so AI/ML techniques are used. For example, the automation of network part resources is possible by extending the network slice orchestration with AI/ML [24].

However, there will be different AI/ML algorithms/tools for the same purposes. But, as noted before, we can only standardize AI/ML frameworks, not the algorithms/tools, because for a given task it is not possible to have a single "smartest" AI/ML algorithm/tool.

6.6 Future AI-based Network Service Provisioning

From the previous parts of this chapter, one could see the approach AI/ML uses in existing and future networks. But the question is why we need AI/ML techniques now and in the future?

Well, at the present time there are ongoing changes in various industries and other sectors by the evolution of emerging technologies represented by cloud computing, Big Data, IoT, and AI. All such technologies will have direct influence on the standardization of future networks. Tens and hundreds of billions of new devices will be connected to telecom and private network over time. The increased number of devices connected to the network increases the cost of network operations. Regarding the network security, the challenges also increase with the number of heterogeneous devices being connected. For support of operation and maintenance in networks, one may note two opposite sides. In that way, business needs drive the network demands, but on the other side exponential increase of connected devices, traffic, and QoS requirement for different services and customers would require more staff for operation and maintenance to response to demand if the legacy approach in network operations is used.

These two opposite sides result in two main targets of telecom networks. One target is development of business models (considering the future network capabilities and possible use cases). The other target is automation of operation and maintenance of the fixed and mobile networks (especially 5G and beyond networks) via automated processes. The first of these two refers to intent-based networks, and AI/ML can be used for it. The second one (which refers to operation and maintenance processes in the networks) is referred in the telecom industry as zero-touch network which is based on network automation via AI/ML approaches applied in network operations maintenance processes.

6.6.1 Intent-based Networks

Future services include heterogeneous services that will result in heterogeneous business intents. For example, to recall, the minimum set of KPIs for monitoring network service provisioning includes IP network service activation time, DNS response time, number of IP network interconnection points, RTT to points for IP network interconnection, one-way IP delay variation, one-way IP packet loss, average data rate achieved in the downlink and uplink (separately), percentage deviation of the average data rate, Internet IP network service availability, and in the case of mobile networks we also had the availability of radio coverage. Why do we refer to these KPIs again in the provision of network services?

Well, because the targets set for the KPIs can be achieved either with manual or manually-assisted automatic configuration (of course, based on pre-written scripts for configuration setup of network nodes) or fully automatic configuration of network resources to meet the targets for the given set of KPIs. The latter case, in fact, is based on the use of AI in network service provisioning, while the current form of AI (as already discussed this week) is the ML.

Is AI magical for telecom networks? First, there is no wizard stick in any AI solution that can provide additional capacity if needed, because it should be deployed first and then used and optimized with the help of AI/ML techniques. For example, if the capacity is maximum 10 Gbit/s in a given network location, and the overall traffic demand from users (e.g. based on their SLAs) is above that 10 Gbit/s (e.g. 20 Gbit/s) then there is no AI/ML solution that can help to add capacity, because the capacity simply does not exist. Then, why will AI/ML be beneficial in future network service provisioning?

Well, there are several possible reasons. In this example, that would be network optimization, which means getting maximum (as capacity) or minimum (as latency) from resources that are deployed in the network for a given set of end users (customers) that simultaneously are using network resources. In the past, in legacy telecom networks, such changes could be done by actions from stuff (working at the telecom operator) based on monitoring of certain network KPIs (e.g. via regular hourly, daily, and other reports), but in the future that may be too slow or not scalable at all. For speedy changes in the network based on different demands (either from the business side or from the network and its customers), the main way forward is adding AI techniques to the network provisioning. Of course, that should be a tested and trusted AI, because "bad" AI will harm the network more than it will help its operations. Using AI does not always mean good, but it can be when it is done properly. And, it is done properly by vendors when they design their solutions themselves, not by telecom operators (which however would be able to set certain configuration parameters in such AI solutions). So, AI solutions are applied by vendors in the equipment that is deployed by telecom operators in their networks.

As noted before, services provided over future networks, including fixed, mobile, and satellite ones, will be configured and operated using diversified ICT infrastructure capabilities based on cloud computing, edge computing, and various types of networking technologies. Then, the main target of AI application is also toward different new verticals, where telecom operators' networks will need to be adapted to the requirements for services provided in different verticals, especially for mission critical services. So, what is the intent in such use cases?

The intent is mapping the business intents onto the network architecture via creation of services that are required by the customers. What is the possible difference in services offered by telecom operators in the future compared with services offered by telecom operators in the past or today?

Well, the main difference is heterogeneity. Until the 2020s, the services of telecom operators can be counted on the fingers of one hand. They include the legacy triple-play, with Internet access service, voice service, and IPTV/TV, over broadband access either fixed, mobile, or satellite. For business users, there were used leased lines in the past, and their present forms today are Virtual Private Networks (VPNs). What are telecom business needs for the future?

Well, telecom operators will need to provide *N* different services, where *N* can be 10, 50, 100, and more different services. Of course, operators may also remain in the offer of the main legacy telecom services, because overall the Internet access services and IPTV will provide the majority of the telecom revenues also in the near future. If one looks into the traffic statistics, it shows that video traffic today constitutes more than 70% of all traffic in all networks, including open Internet network and managed IP networks.

But, if there are needed new services to be offered (and will be needed in the future), then it should be possible to provide them through the same network infrastructure for a short period of time (compared to a network deployment time). That is crucial business intent, to provide new services without the need to build a new network for a new service demand, and have a short time to market – in a similar style of OTT providers (on open Internet) for their services. But, when talking about the telecom operators, the physical network infrastructure is also part of the environment and hence it needs to be flexible and adaptable to existing and future services. In such cases, it becomes important that the business role model is focusing on AI-based network service provisioning (including interactions between various business roles). And then, AI/ML techniques can be used to ease the complexity of the network service provisioning for different businesses.

So, one use case of AI/ML refers to mapping of the business intent (from telecom operators or service providers) to physical or virtual network infrastructure. Networks built on such a principle are called intent-based networks. This can also apply to network slicing. In fact, network slicing was "invented" to provide flexibility in realizing business objectives without the need to rebuild the network infrastructure. Nowadays, we associate network slicing mainly with 5G networks, but the approach is also applicable to fixed access networks, satellite networks, future 6G networks, or simply said – to all future networks.

How do intent-based networks work? The first step is to define business intent. Future high-level business intents are expected to be more complex (even today they are becoming very complex), so one needs to break them down into end-to-end service level goals and choose a set of KPIs for each network domain and each network element. For example, business intent could be design of particular service offer with given QoS constraints. Another example of business could be configuration of a particular network topology (e.g. integrated 5G TN and NTN). Another example of business intent is creation of network slice for building 5G private network service for a smart factory, and so on.

Figure 6.15 shows a high-level view of business intent mapping in telecom networks. Business objectives should be divided into multiple distributed local goals, where AI/ML techniques/tools are aimed at achieving business requirements from such distributed networks.

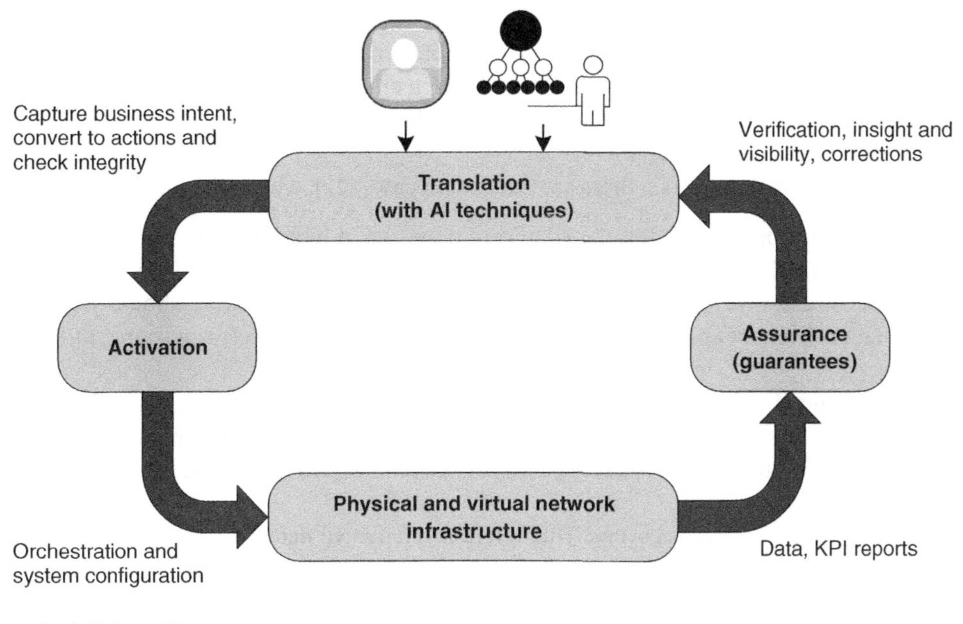

Figure 6.15 Intent-based network.

Finally, with use of AI/ML, there is autonomous execution of business intents on the physical and virtual network infrastructure, which can be done once all service KPIs are set and ML agents in all network elements are consulted to give predictions. Such autonomous actions are based on a closed loop system which uses data analytics, automation, and AI/ML techniques.

6.6.2 Zero-touch Networks

The long-term strategy is to develop plug-and-play telecommunications networks that can be deployed and managed automatically. The term "zero touch" refers to the autonomous allocation of network resources and functions (adding/removing) to scale services up and down as needed. In fact, zero-touch refers to a highly automated network in terms of configurations, its management, as well as operations and maintenance of physical and virtual resources. A zero-touch network that can operate autonomously can be compared to computers and smartphones – when they are powered up, they autonomously operate, update, reconfigure, and people (as end users in such cases) just decide what they want to install or uninstall (as an app/service) and which one to run. However, unlike user equipment, network resources are not packaged into a single box, so the complexity is greater in achieving a zero-touch network considering that includes many network elements in access, core, and transport parts.

Overall, zero-touch network or in other words zero-touch operations (of the network) is based on automation of its operations [25], such as self-configuration, self-optimization, self-healing, etc., targeted to business-oriented scenarios and the automation challenges for telecom operators and verticals. That leads further to end-to-end network and service automation [26]. Considering the diverse use cases, such Zero-Touch Network and Service Management (ZSM) should have a modular, flexible, scalable, extensible, and service-based architecture with open interfaces. The prerequisite for that is softwarization and virtualization of telecom network by using Software-Defined Networking (SDN) and Network Functions Virtualization (NFV), which for example are used in the design of 5G and further mobile networks. The ZSM is data-driven automation based on closed loop and integrated of AI/ML techniques. So, intent-based and zero-touch approaches are usually combined for provision of business sustainable solution for network automation and service management.

Regarding the zero-touch networks, there are three main cases [27], which include:

- Automated radio network slicing: It is aimed toward provision of high QoS with high reliability (e.g. needed for URLCC services in 5G and beyond) while ensuring efficient utilization of radio resources which are always scarce. To achieve such a goal, the mobile network should support continuous data collection, analysis of radio network slice behavior, as well as resource utilization patterns.
- Automatic end-to-end network service deployment: This refers to automatic translation/mapping of service requirements to network parameters/requirements. Therefore, network has to support data models that can be used to specify service requirements and integrate automated network configuration methods.
- Automatic fault detection and recovery: This refers to predictive detection of anomalies and automated recovery decision making, typically based on AI/ML approach. For this purpose, it is required to collect data about network QoS on real-time basis, and then generation of training data for AI/ML tools.

In summary, the goal of zero-touch operations is to make network operations automatic, which means to become more data-driven, predictive, and proactive by using AI/ML technologies and reducing the need for manual interventions in network and service configuration, operations, and maintenance. That will contribute also to better business agility.

In general, network automation through the zero-touch approach is needed first, then to be followed by service management automation. Automating end-to-end orchestration capabilities requires a standards-based approach to network automation and service orchestration (e.g. with standardized service models and APIs). The main benefits of automation are better efficiency in managing physical and virtual network domains, reduction of operational costs, and faster creation and delivery of new services on demand by customers (e.g. vertical industries). Also, automation contributes to improvements in QoS and QoE due to the reduction of human intervention which are more prone to errors than automated processes (assuming such automation is based on proven and functional algorithms).

6.6.3 Discussion

One of the main consumers of zero-touch network solutions provided by telecom equipment vendors are mobile operators. Why? Well, because 5G and beyond mobile networks, when combined with QoS requirements of critical IoT use cases, are possible to be deployed only when applying AI/ML, automation, and data analytics to drive their network operations.

Also, network slicing in 5G requires network automation from both sides, AI-based service provision (intent-based) and AI-based automation of network operations in the created slices (zero-touch).

Of course, these approaches can be applied also to fixed broadband networks (especially considering the shared infrastructure between fixed and mobile networks and their convergence), as well as to NTNs, i.e. satellite broadband, either standalone or integrated with terrestrial 5G mobile networks.

6.7 Blockchain for IoT Data Processing and Management

Future services include many IoT use cases in many different scenarios. IoT is also evolving to penetrate various so-called verticals that were outside the scope of telecommunications/ICTs in the past.

There are different types of IoT service platforms depending upon the IoT service. Overall, there are three main types of IoT service platforms [28]:

- Centralized IoT – In this case, the IoT service platform is deployed in a single location on the server side (e.g. a data center) and managed by one platform provider, which provides centralized services to IoT applications and services.
- Distributed IoT – In this case, the IoT service platform is deployed in multiple locations on the server side and managed by one platform provider, which provides distributed services to IoT applications and services.
- Decentralized IoT – In this case, an IoT service platform is established and maintained by a group of independent stakeholders, where IoT applications/services and IoT devices can be part of the decentralized IoT service platform. Also, they can provide services to each other.

While centralized and distributed IoT platforms can be considered legacy in the telecommunications/ICT world, the new type is the decentralized IoT platform. Also, in this decentralized mode of IoT, blockchain technologies can be used.

6.7.1 Blockchain Definition and Use Cases in Telecom World

Blockchain appeared and became popular with crytocurrencies that is sometimes referred to as "traditional" blockchain. However, we are not interested here about cryptocurrencies (although they also use blockchain), but more about telecom and industrial blockchain. What is the difference?

Well, the industrial blockchain is tailored at solving certain business problems between equal participants in a given business environment. For example, blockchain technology, as a decentralized technology (where different participants in the blockchain have equal rights), gives opportunity to drastically optimize various business processes and that reduces costs via elimination of the intermediators.

What is blockchain by definition? According to the ITU [28], it is peer to peer distributed ledger based on a group of technologies for a new generation of transactional applications which may maintain a continuously growing list of cryptographically secured data records hardened against tampering and revision.

So blockchain by design is decentralized in nature. How can telecom companies take advantage of blockchain? Blockchain system of mobile operators (as equal participants) can provide possibility for even the smallest mobile operators to have access to the global market for telecom services. So, smaller mobile operators will be able to provide services at the same level as large operators, and large operators will be able to expand their bases further to new clients. Regarding the telecom business, the blockchain proponents claim that it gives us the opportunity to cover markets that were simply impossible to cover before. That is mainly because the blockchain system means the absence of any intermediaries or commissions when making mutual settlement of payments, so telecom operators and companies can increase their revenues in that manner.

For example, a mobile operator can publish their own offers in the form of smart contracts (which are written in the code of the blockchain and executed when certain conditions are met and verified, such as releasing funds to a given party, registering telecom service, etc., and blockchain is updated when the transaction is completed), and make such offers available to other operators. Then, the operators choose offers that they would like to provide to their subscribers. A subscriber chooses the wanted offer and pays for it, which triggers creation of a new "request" smart contract for the given offer (which includes the digital identity of the subscriber and payment transaction). Then, based on such smart contract, the user's money is transferred to both the home mobile operator and the visited mobile operator. The idea of the use of blockchain is for operators to interact with each other on an equal basis as partners, as in roaming contracts. So, if this example is applied, then blockchain may replace the roaming agreements and leave them behind.

6.7.2 Blockchain for IoT Services

Blockchain can be used in any market and for any service that has many equal participating parties. So, it may be good for roaming contracts between mobile operators, for adding various parts in a given offered package, but it can also be beneficial for IoT services, in which we have many equal IoT devices deployed on different locations. And, combination of IoT and Blockchain technology resulted in the creation of the terminology (by ITU) of Blockchain of Things (BoT). What is it?

Well, Blockchain of Things (BoT) is a type of decentralized service platform which is based on blockchain-related technologies, and it is targeted to enable various entities from the IoT ecosystem (e.g. IoT devices, IoT gateways, IoT servers, and so on) to participate in the blockchain and perform transactions together.

A given BoT entity can host multiple so-called BoT peers and decentralized applications which connect to BoT peers to execute transactions. BoT peers interact with one another by using decentralized mechanisms in telecom networks including open Internet over them. Such approach can be used in existing networks (e.g. 5G) and future ones (e.g. 6G). Both virtual and physical IoT can be mapped into a BoT through BoT entities (e.g. IoT gateways), as shown in Figure 6.16.

Of course, use of blockchain requires an algorithm to be written by developers and then applied. And there are many different algorithms that can be written for creation of blockchains. Then, the crucial for blockchain use is trust. In fact, "the algorithms of crowding consensus and decentralized storage of the blockchain can make the participants fully trust the transactions and relevant data." But, before that is needed the trust in the "creator" of the blockchain algorithm, so that the party (e.g. telecom vendor) should be a trusted one – that comes first in the trust chain.

Overall, we want to have a trusted and tested blockchain algorithm, which is also scalable (e.g. it does not slow down with higher number of participants in the blockchain or with increasing number of transactions). However, blockchain still needs to improve their scalability.

What are the barriers to blockchain as a technology? Well, blockchain technology is a new technology and that is usually a problem at the beginning. Because it is a new technology, there can be a number of obstacles and also new technologies can result in mistrust.

Then there are certain privacy hurdles. For example, the blockchain is aimed at complete transparency and if some information about the participants in the blockchain needs to be hidden (from other participants), then additional efforts are needed to make the blockchain algorithm.

Also, as with other technologies, there is a need for legal regulation of the blockchain. However, this can be a problem in itself, as traditional regulatory approaches may not be suitable for blockchain technology, so an understanding of blockchain is required for those writing the rules, otherwise it may lead to a lot of misunderstandings. But, since blockchain is a relatively new technology, regulators are not rushing into regulatory frameworks, as the technology first needs to be proven in different use cases in different countries.

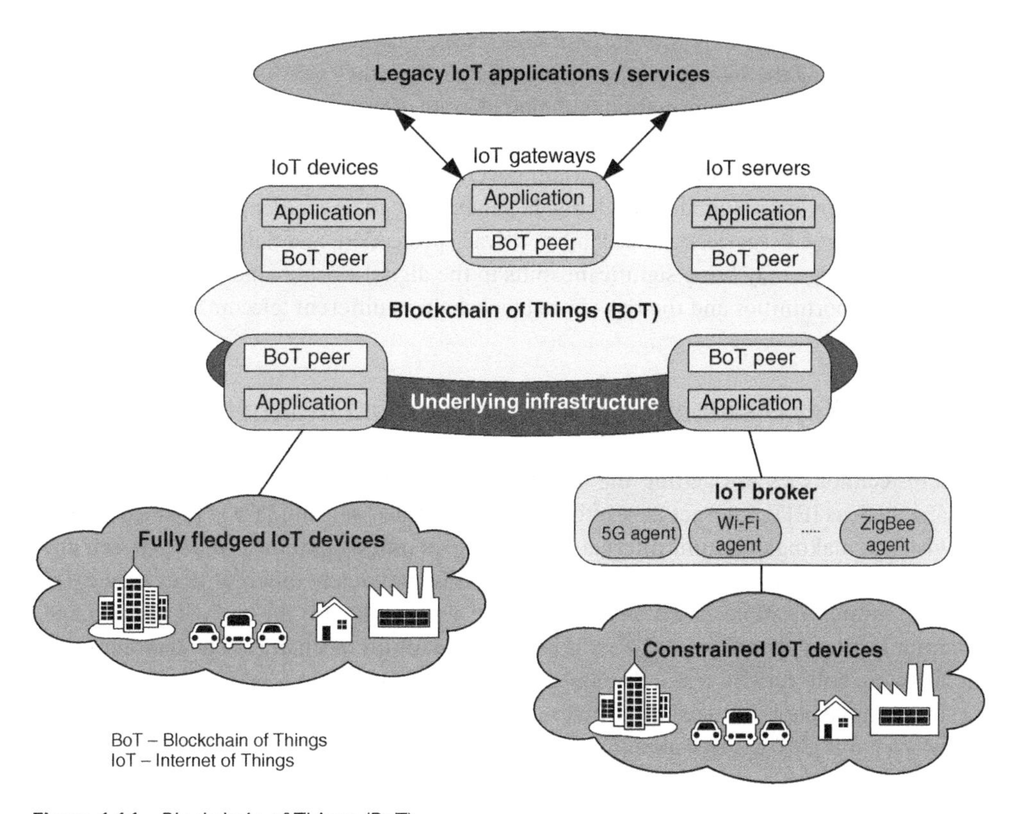

Figure 6.16 Blockchain of Things (BoT).

One of the use cases for decentralized IoT services is smart city services, which involve many IoT devices for different purposes (e.g. measurement, surveillance, monitoring, etc.). So what is the potential of blockchain in relation to IoT and smart city services?

Well, the interest of cities in blockchain (i.e. in the distributed ledger technology) is growing with the advances in smart contracts, cryptocurrencies, digital assets, and other so-called Web 3.0 technologies. The benefit of using blockchain for IoT (and smart city services) is lower costs of connectivity of various IoT devices, keeping balance between storage and data accessibility of such devices, building trust (with the immutable character of blockchain) for IoT services over the untrusted open Internet environment, as well as filling the gaps in lack of standards for IoT authentication and authorization (e.g. IEEE blockchain identity for IoT [29]).

On the other side, there are several challenges with blockchain. For example, the speed with which transactions move in blockchain is slow, and that requires problem solving (by researchers, developers, etc.). The people usually want to understand something with an aim to build trust in it. However, that is just a wish, because for example most of the people do not understand SSL/TLS which is used in https-based Web pages for end-to-end encryption of the application data. However, the trust (of customers, end users) is typically "tied" with the owner of the Web page – so, people build trust in companies and organizations, not in a given technology, because at the end of the day, most of the citizens will not understand new technologies. But they do not need to understand them – the main target is to use their benefits normally in a secure and trusted way.

Another issue with blockchain is that once there is developed a blockchain, it's done with no way of turning back. So, one may say that it is not like AI, because AI can be trained again with new data while blockchain can go only forward, not backwards. However, such feature of the blockchain is good for keeping the history of transactions, but on the other side it could make mistakes around the sharing of personal information or digital identity of people or IoT devices, and then it would very difficult to correct mistakes made in the past. In this regard, for residents in smart cities, a problem could be lost passwords for certain government services or for various wallets on blockchain, particularly for vulnerable residents. In general, the main challenges for use of blockchain in IoT and smart city services are scalability, privacy protection, and data exchangeability.

Blockchain technology may bring significant shifts in the digital world, which requires engagement in both the opportunities and the risks from blockchain in different telecom/ICT use cases.

6.8 Quantum Key Distribution (QKD) for Quantum Internet/IP

The quantum technologies are getting the attention in the telecom world and main standard organizations, such as IETF, Internet Research Task Force (IRTF) [30], and ITU [31], since around 2020. The vision is making quantum Internet by enabling quantum communication between any two points on Earth. However, quantum communication is completely different than legacy (i.e. non-quantum) communications based on bits (ones and zeros), which is the basis of Internet and all telecommunication networks in 2020s. It is based on quantum mechanics. The quantum networks in 2020s are still in early research phases in the 2020s. Quantum communication devices can range from simple photonic devices capable of processing and measuring just one quantum bit (called shortly a qubit [30]) to large quantum computers that could appear in the future.

6.8.1 Qubit for Quantum Communication

A quantum mechanical system is described by its quantum state which is an abstract object that provides a complete description of the quantum system at a particular moment. The quantum system by itself can consist of an atom or photon (as a particle) and the quantum state provides the description of the state of that particle. A string of bits (legacy bits) can be described by one quantum state. Also, one quantum state can also be used to describe an ensemble of many such particles.

Classical communication in the telecom/ICT world uses the binary alphabet of logical "0" and "1," which are represented for example with different pairs (amplitude, phase) in Quadrature Amplitude Modulation (QAM) techniques for each symbol which represents a series of bits depending on the number of points in the QAM constellation (for example, QAM 1024 has the constellation of 10 different pairs of amplitude-phase because $1024 = 2^{10}$). On the other side, a quantum bit, which we called a qubit, although it exists in the same binary space (logical ones and zeros), differs from a classical bit in that its state can exist in a superposition of the two possibilities:

$$|qubit = a|0) + b|1) \tag{6.4}$$

where |X is the Dirac label of a quantum state (which is the value the qubit holds), and in this case binary "0" and "1" as well as the coefficients a and b are complex numbers called probability amplitudes [30]. From a quantum mechanics point of view, such a state can be realized by using electron spin, atomic energy levels, photon polarization, and so on.

To determine the logic value of a given qubit it must be measured, as the qubit loses its superposition in time and irreversibly collapses into one of two possible states, either |0 or |1, which cannot be known in advance, but it is determined by measurement of the used characteristic of the particles (e.g. electron spin). The result of such measurement is classical bit, with logical values "0" corresponding to $|0\rangle$, and accordingly logical value "1" corresponding to $|1\rangle$. In quantum mechanics, we must use the probability for the particle to be in a given state. So, the probability that the qubit (i.e. particle) is in the state $|0\rangle$ is $|a|^2$, and the probability to be in state $|1\rangle$ is $|b|^2$, where it must hold:

$$|a|^2 + |b|^2 = 1 \tag{6.5}$$

Equation (6.5) is based on the fundamental characteristic of quantum mechanical systems.

So far we used only one qubit. What if we need multiple qubits? When multiple qubits are combined into a single quantum state, the space of possible states grows exponentially (with 2 on exponent number of qubits), and all these states can coexist in superposition. Thus, the general form of a two-qubit register can be expressed by the following relation:

$$a|00\rangle + b|01\rangle + c|10\rangle + d|11\rangle \tag{6.6}$$

where the coefficients a, b, c, and d have exactly the same probability interpretation as in the case of the one qubit.

The idea is not to replace the classical networks that are used in fixed, mobile, and satellite communications, but to use them to form a hybrid classical-quantum network that would benefit from the use of quantum communication. Such approach is also referred to as Quantum Information Technology (QIT) [32]. In this decade (2020s), the most known application of quantum communication is Quantum Key Distribution (QKD), which is intended to create and distribute a pair of symmetric encryption keys. Generally, QKD can be used in combination with other advanced technologies from the telecom/ICT world, such as SDN [33] and ML techniques [34].

6.8.2 Quantum Key Distribution (QKD) Technology

The QKD is the main use case of quantum communications and networking in the telecom world in the 2020s. What is the main purpose of QKD as a technology? The simple answer is to increase the security of the data that is transmitted from point A to point B in the network. Why do we need it?

Well, because existing cryptographic key solutions cannot be trusted to be unbreakable for people, organizations, or countries who have almost unlimited computing power (e.g. using distributed computations on very powerful machines in parallel). So, quantum is being developed to be future proof of security regardless of the development of computing technology.

The existing cryptography has several issues that have driven the appearance of QKD. For example, intrusions may remain undetectable. Also, the security is based on the assumption that certain mathematical operations are not possible to be performed in a limited time by any computer (which cannot hold in the future). Further, the keys are irregularly updated and longer key lengths slow down the encryption process on both ends of the communication link. So, these are enough reasons for cryptography going one step further in the future, and that is accomplished with work on QKD, although again this is a completely new technology based on quantum mechanics.

How does it work? QKD technologies provide a means to distribute symmetric random bit strings as a secure key that can be proven to be secure even against an eavesdropper with unlimited computing power. Thus, with the advent of AI and quantum computing, QKD technologies will/may become more important for encrypting critical data. QKD is designed as a technological addition (or you can say, a service), that is, it is added to existing communication networks without the need to build a new network for it.

QKD does not encrypt data, so it is not an encryption technique. What is it? Well, it is a technology that provides two endpoints (QKD nodes) to generate a shared secret key known only to them, and then with that key both ends encrypt and decrypt information. QKD is based on quantum mechanics, which makes it possible to detect any intruder, since any intrusion will perturb the quantum system states at both ends (if the system is perturbed above a predefined level, then this means that there is a potential eavesdropper between them, in which case the connection will not be established at all). But again, QKD is only used to distribute a key used to encrypt data. Then, any encryption algorithm can be used to encrypt the data.

How are keys established between two points/nodes in a telecom network? This is done by establishing a parallel QKD connection between two endpoints that wish to use such a secret key for secure communication. Figure 6.17 shows how QKD can be used for encrypting the data at the start of a VPN tunnel, which is typically used in telecom networks for carrying all types of traffic. Of course, different VPNs are established for different traffic types or different customers (e.g. one VPN for Internet traffic from a given fixed or mobile access network, another tunnel for managed IPTV traffic, a third tunnel for one customer with critical data transfer, a fourth for another vertical industry, and so on). In fact, we have a QKD plane which appears to exist in parallel (as an overlay) to existing telecom networks, which have no changes in their infrastructure in such case, except changes in network nodes (e.g. routers) that will receive and apply a quantum key for encryption and decryption of the traffic on each end of the application data link, respectively. So, a QKD connection between the two ends (for the quantum keys) and also a data connection between the two ends (for transferring user data) is required.

What is a QKD network? A QKD Network (QKDN) is tailored to expand the reach and availability of QKD. Building QKDN as an overlay to existing telecommunications networks and cryptographic infrastructures is challenging, as QKD technology requires specific physical channels (i.e. quantum channels) and operates as a point-to-point connection technology. The important element on both sides of the QKD connection are Key Managers (KMs), because the keys generated by QKD need to be properly managed and transferred in the network, taking into account possible security threats and risks.

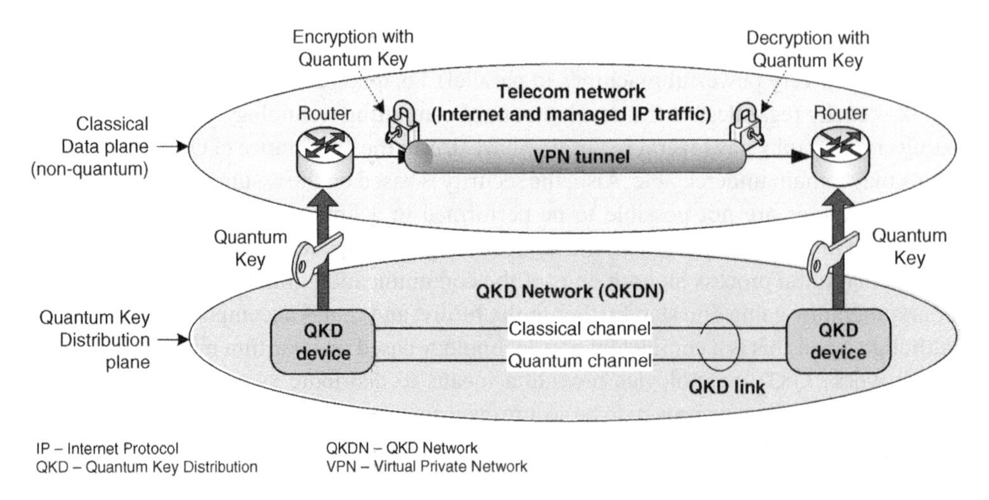

Figure 6.17 QKD use as overlay for securing point-to-point application data link.

What about QKD use cases? There are many use cases. Without naming them all, simply put, QKD can be used in all protocols that use secret keys. While in the traditional approach nowadays we extend the length of the keys with increased computing power of computers (in order to make them difficult to be broken by an attacker), QKD provides unbreakable keys once the communication link is established between the two points using QKD technology.

Is QKD a low-cost technology that can be used everywhere? Well, it is currently an expensive technology (compared to other cryptography solutions) due to the specific equipment required and requiring QKD connections between endpoints. So, today it can be used in cases where the data is very important or critical. From a telecommunications point of view, the first application of QKD is in telecom networks to connect their remote points (e.g. in 5G access network) and later (when the technology is mature and more affordable) it can also be used on end user devices. So, quantum technology is constantly penetrating the telecommunication networks, protocols, and services toward the future.

6.8.3 QKD Application in Telecom Networks

The QKD can be applied in different types of networks, so it is not related only to future ones. When is QKDN important? It is important when encryption of the data is critically important. So, QKDN is not needed for ordinary Web services today, however, it can be used if needed.

What about 5G use cases of QKD technology? The 5G fronthaul is becoming more diverse than 4G fronthaul or before because 5G RAN will include various Remote Radio Units (RRUs) and Distribution Units (DUs) based on decoupled open interfaces to support RRU and DU from different vendors. The security guarantee for 5G fronthaul is an important issue which needs to satisfy high bandwidth, low latency, and high-level security at the same time. In that regard, QKD is considered a promising solution for providing 5G (in the 2020s) and then 6G (in the 2030s) fronthaul network.

Another use case is the application of QKD to the 5G midhaul network to secure confidential data transmission from a smart factory to the cloud. For example, a customer of the mobile operator, such as an auto parts manufacturer, may implement a smart factory using IoT devices and robots which generates a large volume of confidential data such as design documents transmitted over the 5G network to reach the cloud. In this case, QKD is used for securing confidential data transfer over 5G to the cloud from a private 5G network for a smart factory. Since the 5G DU to CU for connectivity uses a fiber-optic network, it is possible to combine QKD with the encryption on the 5G midhaul network.

Overall, QKD and, generally, quantum communication offer certain new possibilities, and then it is about the convergence of its further standardization among different standards organizations and the costs of the equipment versus benefits from its use in achieving business goals in telecom/ICT services.

6.9 Business and Regulatory/Governance Aspects of IoT, Big Data, and AI

Digital transformation (digitalization) is driven by major innovations in the world of telecommunications/ICT that include IoT, AI , and Big Data, as well as cloud computing. All of them in different combinations enable digitization of business processes, increase operational efficiency, and provide opportunities for development and introduction of new business models. Data is collected from IoT and other devices, processed with AI/ML techniques, and stored/accessed in/from data repositories in clouds. The analysis and use of collected data is one of the additional drivers of business growth from a telecom/ICT perspective (in addition to mobile broadband and fixed broadband, as a prerequisite for data growth and further development of new technologies, including use cases of IoT and AI/ML).

Also, the combination of IoT, Big Data, and AI provides the means for telecom operators and other organizations in various verticals to make informed decisions about market trends and customer behavior, interests, or issues, and act accordingly. In a data-driven telecom world, data becomes an asset in itself.

6.9.1 IoT, Big Data, and AI – Opportunities and Challenges

The business potential of IoT, Big Data, and AI is toward creation of new business offers by telecom operators and service providers [35]. Typically, the telecom operator is providing services on a national basis, while we refer to service providers (at least here) to companies that provide OTT services on a global scale (e.g. over the open Internet). Also, the business targets of IoT, Big Data, and AI are services in different verticals, which include all other initially non-ICT related sectors (e.g. education, health, agriculture, and so on).

However, the business aspects of IoT services are different for massive IoT services and for critical IoT services. While massive IoT services (e.g., various sensors, meters, fleet tracking, home surveillance cameras, etc.) are typically provided over the open Internet network (so, they can be provided by global/local service providers or national telecom operators), the critical IoT services can be provided by national telecom operators. Why?

Because telecom operators have nationwide network infrastructure, so they are in a position to provide such IoT services, such as V2X services, Industry 4.0 services, and so on.

The importance of various IoT services is different. In this manner, critical IoT services may need governance more than massive IoT services. And, governance of IoT is also related to Data Processing and Management (DPM). Such DPM governance (or data governance, in short) can be hard or soft, with either an international, national, or local approach. For example, hard approach for international data governance is based on trade agreements, binding standards, as well as cross-border regulation (e.g. establishing cross-border regional corridors for given IoT services that demand such approach, e.g. V2X services, considering that vehicles often move across borders from one jurisdiction to another). On the other side, a soft approach for international data governance is usually based on guiding standards and industry practice. Further, hard approach for national data governance is based on laws (or bylaws), while soft approach for national data governance is based on guidelines and tax policy (or, locally, it is based on procurement, client preference, as well as data management culture).

Further, it is almost impossible for all IoT services (particularly critical ones) to be put into a single regulatory framework. So, each critical IoT service type may need to receive separate treatment in the regulations and governance.

6.9.2 AI Governance

After IoT and data, another important area for governance is AI. It could be problematic if not regulated. But, how to regulate the AI, considering that it is too broad. The first step is understanding of the AI and becoming aware of its presence and its form (e.g. AI means primarily ML algorithms used for various purposes). Then, the driving idea about the AI is to be used for good, which means good of the citizens (e.g. better public services, better healthcare, etc.), for good of businesses (for new innovative services, for higher efficiency in their delivery, etc.), as well as for good of the society (e.g. via services of public interest, such as improved transport, education, security of citizens, energy management, and so on, with application of telecom/ICT, i.e. digital technologies empowered with AI/ML).

One may say that AI in its simplified form can be seen as "data" (which is used for training of algorithms) and ML algorithms (which use the data as the main "food" with the aim to become better). But, on the other side, it would be difficult to regulate AI, because it is about "data" and "algorithms" [36]. Overall, humans determine and program the goals or desired results of all such algorithms. So, the end goal should be regulated, where necessary.

AI is in the early stages of growth in the 2020s. Thus, overall regulation may appear inadequate due to the incremental growth phase of AI over the coming years or even decades. In fact, the development of AI technologies in the telecom/ICT sectors can be expected to be embedded in the development of other telecom/ICT technologies.

But AI use cases in different sectors and applications can result in different levels of risk. For example, European countries have a risk-based approach to managing AI, where the level of risk is the main driver in AI regulations and policies, about what is allowed and what is not allowed (for AI). The considered approach in Europe for AI governance is a risk-based approach, as shown in Figure 6.18, in which AI systems and their applications are grouped into four groups, going minimal risk (at the bottom of the pyramid) up to the unacceptable risk (on the top of the pyramid).

Most of the applications fall into the group with minimal risk, such as the use of AI in video games or email spam filters.

The limited risk group includes AI systems with certain transparency obligations, such as chatbots where users need to be aware that they are interacting with a machine so that they can make an informed decision to continue or withdraw from using that service.

High-risk AI systems will be subject to strict obligations and will need to follow appropriate risk assessment and mitigation systems, have high-quality datasets feeding the system to minimize risks and discriminatory results, traceability of results, detailed documentation of AI system intelligence, clear user information, adequate human supervision to minimize risks, as well as a high level of robustness, security, and accuracy. AI systems in the high-risk group include AI technology

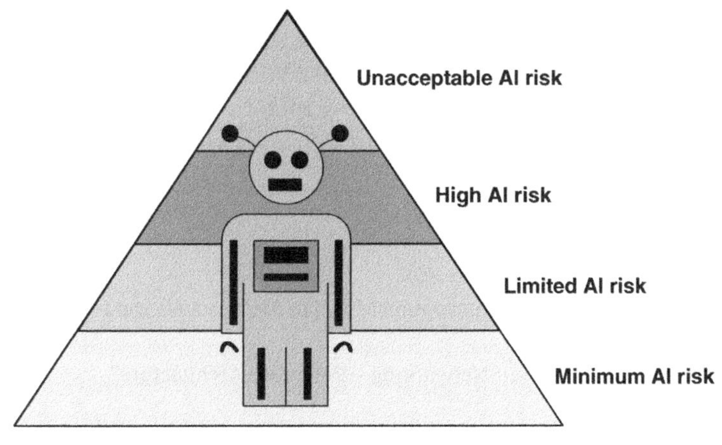

Figure 6.18 Risk-based approach for AI governance.

used in critical infrastructures (e.g. transportation), various safety components of products (e.g. in robot-assisted surgery), employment (e.g. resume sorting software for recruitment procedures), law enforcement, and basic human rights (e.g. evaluating the reliability of evidence), justice, and democratic processes (e.g., applying the law to a particular set of facts) and so on.

There is always the question of whether AI/ML algorithms can be controlled in the same way as humans, since they are originally created by humans, but can then change autonomously and result in unexpected behavior in certain untested or unforeseen scenarios (it is not theoretically possible to test in advance all possible scenarios that may occur, but only the most likely ones).

Finally, unacceptable risk covers all AI systems that are considered a clear threat to the safety, livelihood, and rights of people. Such AI systems will be prohibited (e.g. toys that use voice assistance that encourages dangerous behavior, etc.).

In general, the management of any technology, including AI/ML, is aimed at the future improvement of telecommunications/ICT systems and services (through greater automation, greater efficiency, flexibility, etc.), so it should be positioned to deliver benefits from use against possible risks.

By their very nature [37] AI strategies are a response to the moment in which they exist. This means that additional activities will be needed. Quantitative indicators will be needed in order to assess the future impact of AI on the economy and society. However, such strategies cannot be limited to the telecommunications/ICT sector alone, given the cross-sector nature of AI, so collaboration across a wide range of sectors will be necessary. Also, it will be a never-ending process, as AI will continue to evolve with increasing computing use for developing AI systems, and this may drive future technology development to be faster than ever before.

Finally, one may note that AI offers a huge potential to citizens, telecom/ICTs, and other businesses and society in general, but also creates many challenges about how to deal with it. That will require continuous and gradual work on AI policies and regulations, with an aim to keep their developments with the AI developments, and not to kill or lock the AI, but to "train" it (with trusted data) to serve the humans the best it can, especially considering the never-ending process of digitization of everything which will constantly develop in the future.

References

1 Janevski, T. (2019). *QoS for Fixed and Mobile Ultra-Broadband*. USA: John Wiley & Sons.
2 Janevski, T. (2015). *Internet Technologies for Fixed and Mobile Networks*. USA: Artech House.
3 Janevski, T. (2014). *NGN Architectures, Protocols and Services*. UK: John Wiley & Sons.
4 ITU-T Rec. Y.4000/Y.2060, "Overview of the Internet of things", June 2012.
5 3GPP TS 23.501 V18.2.2, "System Architecture for the 5G System (5GS); Stage 2 (Release 18)", July 2023.
6 Ericsson, "RedCap—Expanding the 5G Device Ecosystem for Consumers and Industries", February 2023.
7 ITU-T Rec. Y.3605, "Big Data – Reference architecture", 2020.
8 ITU-T Rec. Y. 3603, "Big Data—Requirements and Conceptual Model of Metadata for Data Catalogue", 2019.
9 ITU-T Rec. Y.3650, "Framework of Big data-Driven Networking—Reference Architecture", September 2020.
10 ITU-T Rec. 3854, "Big Data Driven Networking—Machine Learning Mechanism", February 2022.
11 Alan Turing, "Computing Machinery and Intelligence", MIND—a quarterly review of psychology and philosophy, Volume LIX, No. 236, October 1950.

12 United Nations, "United Nations Activities on Artificial Intelligence (AI)", 2019.

13 Moore, G. (1965). Cramming More Components Onto Integrated Circuits. *Electronics* 38 (8).

14 ITU-T Rec. Y.3171, "Architectural Framework for Machine Learning in Future Networks Including IMT-2020", June 2019.

15 Ericsson, "How Machine Learning-Powered Demand Forecasting can Boost Customer Satisfaction", https://www.ericsson.com/en/blog/2022/7/machine-learning-demand-forecasting, accessed in August 2023, 2023.

16 3GPP TR 22.874, "Study on Traffic Characteristics and Performance Requirements for AI/ML Model Transfer in 5GS (Release 18)", December 2021.

17 ITU-T Rec. Y.3172, "Architectural Framework for Machine Learning in Future Networks Including IMT-2020", June 2019.

18 ITU-T Rec. Y.3174, "Framework for Data Handling to Enable Machine Learning in Future Networks Including IMT-2020", February 2020.

19 ITU-T Rec. Y.3176, "Machine Learning Marketplace Integration in Future Networks Including IMT-2020", September 2020.

20 ITU-T Rec. Y. 3177, "Architectural Framework for Artificial Intelligence-based Network Automation for Resource and Fault Management in Future Networks Including IMT-2020", February 2021.

21 3GPP TR 37.817 V17.0.0, "Study on enhancement for Data Collection for NR and EN-DC (Release 17)", March 2022.

22 ITU-T Y.3183, "Framework for Network Slicing Management Assisted by Machine Learning Leveraging Quality of Experience Feedback from Verticals", January 2023.

23 ITU-T Rec. Y.3178, "Functional Framework for Artificial Intelligence-Based Network Service Provisioning in Future Networks Including IMT-2020", July 2021.

24 ITU-T Y.3182, "Machine Learning Based End-to-end Multi-Domain Network Slice Management and orchestration", September 2022.

25 Ericsson, "Operations of the Future", 2021.

26 ETSI GR ZSM 004 V2.1.1, "Zero-Touch Network and Service Management (ZSM): Landscape", January 2021.

27 Omdia, "Achieving a true Zero-Touch Network Vision", March 2020.

28 ITU-T Rec. Y.4464, "Framework of Blockchain of Things as Decentralized Service Platform", January 2020.

29 IEEE, https://blockchain.ieee.org/standards, last accessed in August 2023, 2023.

30 IETF RFC 9340, "Architectural Principles for a Quantum Internet", March 2023.

31 ITU-T Rec. Y.3800, "Overview on Networks Supporting Quantum Key Distribution", October 2019.

32 ITU-T Technical Report, "Quantum Information Technology for Networks Use Cases: Quantum Key Distribution Network", 24 November 2021.

33 ITU-T Rec. Y.3805, "Quantum Key Distribution Networks – Software-Defined Networking Control", December 2021.

34 ITU-T Y.3814, "Quantum Key Distribution Networks—Functional Requirements and Architecture for Machine Learning Enablement", January 2023.

35 ITU-D, "Emerging Technology Trends: Artificial Intelligence and Big Data for development 4.0", 2021.

36 European Commission, "On Artificial Intelligence—A European Approach to Excellence and Trust", 2020.

37 UK Government, "National AI Strategy", September 2021.

7

Cloud Computing for Telecoms and OTTs

The emerging trend of all Internet applications and telecom services that include transfer of data is toward the use of cloud computing either with dedicated QoS or via open Internet access [1]. The cloud computing has emerged with the spread of Internet technologies in the telecom world [2], which provided the possibility for any two computers (client and server) to connect (client connects to server) based on Internet addressing and naming (using domain name system and IP addresses). That has become a paradigm that draws much attention including standardization organizations, service providers, and end users within the standardization frameworks [3]. However, the idea about distance computing is not really new, but in this 21st century it took the standardized framework under the name of cloud computing.

7.1 Cloud Computing Architectures

The approach to data storage on servers began with the penetration of Internet client-server communications, where servers are located in data centers. Typically, a single server machine has more processing and other capabilities than client machines (e.g. computers and smartphones) because a single server typically serves many clients simultaneously. So, the resources of the servers are shared between multiple simultaneous connections made to them by clients in different locations. From the concept of data centers (with many servers), over time, we have moved to the so-called "cloud computing."

Let's define cloud computing – what does it refer to? Well, according to the ITU definition, cloud computing is a paradigm for enabling network access to a scalable and elastic pool of shared physical or virtual resources by providing self-service and on-demand administration [4].

The main idea of cloud computing is to use resources at a distance, such as storage and processing power, therefore, the key requirements for its deployment are bitrates and end-to-end delays. So, for better performances, the cloud computing approach requires high bitrates on the end users side, including residential and business users. So, broadband access (including broadband Internet access) is a major requirement for use of cloud computing for many services, from data storage and updates of operating system (OS) and applications, to Web services and real-time services such as video on demand, AR/VR (Augmented Reality/Virtual Reality) services, and so on.

Overall, higher bitrates also contribute to lower latency, since latency is inversely proportional to the sustained end-to-end bitrate (e.g. from the cloud application on the end users' side to the cloud servers on the cloud service provider's side). But one should not forget the transmission delay (of IP packets) including all the buffering and processing of the network nodes on their way.

Future Fixed and Mobile Broadband Internet, Clouds, and IoT/AI, First Edition. Toni Janevski.

To provide fewer delays, it is also required for cloud servers to be closer to the users whom they serve, with the aim to have a lower round-trip time (RTT). Why?

That is required with the aim to have high end user experience (e.g. for human end user) similar to access to locally available resources on the end user equipment (e.g. personal computer, smartphone, and so on).

The cloud computing framework includes different functions which support the cloud services. What is included in the cloud computing functional architecture?

The cloud computing functional architecture includes four main functional layers, as shown in Figure 7.1.

Going from the user toward the cloud physical resources, the top layer is the so-called user layer, which includes user functions such as user interface to cloud services (e.g. Web-based interface via a browser), business functions, and administration functions. The business function includes selection and purchase of cloud services, as well as accounting and financial management regarding them. The business capabilities are offered through the cloud services, so they are related to the cloud service provisioning.

Below the user layer (in the functional cloud architecture) is placed the access layer, which provides access mechanisms for displaying cloud service capabilities. This layer is responsible for the distribution and interconnection of cloud functions. It also provides administrative and business capabilities of cloud services. This layer also implements QoS enforcement functions for cloud services, as well as cloud security aspects such as data encryption and integrity checks.

The third layer (going from top to bottom) is the service layer, which includes service, business, and administrative capabilities. It contains the implementation of cloud services and their capabilities. This layer relies on the underlying resource layer and its capabilities, which directly affect

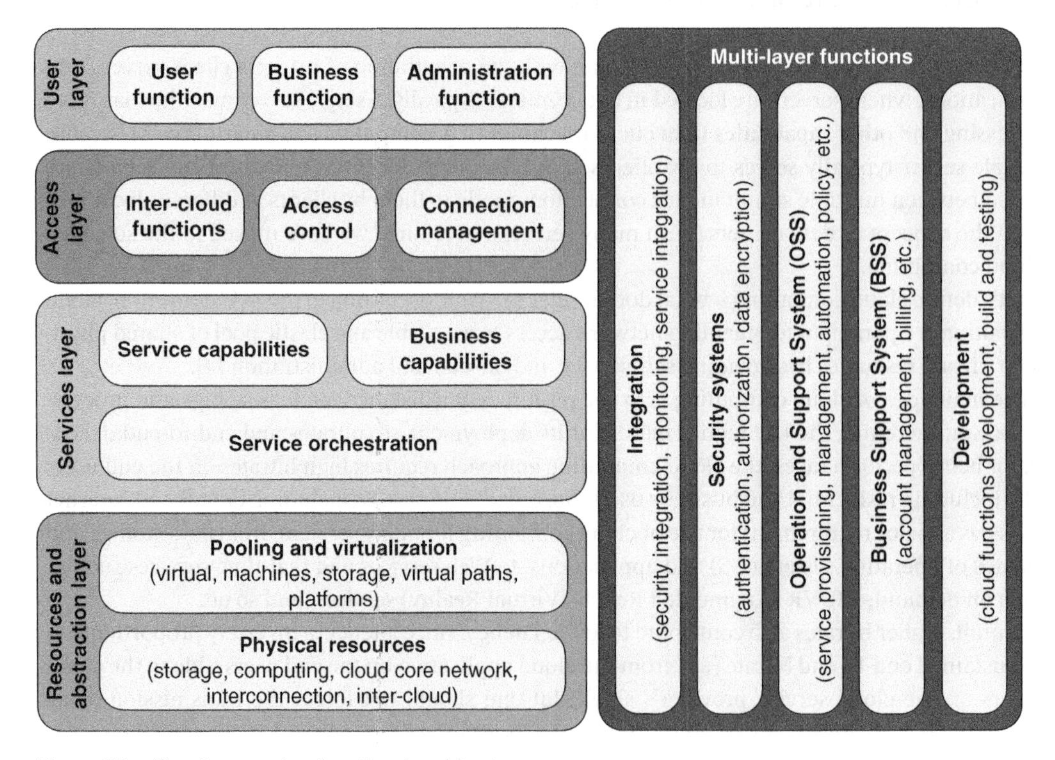

Figure 7.1 Cloud computing functional architecture.

the QoS specified in the service level agreement (SLA) between the cloud provider and the cloud customer. The service layer includes all the necessary software components for the implementation of services and service interfaces to cloud clients. Business capabilities are related to the business functions of providing cloud services, where such functions are embedded in the business and support system (BSS), shown in the multi-function segment of Figure 7.1. Further, the service layer also includes the administrative capabilities that provide functions related to the operation of the operation and support system (OSS) and BSS. Finally, service orchestration refers to the coordination, aggregation, and composition of multiple service components required for delivery of cloud services.

Finally, at the bottom of the functional layering architecture for cloud computing is the resource and abstraction layer. It contains physical resources, such as servers (in data centers), switches and routers for the networking, and storage devices (e.g. disks). Further, it includes functions for pooling and virtualization, which are used for configuration and operation of virtual machines (VMs), virtual storage, and software platforms over the underlying physical resources. All physical and virtual resources of the cloud system are managed via the OSS and its functional components which run on parts of the physical or virtual resources.

In parallel with the four layers of the cloud's functional architecture, there are so-called multi-layer functions (Figure 7.1) that operate across different layers. Their integration includes functional components that are required to create unified cloud architecture, including security integration, monitoring of functional components, and service integration. Multi-layer functions also include security systems for user authentication and authorization, as well as data encryption/decryption and privacy (which should be in accordance with the regulation and legislation in a given country or region, especially for "sensitive" cloud services, such as those used by governments, banks and various enterprises). Furthermore, multi-layered functions include BSS (e.g. account management and billing) and OSS (e.g. service provisioning, governance, automation, and policy). Part of the multi-layer segment in cloud framework is the development of cloud functions, including the building of cloud services and their testing.

But which application and services use cloud computing? Well, almost all applications/services provided over the open Internet involve the use of cloud computing. In other words, today's Internet/IP service model is based on the cloud computing paradigm, especially over-the-top (OTT) services provided through open Internet access. Services that do not require or use cloud computing are real-time conversational services such as voice (i.e. telephony) and video telephony services. But in the OTT application space, even such voice services or live streaming video are combined with other data-based services such as file sharing (e.g. sharing photos, videos, and files) or social networks, which all involve cloud computing. Also, new services such as various forms of Internet of Things (IoT), AR/VR services, AI-based services, and many others require cloud infrastructures for their provision.

7.2 Cloud Ecosystem

For cloud computing services, we have cloud providers and cloud customers as well as third party players for development, provision, or operation of various aspects of the cloud services. Based on such analyses all cloud computing activities can be categorized into three main roles, which include (1) activities that use services, (2) activities that provide services, and (3) activities that support cloud services. Also, one should note that a party may play more than one role at any given point in time, but also may be engaged only in a limited subset of activities of that role. The noted

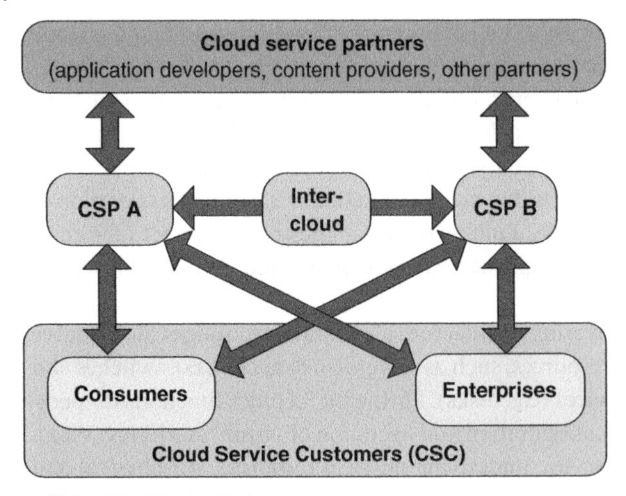

Figure 7.2 Cloud computing ecosystem.

roles of cloud computing services and their relations define the cloud computing ecosystem which is shown in Figure 7.2.

According to such cloud ecosystem, the major cloud computing roles include the following:

- Cloud service customer (CSC) is an entity that has a business relationship with a cloud service provider (as well as a cloud service partner) to use cloud services. There are several sub-roles defined for the CSC role, which include cloud service user (a customer-side individual or entity that uses cloud services), service administrator (who has the goal to ensure the smooth operation of the cloud service), service business manager (responsible for efficient acquisition and use of cloud services), and service integrator (responsible for the integration of the cloud service with the other ICT services of the cloud customer).
- Cloud service provider (CSP) provides the cloud computing services and ensures their delivery to cloud customers. It includes many activities such as provision of cloud service, deployment and monitoring, business management, audit of data, cloud operations management, customer care, inter-cloud provider services, cloud security and risk management as well as network interconnection (e.g. connecting clouds to telecom infrastructure regionally or globally).
- Cloud service partner (CSN) is a third party (different than CSC and CSP) which is engaged to support either the CSC or CSP. So, depending on the partner and its target role its activities may vary. The partner can also be engaged in the cloud service development (that is, for the design, development, testing, and maintenance of the cloud service), audit of the cloud service, acting as cloud service broker between the CSC and CSP, performing customer assessment, and setting up an SLA for the cloud service.

So, the CSP uses cloud infrastructure which consists of compute, storage, and network resources to deploy and deliver any kind of cloud services. The CSC accesses and uses cloud services deployed in and delivered by cloud infrastructure and it is an essential actor of the cloud ecosystem.

In many cases it is required for one CSP to use resources or applications from other CSPs. In Figure 7.2, CSP-A is connected to CSP-B via inter-cloud, which is a paradigm used for enabling the interworking between two or more CSPs [5].

7.2.1 Cloud Deployment Models

Cloud deployment models represent how cloud computing can be organized based on the control and sharing of physical or virtual resources, either public or private or combined. So, cloud deployment models include:

- Public cloud: In this cloud deployment model, cloud services are potentially available to any CSC (e.g. if we have an OTT cloud service, then it may be available to any customer with unrestricted access to the open Internet) and resources are controlled by the CSP.
- Private cloud: This is a deployment model where cloud services are used exclusively by a single CSC and also cloud resources are controlled by that customer.
- Community cloud: This is a cloud deployment model where cloud services exclusively support and are shared by a specific group of CSCs who have shared requirements and have a relationship with one another. In this case the cloud resources are controlled by one or more members of the community.
- Hybrid cloud: This is a cloud deployment model which is based on a combination of at least two different models. The entities are bound together by using technologies that enable interoperability as well as data and application portability.

Finally, cloud computing is a paradigm/framework targeted to enable convenient, on-demand network access by different parties (provider, customers, and third parties) to a shared pool of configurable resources which include networks, servers, storage, applications, and services. The goal is for cloud computing services to be rapidly provisioned and released with minimal engagement and interaction of the CSP. Also, cloud computing customers (CCCs) can use resources to store data, but they can also use cloud services to develop, host, and run their own services and applications on demand based on their business needs. The cloud computing framework provides a flexible approach for delivery of cloud services to any appropriate device (e.g. computer and smartphone), anytime and anywhere in the cloud computing environment, where the access to the services is normally dependent upon the cloud deployment model (e.g. private cloud services are not publicly available to everyone).

But, considering that there are different targets in the cloud computing approach, i.e. different service types, in Section 7.3 we provide details about cloud service models and their characteristics.

7.3 Cloud Service Models

The cloud capabilities can belong to three main types, which include application, platform, and infrastructure. Based on such capabilities are defined multiple cloud service categories. They may use one or more of the three main types of cloud capabilities. The cloud service models that use only one of the three main capabilities (i.e. application, platform or infrastructure) are in fact the main three cloud services models (Figure 7.3), given as follows:

1) Infrastructure as a Service (IaaS) – uses infrastructure capabilities;
2) Platform as a Service (PaaS) – uses platform capabilities;
3) Software as a Service (SaaS) – uses application capabilities.

Infrastructure as a Service (IaaS) is a cloud service model that provides on-demand access to infrastructure components (physical and virtual) such as storage, servers, and networking. This type is aimed at enterprises and provides cost-effective and rapid deployment of their services

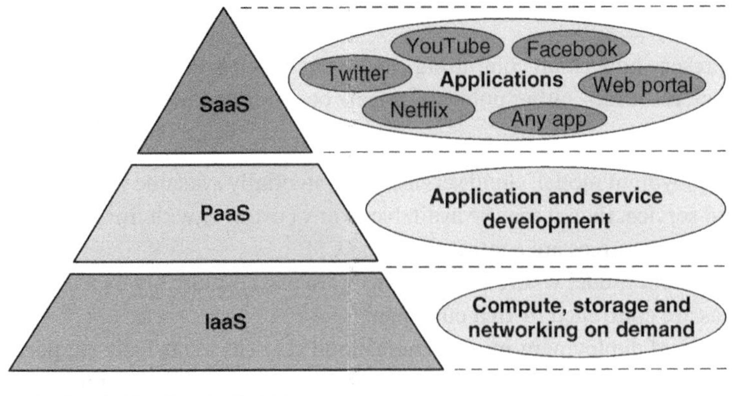

Figure 7.3 Main cloud services models.

using CSP infrastructure instead of creating their own data center and network, starting from procurement through deployment and testing to operation and maintenance with their own ICT works. Examples of IaaS are Microsoft Azure, Amazon Web Services (AWS), Google cloud platforms, Alibaba cloud, and others.

Platform as a Service (PaaS) is a cloud service model that provides platform in a scalable manner which is used as development and deployment of various applications in the cloud. It includes hardware as IaaS (storage, servers, and networking), but additionally it provides middleware with software tools (for development of applications/services) and OSs, business intelligence, and databases. In this way, PaaS is suitable for application and solution developers by reducing the need for local installation of hardware and software required for such purpose and using pay-as-you-go PaaS offerings from CSPs (e.g. Google, Microsoft, Amazon, and others).

Software as a Service (SaaS) is a cloud service model in which cloud customers use cloud applications typically over the Internet. Typical examples include websites, email, social networking (e.g. Facebook or Twitter), video sharing and streaming sites (e.g. YouTube and Netflix), online gaming, online office tools (e.g. Office 365), online cloud storage services (e.g. Google Drive), and so on. In this case, the application software and data are hosted on remote servers (in the cloud); they are maintained by a cloud provider or service provider (which has business relation with the cloud provider) and used by end users (CSCs) via defined interfaces such as legacy Web browsers (for various end user devices, such as desktops, laptops and smartphones) as well as proprietary apps for access to SaaS-based services via mobile handsets (e.g. smartphones). SaaS provides access to services by CSCs without the need of installing specific software or hardware. For example, online video storage and video services (on demand) require only updated browsers (or apps in mobile devices), without a need for replacing the hardware such as VHS cassette players (in the 1990s) with DVD disc players (in the 2000s) and then Blu-Ray disc player (in the 2010s). Of course, SaaS-based services also change over time with increasing bitrates in broadband networks and increasing processing capacities of end user devices (such as computers and smartphones), so video resolutions and video frames per second (fps) are increasing requiring end users to once again replace their PCs and smartphones with newer ones (with higher capabilities required for increasingly demanding services such as 2K/4K/8K/. . ./2^NK videos and eXtended reality – XR services).

If we want to provide Network as a Service (NaaS) [6] as a cloud service model, then we need to provide all cloud capabilities including infrastructure (storage, servers, networking), platforms,

Table 7.1 Cloud service models versus cloud capability types.

	Cloud capability types		
Cloud service category	**Application**	**Platform**	**Infrastructure**
Infrastructure as a Service (IaaS)			X
Platform as a Service (PaaS)		X	
Software as a Service (SaaS)	X		X
Network as a Service (NaaS)	X	X	X

and services (applications), which means we need to have all three main cloud service capabilities, as shown in Table 7.1 (where NaaS is compared to IaaS, PaaS, and SaaS).

Besides the main cloud service models given in Table 7.1, there are also other cloud service models that are used for various services (e.g. OTT services), such as Email as a Service, Desktop as a Service, Databases as a Service, and others. With the development of new technologies and their use cases in telecom/ICT world also appear new cloud service models. In that manner, with the emerging blockchain and AI/ML technologies, new service models in the cloud environments are Blockchain as a Service (BaaS) and Machine Learning as a Service (MLaaS).

7.3.1 Machine Learning as a Service (MLaaS)

The development of ML as a technique to enable machines or computers to learn how to perform tasks without being explicitly programmed resulted in the development of ML models, learning algorithms, and training data. As we already discussed in Chapter 6, an ML model uses input data to do tasks based on defined mathematical representations. The learning algorithm trains an ML model using training data for the task.

MLaaS is different than SaaS because the CSPs provide to cloud customers (typically enterprises) access to ML algorithms, computing power, storage, and certain data (for the ML algorithm), and the customers can build and deploy their own models [7]. The ML developer configures the ML model and learning algorithm and collects training data based on a task setting. Use cases of MLaaS include (but are not limited to) predictions/forecasting, image recognition, understanding large datasets (data exploration), adding AI/ML uses in business applications, as well as the possibility to build, train, and deploy own ML models (by using the resources of MLaaS provider).

MLaaS is a cloud service category in which there is highly demanding ML processing with a large amount of computing power and resources that are needed for training of the ML models. Such computing power and also storage capacity is required for large datasets needed for ML training and the required complex computation. However, MLaaS resolves such requirements by provision of on-demand elastic computing capabilities and resources in the cloud.

There are several cloud computing sub-roles that are mapped to MLaaS roles, as given in Table 7.2, which includes the sub-roles of the CSP, CSN, and CSC mapped to the MLaaS ecosystem. The relation between different roles in the MLaaS ecosystem are shown in Figure 7.4.

Figure 7.4 shows the relations between the players in the MLaaS ecosystem. The ML data provider from CSN provides the data to ML model developer which is also CSN to CSP (both roles data provider and model developer can be done by a single third party – CSN, or different parties). The CSP for MLaaS service model is, in fact, the ML service provider, who provides ML model training

Table 7.2 MLaaS ecosystem and roles.

MLaaS ecosystem	MLaaS roles
ML data provider	CSN: MLDP (ML data provider)
ML model provider	CSN: MLMD (ML model developer)
ML framework provider	CSP: MLSP (ML service provider)
ML framework customer	CSC: MLSU (ML service user)

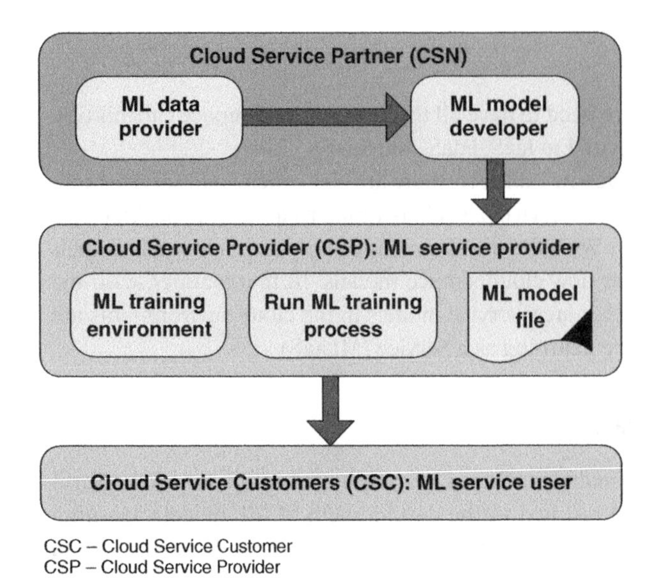

Figure 7.4 MLaaS ecosystem.

environment and then operating environment for running of the ML model by the CSC who is the ML service user.

There can be different applications of MLaaS services. One use case example of MLaaS is an application for traffic speed information gathered from sensors on streets or in vehicles. The data from traffic measurements (of cars on streets) is used for training of the ML model, which further provides predictions about the traffic (e.g. on time intervals such as 5, 10 minutes) with the aim to solve traffic congestion or to optimize traffic light settings for minimization of the congestion under the given vehicle traffic load on city streets. Another use case is using the traffic measurements on links in a transport network and loading it into the ML model in the cloud for training the ML model for accurate traffic predication, which is used for configuring the load balancing in the transport network based on given set of QoS/QoE parameters (i.e. Key Performance Indicators – KPIs). Besides network traffic, the MLaaS can be used for business analytics and predictions in different verticals (i.e. not only in the telecom/ICT sector).

7.3.2 Blockchain as a Service (BaaS)

Another new cloud service type is related to emerging blockchain technology. As already noted in Chapter 6 , blockchain is a type of distributed ledger technology (DLT). It is composed of digitally

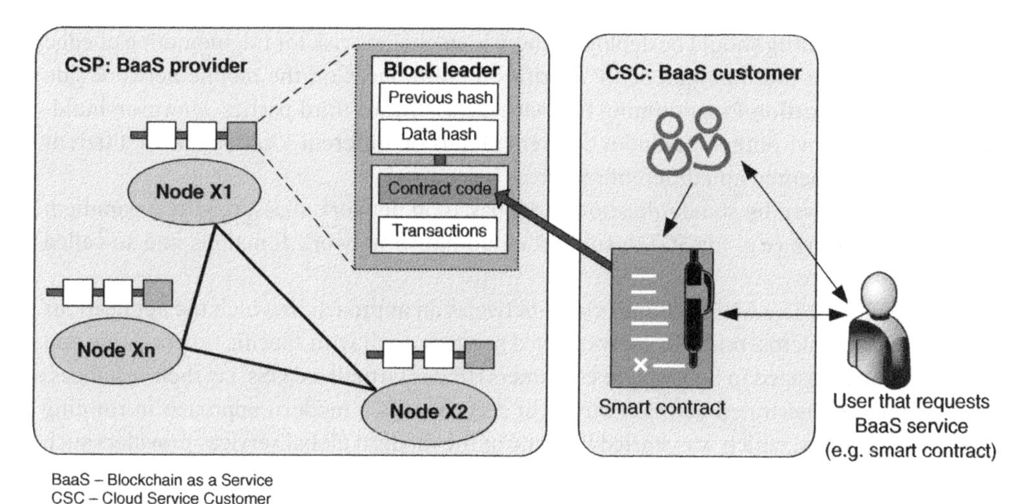

Figure 7.5 Blockchain as a Service (BaaS).

recorded data which is arranged as a successively growing chain of blocks. Each such block is cryptographically linked and hardened against manipulation and malicious changes. So, the blockchain platform in each node creates ledgers consisting of immutable records of transactions. If such a block is generated, then it is transmitted to all other nodes in the blockchain network. Finally, the blockchain platform validates the generated block, and if everything is okay then the block is stored in storage of each blockchain node. In a blockchain network, the data transfer between blockchain nodes (e.g. transactions) uses peer-to-peer connections.

BaaS is a cloud service model in which the main capability provided by CSP to CSC is the ability of setting up a blockchain platform, which is used for developing the application that uses blockchain approach [8]. The CSCs use BaaS to create and operate the blockchain network and to develop (and use) blockchain-based applications on the blockchain platform in the cloud.

The BaaS cloud ecosystem provides additional sub-roles and activities (specific for blockchain) based on the architectural user view defined for the cloud computing, which are shown in Figure 7.5.

Regarding the BaaS, example use cases can be found in different OTT cloud platforms (e.g. Microsoft Azure, AWS, and so on). In many cases open source frameworks for the blockchain are used. One of the use cases of blockchain (via BaaS) is smart contract, which represents a business logic or predefined criteria represented by a computer code managed by a blockchain. Such smart contract is automatically enforced when predefined rules (agreed by all parties) are met. What are the other use cases of BaaS? Well, there are horizontal use cases (e.g. identity and security management, data management, and so on), and vertical use cases such as voting (by citizens), healthcare (e.g. access to health records), and energy (e.g. micro grids).

7.4 Cloud-native and Microservices for OTT Providers and Telecoms

Edge computing is being specified in many standards and also by open source organization. For example, 3GPP has accelerated its activities toward edge clouds, but also International Telecommunication Union (ITU), European Telecommunications Standards Institute (ETSI), Cloud Native Computing Foundation (CNCF), and others. When many organizations work to specify how edge computing should be deployed, there is an obvious risk for fragmentation of edge computing solutions. In that manner, 3GPP specifies the interfaces on the mobile network side (e.g. 5G), such as Application Programming Interface (API) toward third parties. However, building distributed cloud environment includes different clouds on different locations, with different functions being implemented and different resources being used.

The new services, driven by standardization of 5G based on network slices, require a paradigm shift in the core network (e.g. for 5G) to support cloud-native network functions and so-called container-based software.

What is cloud-native? The CNCF defines cloud-native as an approach in which the applications are broken down into microservices (loosely coupled services, each with specific functionality and self-contained) and packaged in lightweight containers (that is virtualized OSs, i.e. their instances) to be deployed and orchestrated across a variety of servers. It is a modern approach in running cloud-based applications, which was started by some of the modern global services providers such as Netflix, Twitter, Alibaba, Uber, Facebook, and others.

So, in this case the global OTT service providers (on the open Internet), which by default provide services via distributed clouds (which they break into smaller pieces and that is called cloud-native) influence the telecom vendors who provide equipment for mobile and fixed operators. The main target is to support new business models for existing and new services in telecom networks by using the once deployed infrastructure.

In general, the cloud-native approach is driven by softwarization and virtualization of networks, in particular in 5G mobile networks as the main driver.

7.4.1 Cloud-native for Telecoms and OTTs

The move to 5G NR standalone architecture (with 5G access and 5G core network) was needed to support new advanced 5G and beyond services. However, such transition requires a shift to cloud-native technologies and a cloud-native infrastructure to carry out cloud-native functions and cloud-native services/applications. Such cloud-native infrastructure may be comprised of various microservices, some of which may be based on open source software, which is vendor-dependent. The main line is that all building blocks in a cloud-native approach are called containers.

What is a container? Well, a container is composed of a complete runtime environment that includes the application, all dependencies built into it, libraries, executable files (binaries), as well as configuration files, placed in a single "package" called a container. Multiple containers can run on the same machine (i.e. on the same host computer) and all containers use the same kernel on that computer, with each container serving an isolated process (application) in the user space. At the same time, containers take up less memory than VMs (a container image usually has a size of several tens of megabytes – MBs).

What is the difference between containers and VMs? VMs are an abstraction of physical hardware, so one physical machine (computer) can run multiple VMs. Unlike VMs, the container does not necessarily contain an OS on a given host, which means it is simpler than VMs because containers share the host's OS (kernel) and build containers on top of it as needed. So, containers take up fewer resources (e.g. less memory) and provide flexibility in building services and applications. One typical example of such implementation is Docker [9]. Let us explain in a little detail how Docker works (with an aim to get a picture about containers). The developer first creates a Dockerfile, in which he defines all applications and features (dependencies) that are used to create Docker images. In a Docker image, all applications and their features are defined and with that, in fact, establish the Docker containers that are running instances of the Docker images. Such Docker images can be placed in the cloud which gives a Docker Hub.

Let us define what does orchestration do in Docker? Orchestration can best be described through an example. For example, let us have an application that has both high-traffic and high-availability requirements. Because of these requirements, multiple machines are usually needed (for higher application availability due to heavy traffic) so that in case one machine fails to serve the application, the application is still available through at least one, two, or more other machines. Without orchestration, each machine must be accessed individually for configuration, control, and setup, which is typically time-consuming and error-prone due to the manual work on each of the machines on which a particular application needs to run. Orchestration (via programmed orchestration tools) allows offloading of most such manual work via synchronization of machines in terms of load sharing, which in a single word means automation. A typical orchestration in Docker uses so-called Docker Swarm, which can deploy an application to many host machines with just one command. Also, in case of failure (crash) of one of the machines (each machine is a node in the swarm), the other machines (i.e. nodes) will automatically take over the load so that the application continues to function smoothly.

Thus, orchestration in Docker is an automatic process of managing individual application containers based on the so-called microservices (which is part of the Control Plane in the cloud). Various microservices solutions are used in the cloud by global OTT service providers of cloud solutions, such as Google, Microsoft, and Amazon, which include the following examples:

- The Google Cloud platform uses the so-called Kubernetes [10] that are open-source container orchestration systems.
- Amazon Elastic Container Service ([11] is an Amazon web service that enables the Elastic Compute Cloud (EC2) service to run and manage containers in the Amazon cloud.
- Microsoft Azure uses Azure Container Service [12] as a solution for hosting web services, similar to Elastic Container Service on Amazon.

Regarding the telecom operators, containers such as Kubernetes and Docker in cloud-native implementations should be "hidden" from the telecom operator via defined interfaces (with certain configuration options).

7.4.2 Cloud-native in 5G Mobile Networks

One of the reasons for cloud-native introduction in the telecom sector is the need for transition from 4G to 5G via coexistence of two core networks, 4G EPC and 5G Core. A similar approach may be expected to be used for transition from 5G to 6G. This means that orchestration needs to work with a hybrid environment of multiple mobile networks, and network infrastructure as the mobile service will work seamlessly across them.

There is a possibility that some network functions will be virtualized with VMs and some with containers (e.g. Docker), while some functionality still realized as physical equipment.

Figure 7.6 shows 5G deployment stack for cloud-native, which incorporates the existence of IaaS consisting of physical resources (processing power, storage, networking), PaaS [13], and shared data environment, which may be an open source solution or proprietary solution, and standardized 5G network functions on the top. Overall, 5G network slicing can be used for provision of different services to different verticals (with different requirements) by using edge computing in 5G access and cloud-native approach for 5G CN.

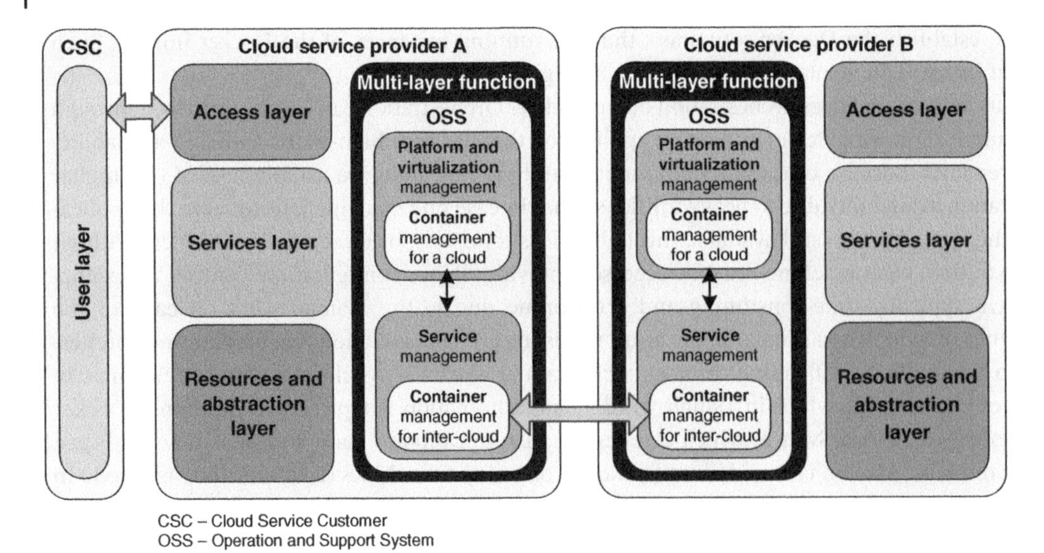

Figure 7.6 5G deployment stack for cloud-native.

Finally, the future will certainly prove whether the emerging cloud-native approach will bring the expected flexibility and agility to 5G and beyond mobile operators for provision of expected new services and new business models to different verticals besides the enhanced legacy mobile broadband services.

7.4.3 Cloud-native IoT

Cloud-native is natively related to IoT services, considering that IoT itself is all about connecting devices, and one of its key requirements is automation. Connected IoT devices and applications are being integrated into different existing and future business processes. However, IoT is also a source of new revenue streams obtained through enabling IoT applications/services and via monetization of the obtained data from various devices. Due to such heterogeneity in IoT devices, services, and potential verticals, the cloud-native approach can bring benefits the IoT ecosystem.

What are the most important attributes of cloud-native IoT? The first attribute is horizontal scalability. The IoT software is typically designed to support load balancing (because new IoT devices appear and older ones may disappear over time). The number of end IoT devices is huge (just to remind that the requirements of each new mobile generation increases the number of supported devices per square kilometer) and usually their traffic is directed to/from many IoT gateways. The sink node (or source node, e.g. for configuration purposes) of IoT traffic are clouds the where IoT platform are located. There can be a large number of computing instances that are allocated by the cloud infrastructure to different IoT services and end devices. In that way, cloud-native IoT allows automated scale-up and scale-down computing and storage capacity, according to a given set of policies and business goals.

The second attribute is high-throughput, as IoT software needs to handle large amounts of data without any impact on system performance. Why? Because IoT traffic can be "bursty," it is a high ratio of peak to average bitrate. On the other hand, throughput is not always an issue, as there are IoT services with very low data throughput generated by limited, low-power devices.

The third attribute important for critical IoT services is low latency (e.g. needed for applications such as autonomous driving or industrial automation). Compute by design reduces latency by moving computing resources closer to end IoT devices. However, this must be matched with system architecture designs that are capable of supporting low latency. With cloud-native, developers and system architects have the ability to mix and match tools to meet the different requirements and traffic patterns associated with IoT applications.

Finally, cloud-native provides no single point of failure. The software components of an IoT server could be optimized around tasks performed for specific purposes – microservices that communicate using message brokers. Then, the design, orchestration, and administration of the microservices that are part of the IoT platform, as well as their broker, must be done in such a way that avoids single points of failure.

So, cloud-native is required for telecom IoT applications that have strict performance requirements and need separate network slices (e.g. 5G network slices for URLLC services). In fact, network slicing on demand and multi-tenant (based on various systems produced by different vendors) is possible or easier with the cloud-native technologies.

For example, different vertical applications may have different requirements on data being collected. Also, charging and billing mechanisms can be customized for a given network slice that is used by the particular customers. In order to adapt to heterogeneous requirements by different vertical industries that use telecom IoT services, network slicing as an end-to-end system architecture solution applied to all domains including access and core networks, transport network, and cloud infrastructure. Network slicing could also extend beyond a single location of the IoT service, such as cross-site communication on a global scale. Such approach requires cloud-native design for telecom networks (e.g. 5G/6G mobile networks) by telecom equipment vendors.

7.4.4 Discussion on Cloud-native

Cloud-native solutions that exist today are not fully open and vendor-neutral. Also, in practice, the ability to monitor, optimize, and modify systems is not ubiquitous. Performances for telecom cloud-native services are not yet proven at mass scale (at least, by the mid-2020s). Examples of truly innovative services built on telecom cloud platforms are only a few. However, it can be expected that such large-scale cloud-native IoT applications will be offered by telecommunications operators in the future (in the 2030s and beyond).

The main goal of cloud-native is to speed up the application development and to optimize its operation in the cloud, to make it easier to build up and start and to make it easier for developers and staff engaged in its maintenance. As already typical, cloud-native applications are used primarily by the OTT global service providers because they cover very large geographic areas (e.g. the whole world) with multiple clouds on a global scale. On the other side, telecom operators try to follow the same approach as OTT service providers with cloud-native services; however, their business cases are more limited because they are national operator with services provided in a single country, which can be a small country with a small population (e.g. a country on a Pacific island).

7.5 Edge Computing

Cloud computing has emerging requirements for support of real-time services, such as AR/VR services, IoT services, and new innovative use cases in mobile networks (e.g. V2X services). The main challenges for provisioning of real-time services by cloud computing are (1) end-to-end latency between CSP and its customers and (2) the load congestion at the data centers of the CSP. Therefore, further development of cloud architectures includes the so-called distributed cloud concept [14]. According to it, cloud capabilities (for provision of cloud services) are distributed via core, regional, and edge clouds. Core cloud has the largest computing and data storage capabilities, while regional clouds (between core and edge clouds) are optional for load sharing as well as QoS/QoE enhancement of cloud services. The edge clouds are clouds with limited resources (e.g. limited processing power and limited data storage) deployed at the edge of network accessed by the CSC. The main advantage of introduction of the edge computing is lower delay, considering that the edge cloud is much closer to the end customers compared to the core clouds. However, the cloud management and orchestration systems should be intelligent to detect which tasks should be executed at the edge and which tasks are less sensitive to latency and can be transferred for execution to regional or core clouds in the architecture.

Edge cloud is especially important for the mobile network, considering that mobile users move in different locations (so they have variable distance from centralized clouds, and also users are closer to different network edges at different times), while on the other hand their devices are limited in terms of battery power, processing capabilities, and storage (when compared to desktop computers).

7.5.1 5G Core and Edge Computing

The advances in 5G and beyond mobile networks include the network slicing approach. As already noted in Chapters 4 and 5, network slices in 5G core can be created for different purposes. For example, to serve different traffic types, a slice designed for enhanced Mobile Broadband (eMBB) traffic is able to handle very high-throughput per user. Another slice, for massive IoT (mIoT), rather serves a large number of subscribers that transmit small data infrequently. Slices can also be created to serve subscribers belonging to different enterprises, e.g. a slice dedicated to subscribers for each mobile virtual network operator (MVNO).

Let us recall what the 5G Core design approach looks like? It can be compared to the design of the OS, which has a library of functions and various third-party applications can be installed through an API. In a similar way, the 5G network works through the network functions and the external interface (API) to create services, and from third parties by using the given network functions in 5G. That way, a mobile operator that has a 5G network will be able to (theoretically) "install" and "uninstall" various applications from the network, in a similar way (metaphorically) to how we install and uninstall various applications on our smartphones and laptops today. 5G Core is also related to further advances in cloud computing toward the edge between the core network and the 5G RAN. What is the target of mobile edge computing?

Well, the goal is to reduce the delay for critical services. Another goal is to reduce traffic (when possible) for the high-demand services expected in 5G and wider networks in the future. One such service is VR, which has very strict throughput and latency requirements for an ideal user experience with strong interaction. Of course, each service with such requirements will have different levels of service experience, similar to the different resolutions that can be selected when viewing video content over the public Internet.

What else could be the potential uses for edge computing in 5G? Overall, edge computing can be used by many future services in different verticals (e.g. IoT services). They will be the new businesses in the world of mobile telecommunications that will promote the further decentralization of cloud data centers. The decentralization of data centers pushes them (the clouds) to network edges (in parts – the central clouds will remain with higher processing and storage capabilities). Also, it was primarily targeted to be used by 5G services (e.g. critical IoT services over 5G). In that manner, one gets transformation from the centralized to distributed cloud environment.

What is involved in such a distributed cloud concept? Well, on one side, there is the traditional core cloud (physically, it is a data center deployed in a specific location and connected to the

Internet or a specialized IP network with sufficient bandwidth). On the other side (closer to the end user side) are edge clouds, deployed at the edges of the network, which are typically limited in terms of space for them and the power they can use (when compared to "legacy" clouds).

As for edge computing (or edge clouds), 5G and future mobile generations are inextricably linked to it. Why? Well, because 5G actually increases the need for edge computing by delivering critical IoT services, then demanding AR/VR services, V2X services, and other future services. To this end, 5G mobile operators may also require third-party 5G applications to be hosted on their 5G network and in parallel with their own services. However, third-party application providers and OTT content providers may be concerned about being locked in (or also locked out) by the mobile edge. Thus, such third-party vendors may prefer arrangements for fast access to the mobile edge while their applications run in their own data centers.

But, what does the edge cloud do? It extends certain cloud capabilities, such as storage, computing, network, AI algorithms (e.g. for orchestration), and so on, to the network edge. Then, the core cloud and edge cloud cooperate with each other to realize the cloud to thing continuum. With such edge cloud approach, the problem of latency (between the cloud and end device/machine) is solved. Also, the bandwidth utilization is reduced between the edge and the core cloud (i.e. data centers) due to offload of such traffic on the shorter distance between the devices and edge cloud.

Which future services can use edge computing? There are many foreseen services. Examples include the already noted AR/VR, public safety, remote operations, IPTV, and video delivery via different access networks, demanding services at venues (e.g. stadiums and exhibition halls), vehicle-to-everything (V2X) communication, industrial IoT and smart factories, content delivery networks (CDNs) caching at the edge, application computation offloading, camera as a service, and so on. Is there any negative implication of edge computing?

Well, normally, edge computing will require some additional capabilities (e.g. storage and processing power) which will increase the price of the edge network equipment. But, for many critical services (e.g. 5G/6G URLLC services), edge computing will be the fundamental approach with the aim to achieve very low latencies, in range of a few milliseconds or less. Also, edge clouds are not necessarily placed at fixed locations; they can be mobile units (i.e. embedded in moving stations). We have such examples in vehicle-to-network (V2N) communications where edge clouds are located in the vehicle (which is mobile), and also edge clouds are placed on the edge platform at fixed locations in the mobile network (e.g. 5G/6G fronthaul), which can be collocated/integrated with the base station or standalone. So, there can be multiple edge clouds on different sides of the connection in the access networks, which are used for provision of the given service.

7.5.2 Telecom Edge Clouds

The edge clouds are typically a possibility for telecom operators to extend network capabilities to network edges, primarily because they own the physical network infrastructure. The exceptions are edge clouds that are deployed by third parties (e.g. vehicle vendors for clouds deployed by design in the vehicle). Telecom operators typically base their infrastructure on standards, considering that they need interoperability with other telecom operators, especially considering the control plane traffic (i.e. signaling), which is present in the telecom world since the introduction of voice services in the 20th century onward. Different standards development organizations (SDOs) develop their own standards and frameworks for edge computing, targeted for their use by telecom operators, particularly mobile operators. In that way, the ETSI has standardized edge framework called Multi-access Edge Computing (MEC) targeted for use of edge computing in any type of network (although, initially, it was defined as mobile edge computing, which further evolved due to integration of different access networks with the same core and transport networks) [15]. Also, ITU specified the functional requirements for edge cloud management [16] targeted mainly to telecom operators. The Global System for Mobile Communications Association (GSMA) tries to define end-to-end the telecom operator platform that will be a common solution for edge computing environments for exposing network capabilities and edge computing resources [17].

The initial focus in edge computing frameworks was the focus on edge computing with Network-as-a-Service approach (e.g. for QoS control). Future developments of the edge clouds are toward their support for end-to-end network slicing (e.g. in 5G-Advanced and 6G mobile networks).

Given the importance of edge computing for future IoT services, there have been recent developments in the telecom operator sector in that direction. As such, the Telecom Edge Cloud (TEC) is a global platform solution for exposing, managing, and marketing edge computing, network resources, and capabilities across different telecom operators and across national borders, which may use existing and future network assets. The TEC platform is targeted to be a digital platform that will provide packages based on assets (computing and network resources) of the mobile network operators (MNOs), which will be used by TEC platform customers (e.g. service providers, which include also the telecom operators as such, but also other parties such as different vertical industries) to deliver services based on the use of edge computing. On the other side, the end users will be able to consume such services across different networks in a seamless manner. The scope of TEC is shown in Figure 7.7.

The main target of TEC is to enable MNOs to use their edge resources for different services and with that to monetize them additionally (via provision of new services, such as critical IoT services to verticals). For such purpose, mobile operators need a platform based with CSCs for TEC services to deploy them easily with the aim to deliver new services (e.g. URLLC services in 5G, 6G, and beyond) and use experience for bandwidth demanding and latency sensitive services such as cloud XR gaming, holographic videoconference, and so on. A specific example is V2X across multiple operators and different countries which may be practically feasible on the basis of edge federation approaches.

An application is considered cloud-native when it seeks and leverages the benefits of telecom edge computing, such as very low latency and high performance, that can be achieved through a

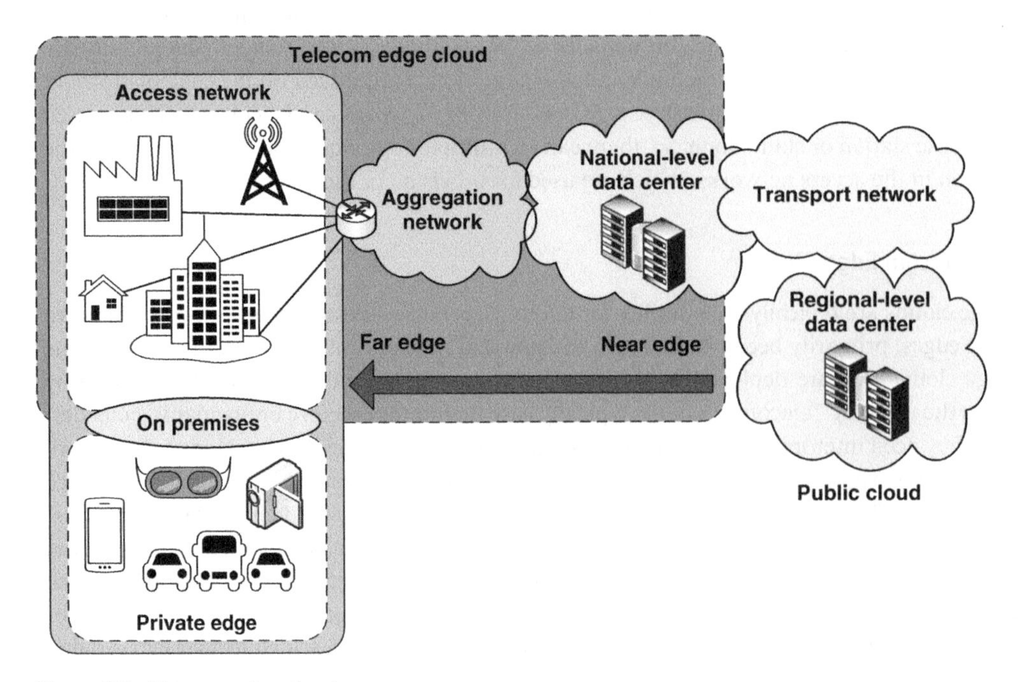

Figure 7.7 Telecom edge cloud scope.

distributed network of edge nodes. The use of edge computing requires edge applications to have tighter integration with network elements (which is different from OTT cloud services in which there is no mutual cooperation between the telecommunications operator that has the physical infrastructure, including the edge resources, and on the other hand, OTT provider services that use the public Internet in which net neutrality is applied as a policy). Such integration also applies to network assets, not only to edge clouds in user premises equipment, such as the local disk on smartphones or the memory in a home surveillance camera.

There are several specific characteristics that cloud-native applications and services need to have to use various edge resources (hardware and software). They are expected to provide low latency and high-performing services to the end users, which requires flexible and efficient architecture with real-time characteristics. Unlike legacy cloud computing services (based on use of centralized cloud resources in regional data centers), which typically serve a wide geographic area and the population in it, edge-native services provide the intended services only over a limited area (e.g. downtown city area, factory floor, and logistic warehouse). Various devices can be connected to the edge computing facilities, including smartphones, various cellular IoT devices, vehicles (cars, buses, and trucks), trains, drones, and so on, which all can move across different regions and countries. For services that include such mobile end user devices (e.g. connected via 5G mobile network, or 6G in the future, and so on), the edge-native applications may need capabilities for handling end user mobility with the aim to ensure service continuity, by performing handing over from one edge cloud node to another one in the same telecom network or in another (adjacent) one. One may make a parallel between handovers between cells in mobile networks (which function well since the 1990s from 2G onward) to edge-cloud handovers for cloud-native applications (which, again, include the most demanding ones in terms of latency, reliability, and bandwidth).

Considering that edge clouds have limited resources when compared to legacy clouds, security has an increased importance. Considering that telecom operators have more capacity (in assets, stuff, and capital) to invest in security solutions for all parts of the network (hardware and software), the provision of higher security typically gives an advantage to telecom edge clouds. On the other side, by keeping the data at the edge devices (at edge clouds), which is normally closer to the end user premises, the cloud edge applications generally provide better data privacy for all users, including business users (enterprises) and residential users. However, one may note that the edge cloud services offered to residential and business customers may differ, because business customers typically have the capability to include or host more edge cloud resources considering that on average companies are deployed in larger areas than homes.

Edge-native applications for enterprises typically require high reliability. Typical examples are URLLC (i.e. critical IoT services) services in 5G and beyond mobile networks (e.g. motion control of machines in smart factories, control of driverless cars and drones, smart grid, and critical smart city services). For such service, high reliability is required for edge-native services, because an error can result in high loss or damages at the customer end (e.g. dangerous movement of machines in a smart factory and accident of the driverless car or drone).

7.6 Future OTT Cloud Services

Both telecom operators and OTT service providers (local or global) benefit from cloud computing. Telecom operators have network infrastructure, so they can also provide cloud services to the edge of the network in different locations and using different network nodes or data centers. But also, OTT service providers have the opportunity to use edge computing, even without renting it from telecom operators. For example, end user equipment, such as a home surveillance camera or a connectivity device installed in a car, has a specific data store that is part of the end user equipment (not the telecom infrastructure).

In this book, as a reminder, OTT refers to all applications and services that are available over the open internet on the basis of net neutrality. However, the Internet access services in all countries is provided by telecom operators who are typically national, or they can be international such as satellite Internet providers which again need to obtain national licenses in countries where they provide such satellite Internet access service. So, all services offered through the public Internet can be treated as OTT services or applications. This includes all websites on the Internet, email service, online social networks (e.g. Facebook and Twitter), search engines (e.g. Google and Yahoo), video aggregation/sharing sites (e.g. YouTube), picture sharing sites (e.g. Instagram), voice services over Internet (e.g. Skype, Viber, and WhatsApp), torrent services (e.g. BitTorrent), video streaming (e.g. live streams over the Internet) and audio streaming (e.g. radio over the Internet), OTT cloud computing services (e.g. Google Docs), gaming over the Internet (e.g. Steam platform), governmental services over the Internet (which increase over time with their digitalization), and all other existing and future services provided over the open Internet network on the basis of network neutrality. All such services today and in the future are based on cloud computing, because in all cases the data from users and businesses is stored in databases and in memory units in data centers (which are, in fact, legacy clouds) and other network nodes and hosts between the legacy cloud and user equipment which offer data storage and processing capabilities for use by the OTT service provider over the open Internet.

OTT service providers, especially larger global providers (e.g. Google, Facebook, Amazon, Microsoft, and Apple) are taking the full benefit of telecom operators' network infrastructures to grow their businesses. On the other side, OTT service providers increase the appeal for broadband on the side of the telecom customers, which increases the revenues (of the telecom operators) for selling broadband Internet access service. Why?

This is because without OTT the demand for broadband access will be much lower than it is currently. So, both OTT providers and telecom operators are already highly dependent on each other. In the future, this may also apply to edge cloud services, since net neutrality does not prevent telecom operators from leasing to OTT service operators a part of their edge cloud infrastructure that will clearly become available in the future, in the 2030s and beyond, as in the first decades of the 21st century, telecom operators use cloud services from large online service providers for their purposes related to business analytics and backup.

What is the future of OTT cloud services? Well, the increased use of digital services and application (i.e. digitalization in many vertical sectors and industries) increases the number of different platforms and complexity of the networks. The main novelties for future OTT clouds include (but are not limited to):

- Multi-cloud –based on the public cloud from two or more independent CSPs at the same time, which are used for the customers' businesses. It differs from hybrid cloud (which includes two or more of cloud deployment models, such as public, private, and community cloud), inter-cloud (which refers to interworking between CSPs), and federated cloud (which refers to interworking of two or more CSPs for federated use of the cloud resources, based on federations agreement between CSPs) [18].
- Quantum cloud computing – can be used for increased security for future OTT services which use clouds and, at the same time, require higher security for encrypting sensitive data (e.g. from governments and enterprises), for example, using the quantum key distribution – QKD (as given in Chapter 6 of this book). With the emergence of quantum technologies, the OTT cloud providers add them as offers to their cloud's portfolios [19].

- Edge computing – future OTT clouds will largely use the edge cloud resources available in user equipment. Use of end user resources is not new in the OTT space, for example, since the 2010s, updates for OSs (e.g. Windows 10) downloaded by one computer could be shared with other computers in the local environment, so the first computer in fact serves as an edge cloud for the others. A similar example can be peer-to-peer file sharing services which in fact rely on a combination of memory and bandwidth resources of end users as well as servers (e.g. BitTorrent) [2]. However, use of telecom operator edge resources owned by telecom operators by OTT service providers in the future, without jeopardizing the network neutrality of open Internet, may contribute to new "killer" OTT applications, such as OTT XR services and OTT critical IoT services. The future is likely to be based on massive use of edge cloud computing and for OTT services, which are mainly based on legacy clouds (data centers) and CDNs.
- AI/ ML - the introduction of AI penetrated OTT cloud services much faster than services from telecom operators (due to the shorter time from innovation to offering services in the digital market), offering a higher level of automation than before (before the increased application of AI in the ICT world), real-time data analytics, and improved user experience that are redefining the digital experience for businesses and individuals [20]. Also, AI/ML is offered by CSPs and as a service for use by companies, including startups and entrepreneurs.

The OTT service providers, especially global ones, already have and use distributed CDN. Also, large OTT service providers deploy their own submarine cables for connecting their data centers in different locations around the world (as we have noted in Chapter 3 in this book).

However, constantly increasing the use of OTT cloud computing (in all forms, from the legacy clouds to edge clouds) will result in increased needs for scenarios in which CSC cooperate with the multi-cloud manager (as cloud service partners to CSPs) for simplified use of different cloud services on a global scale (as shown in Figure 7.8). This is particularly important in cases when the CSC is an enterprise that targets development of a global business by using cloud. In such cases, the cloud business customers need a global-scale computing infrastructure which will consist of

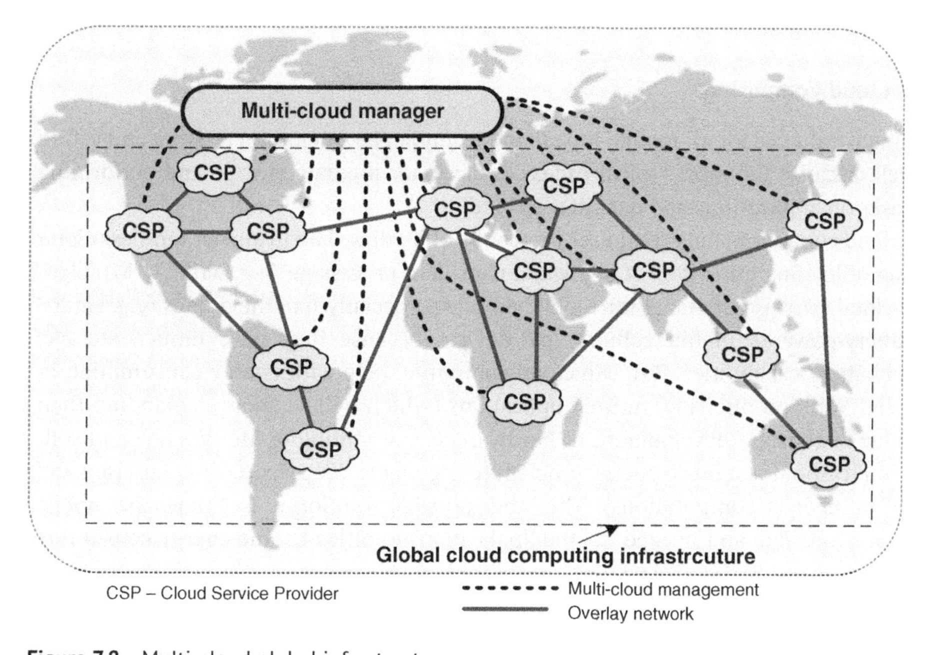

Figure 7.8 Multi-cloud global infrastructure.

different cloud services in terms of virtual machines distributed globally. If one assumes that it is not possible for a single CSP to provide all the required cloud services considering the required capacity, geographic location, and cloud service type , then use of the noted multi-cloud will be beneficial for the cloud service customers.

7.7 Future Telecom Cloud Services

Future telecom cloud services include all future possibilities for OTT cloud services, such as multi-cloud, quantum cloud computing, edge computing, and AI/ML cloud services. However, telecom infrastructure is typically limited to the territory of a given country which may vary from one country to another (e.g. there are larger and smaller countries in terms of both territory and population, which on the other hand influence the potential customer base). However, there are several specifics to future telecom cloud services, considering that telecom operators have the physical infrastructure including networks and data centers, i.e. core clouds (e.g. collocated with the operator's core network nodes).

The main drivers of future telecom cloud services are related to bandwidth demanding services (e.g. XR services) and mission critical services that are sensitive to bandwidth and latency, respectively. Considering that mobile broadband access is the main access to Internet and, generally, digital services in many developing telecom markets worldwide, and additionally mobile communication is a personal communication (each human has their own smartphone and carries it all the time), the focus for future telecom cloud services is their use for future mobile services provided by telecom operators. The other important aspect regarding the future telecom clouds are edge clouds (particularly in terms of their use cases in future mobile services, in 5G-Advanced and further in 6G during the 2030s) and their federation among multiple mobile network operators. Besides the use of mobile cloud computing and edge cloud federation, cloud-native approach in design of future telecom networks is expected to make them more open to innovations by different parties for services needed in different verticals.

7.7.1 Mobile Cloud Computing

Cloud computing is valuable for all mobile broadband technologies, including existing and future ones. Why? Well, because the most typical use cases of clouds are data storage and performing computation tasks on applications and data in the cloud.

In that way, cloud computing helps end user devices by offloading data to almost limitless cloud storage and also offloading execution of applications or data processing (e.g. with AI/ML algorithms) in the cloud. This is beneficial for mobile devices, especially handheld ones (e.g. smartphones) or battery-powered mobile/cellular IoT devices because they have limited storage, processing, and battery capabilities. But, can cloud computing help with energy consumption in mobile terminals? Well, on one side, future applications (which will be more demanding than existing ones, due to constant development of hardware, e.g. according to Moore's law and software that runs on it) should be made energy efficient (e.g. going to deep sleep when not used and have optimized number of computations). The other possible solution is to enable the mobile devices (where it is possible and needed for the application) to offload some energy consuming tasks to the cloud.

Can offloading of computing demanding applications to the cloud save energy for mobile devices? One may do a quantitative study to find the answer [2], [21] by using Eq. (7.1).

$$Cloud\_efficiency = P_c \frac{C}{M} - P_i \frac{C}{S} - P_{tr} \frac{D}{B} \quad (7.1)$$

where S is the speed of cloud to compute C instructions, M is the speed of mobile to compute C instructions, D is the data need to transmit, B is the bandwidth (bitrate) of the mobile Internet connection, P_c is the energy cost per second when the mobile device is doing computing, P_i is the energy cost per second when the mobile device is idle, and P_{tr} is the energy cost per second when the mobile is performing transmission of the data. Typically clouds have many times higher processing power than mobile devices. In that manner, if we assume that the cloud is N times faster than the mobile device, i.e. $S = N \times M$, then one can rewrite (7.1) in the form given with Eq. (7.2).

$$Cloud\_efficiency = \frac{C}{M}\left(P_c - \frac{P_i}{N}\right) - P_{tr} \frac{D}{B} \underset{N \to \infty}{\approx} P_c \frac{C}{M} - P_{tr} \frac{D}{B} \quad (7.2)$$

Typically, the processing capability of clouds is many times higher than the processing capability of a single mobile device (such as smartphone); there we can use that N tends to infinity approximately. If the resulting value from Eq. (7.2) is positive then offloading the task to the cloud is justified because energy is saved, otherwise, it is not.

If we continue the analysis further, due to user mobility in the mobile network (either 5G or 4G or 6G) there are variations of energy consumption and the latency experienced by end user devices for communication with the clouds inside the mobile network (e.g. edge clouds) and outside the RAN (e.g. public clouds). In typical public mobile networks, different mobile users are located at different locations of the serving cell. Then, cell size can range from very small (e.g. micro cells and pico cells), tens to hundreds of meters in radius, to macro cells that are primarily used for coverage of rural and less populated areas (up to a couple of tens of kilometers in radius). In such heterogeneous cellular (i.e. mobile) environment, end users that are located at the edge of the macro cells will need to use more power for uplink transmissions (due to bigger distance to the base station), because parameter B (the bitrate) in Eqs. (7.1) and (7.2) can become significantly lower (due to more robust modulation and coding scheme used when signal-to-noise ratio is worse, and it is worse as the distance increases) and cloud computing in such cases will not save the energy of the mobile device [22].

Due to mobility of end users, it is hard to control the battery (i.e. power) consumption in a macro-cellular mobile network, therefore, future mobile networks (further developments of 5G/5G-Advanced and future 6G networks) need to use small cells in future mobile networks.

A similar analysis can be applied to the latency, which is very important for mobile services that use clouds. The delays in 5G mobile networks can be pushed down to 1ms (meaning lower than 1 ms delays separately in the access and in the core parts) and below in 6G access networks; however, the speed of the light is the maximum (approximately) speed for all signals on Earth (according to the current knowledge of the human kind) so transmission delay also matters a lot, because with the speed of light of 300000 km/s it takes 1 ms to transmit signals over 300 km distance without any delay due to processing and buffering in network nodes on the path. In that way, to have very low delays (e.g. for URLLC services in 5G and URLLC+ in 6G mobile networks) the clouds (where the data is stored and processed) need to be as close as possible to the network edge (that is, close to the access network). Also, future XR services (e.g. AR, VR, or Mixed Reality – MR) require the support of edge clouds for rendering the XR content that is shared or used by end user devices.

7.7.2 Future Telecom Edge Clouds Federation

We already know what is edge computing and what are its advantages. However, for practical implementation of edge computing with edge cloud resources owned by telecom operators, there are required standards. On the other side, edge clouds have higher potential in mobile networks, such as 5G, 6G, and beyond. The standardization of edge computing for mobile networks seems to be easier than before considering that there is primarily a single 5G standard that is deployed globally, that is 3GPP standard; either that is 5G RIT (with 5G New Radio) or 5G SRIT (which includes 4G LTE and 5G RIT in the set of radio interface technologies, however, both are standardized by 3GPP). So, standardization of mobile technologies from 5G is unified (which happened before in the local area technologies, with Ethernet for LAN, i.e. IEEE 802.3 standard and Wi-Fi for Wireless LAN (WLAN), i.e. 802.11 family of standards).

As noted before, the main advantage of using edge computing was the reduced latency, but also enhanced reliability and resiliency of the telecom network, because a failure at the edge does not result in failure of the whole system (as in the case of centralized servers, i.e. legacy clouds). But, such requirements of low latency and high reliability (e.g. URLLC) are characteristic of critical IoT services such as autonomous vehicles, Industrial IoT, critical medicine services, etc., that are provided via non-public networks (NPNs), which in such cases normally are not part of the public Internet. In general, edge computing is crucial to implementation of many future services in NPNs in various sectors and across different industries (which are using ICTs).

Although edge computing is applicable to both fixed and mobile networks, here, we focus on its application in 5G and beyond mobile network from 3GPP, which are the initial target, especially because edge computing in a standardized manner in 3GPP mobile networks will require also roaming capabilities. Also, roaming is a typical feature of 3GPP mobile networks that started with 2G (i.e. GSM), and expands from voice and SMS to Internet access service in 3G and 4G, and further it will expand to IoT roaming (which includes edge roaming and federation) in 5G and 6G mobile networks. All roaming capabilities introduced in a given mobile generation continue to live in future ones (with enhanced standards over time).

The overall future goal regarding edge computing is defining a holistic view on the cloud-native transformation of telecom infrastructure toward hybrid scenario of edge clouds and legacy core network. The OTT service providers have established their PaaS and SaaS solutions over the Internet in the 2010s and 2020s as legacy ones, which are used for certain processes also by telecom operators.

The standardization of edge computing in telecom networks was initially realized by ETSI's MEC [23], which addresses various new vertical markets and enterprise customers (e.g. V2X, video analytics, location services, IoT, AR, content distribution, and data caching) by offering cloud computing environment to developers at the edge of telecom networks. Also, the already addressed in this chapter GSMA TEC [17] was another proposal in that regard for telecom operators. However, mobile standardization is done by 3GPP, considering that the only deployed 5G standards in the world in the 2020s are those from 3GPP. Hence, standardization efforts of 3GPP in 5G-Advanced releases (Release 18 and onward) focused on edge computing [24].

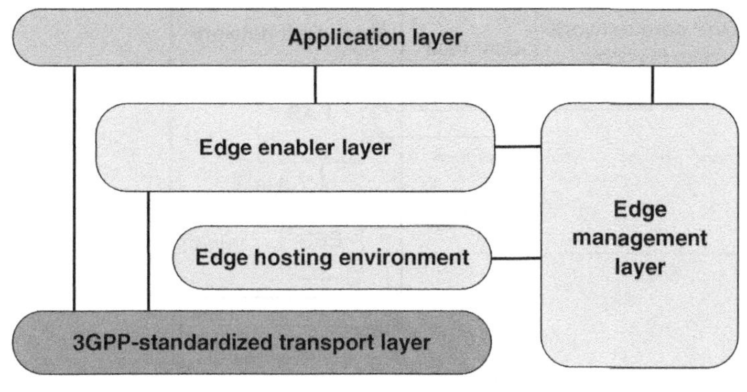

Figure 7.9 3GPP edge computing for telecom operators.

The 3GPP edge computing work is shown in Figure 7.9. It has five layers which include the following:

1) Application layer: It is the layer where the application (e.g. from the telecom operators or a third party) is defined to consume edge computing capabilities specified in 3GPP standards.
2) Edge enabler layer: It contains the overall functionalities in edge entities (e.g. base stations and other network nodes) which support the consumer applications [25]. This layer provides APIs that are used by developers for creation of applications that use edge capabilities.
3) Edge hosting environment: It refers to an environment for execution of edge Application Server (AS), which is typically not specified by 3GPP.
4) Edge management layer: It refers to management of the edge system, including the network and platform for edge hosting, and the application on top [26]. The management of edge computing may be done by three types of providers which are: (1) mobile telecom operators, (2) edge computing service providers (ECSPs), and (3) application service providers (ASPs).
5) 3GPP transport layer: It includes overall functionality provided by different 3GPP networks, such as 4G, 5G, and beyond, in support of the applications that use the edge system of the mobile network.

One should note that initially 4G systems were not designed for edge computing (although it is possible to do it in Evolved Packet System (EPS) by applying Control Plane and User Plane separation). On the other side, the 5G system specified edge computing (3GPP Releases 15–17), enhanced it in 5G-Advanced (started with 3GPP Releases 18), and further edge capabilities continue to be an integral part of 6G systems.

However, there can be multiple Edge Application Servers (EASs) in the network. Therefore, the 5G system supports also Edge Application Server Discovery Function (EASDF) which is used for discovery of IP addresses of the EAS.

Figure 7.10 shows the architecture for support of federation of ECS (Edge Computing Systems). The ECS network architecture by 3GPP includes several interfaces labeled as EDGE-n ($n = 1, \ldots, 10$) between the ECS and 3GPP Core Network (e.g. 5G Core, with UPF, PCF, NEF, and other network functions defined for it) as well as between ECS and User Equipment (UE), as shown in Figure 7.10. In the edge federation architecture is introduced EDGE-10 interface (for exchange of configuration information for Edge Data Networks – EDNs), between different ECSs, which can be provided by the same or different ECS providers. The data at the edge is stored in the Edge Repository (ER) as

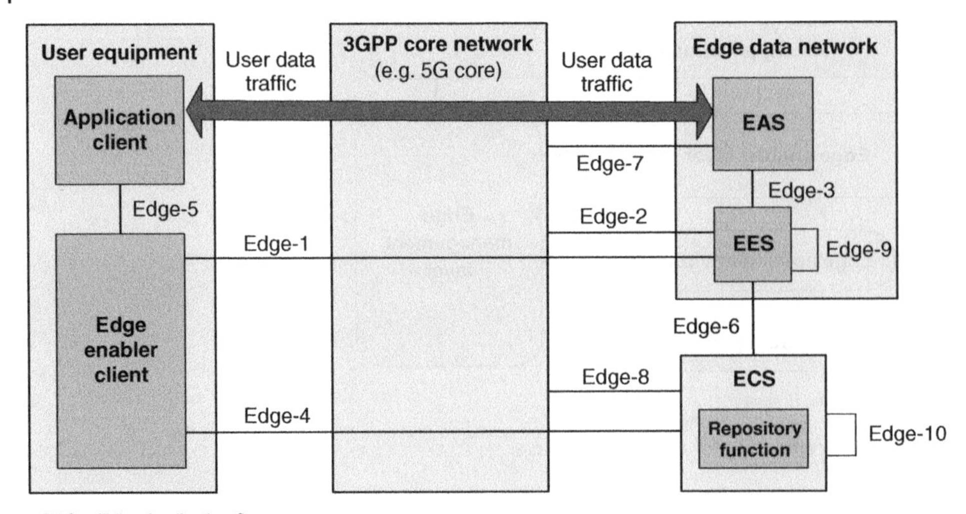

Figure 7.10 Edge system architecture for federation support.

parts of the ECS, therefore, it is also referred to as ECS-ER. This repository (ECS-ER) receives and stores information about edge deployments from other ECSs. In such edge federations, EDGE-10 interface is used both for exchanging of ECS profile during ECS discovery and for exchanging EDN configuration information during service provisioning information retrieval. In Figure 7.10, EAS and EES can be either trusted or non-trusted application functions (AFs). In case of 5G CN, when the EDN is a trusted one, it connects directly to Policy Control Function (PCF) or via Edge-2 and Edge-7 interfaces. On the other side, when EDN is untrusted (in respect to the 3GPP core network, in this case, the 5G core) it has interfaces toward Network Exposure Function (NEF) via Edge-8 interface. The ECS management system requests the management system of the mobile network (in this case, the 5G network) to identify the interface of PCF or NEF to which it has to connect. For that purpose, it has to specify the EDN identifier (e.g. EAS identification and EES address). The goal is for the ECS management system to connect the deployed EAS to User Plane Function (UPF) in the 5G CN for transfer of data traffic of edge-native applications that use EDN [26]. The EDNs appear as local data networks to UPF in the 5G core.

Overall, 5G systems natively support edge computing, so future applications/services provided by telecom operators or third parties using standardized edge system interfaces are called edge-native applications. Such applications are targeted at various sectors, including entertainment and gaming (e.g. XR), industrial IoT, V2X services, unmanned aerial vehicles (UAVs), and a number of native applications developed in 6G and beyond.

7.8 Business Aspects and Regulation of Cloud Computing (Including Security and Privacy)

7.8.1 Business Aspects of Cloud Computing

Cloud services are different and they are becoming more differentiated, so the markets they supply are different, from wholesale to retail, business, and public sector as well as consumers. An increasing number of ICT companies are either establishing new cloud services or converting existing services to the cloud. The most common categorization of cloud services is into three main services (as noted in this chapter): SaaS, PaaS, and IaaS; although the label "X as a Service" is also used for a range of different services.

SaaS primarily involves the use of remote applications by end users, including productivity-related applications such as Google Docs and Microsoft Office 365, social networking such as Facebook and Twitter, and OTT content delivery such as video-on-demand services (e.g. YouTube). PaaS are typically developer-centric, enabling collaborative application development, such as open source software communities. IaaS generally involves the provision of virtual machines, which offer processing and storage capacity.

There are various scenarios of cloud adoption which result in various cloud service provider categories [27] (as shown in Figure 7.11):

- Cloud technology providers: Some public and private entities, such as public administration and enterprises, implement the cloud in their own ICT infrastructure either for their own use or to become CSPs to provide cloud services for others or on behalf of others. For example, some government agencies (e.g. agencies for digital transformation) have established shared infrastructure delivered as a cloud to other government agencies. That requires an ecosystem which includes different industry players including vendors and ICT companies that are deploying data centers, hardware and software vendors, network players, security solution vendors, and energy supply.
- CSPs: The CSP can be OTT or telecom entities, which typically deliver multipurpose cloud services in the form of public cloud services. Such services can be IaaS, PaaS, SaaS, or their combinations. Also, one OTT cloud provider may use cloud service from another OTT cloud provider, for example, Apple's iCloud SaaS can be hosted on Amazon IaaS [27]. Telecom CSPs are capable of providing customized cloud solutions to different industries (e.g. industrial IoT, V2X, and smart city), which may include also network infrastructure (e.g. NaaS service). Also, the future is expected to bring many new edge-native services from telecom operators that have their own access and core network infrastructure.
- ASPs are entities that typically run applications on multipurpose cloud infrastructure provided by OTT providers (as regional or global ASPs) or telecoms (e.g. for edge-native applications). This category includes software vendors who develop cloud services for others to use. But to deliver such services, ASPs should use cloud infrastructure to build and manage their cloud services instead of building their own. This category is the most dynamic in the cloud services

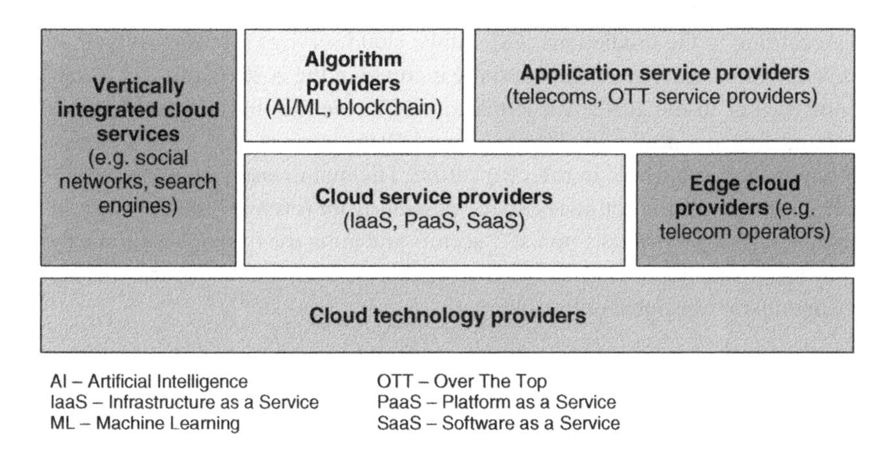

Figure 7.11 Cloud service provider categories.

market, as many enterprises (from small to large) can use ready-made cloud services to provide their own digital services to their customers.

- Vertically integrated OTT clouds are typically OTT global services (e.g. social networks and search engines on Web) that are running on the proprietary developed cloud infrastructure (e.g. developed by the vendors or a third party). In many cases, cloud infrastructure for this category is purpose-built for the specific needs of the services they provide.
- Edge CSPs are typically telecom operators who provide edge data networks and edge enablers for building edge-native applications and services by telecom operators or third parties (e.g. edge-native service providers).
- Algorithm providers are a new dynamic category of cloud-based service providers, which is emerging as a necessity in the future with the further development of AI/ML and new technologies such as blockchain and quantum computing. In this case, instead of providing fully developed cloud-based applications/services, algorithm providers provide ready-to-use niche algorithms that are natively developed and intended to run in a cloud environment and to be offered to other application developers through a suitable API. This is how most AI models are delivered, as well as through the development of other newer technologies (for example, blockchain).

Furthermore, recent developments in cloud-based solutions are transforming cloud-oriented business models. The end users move from one cloud to another when using online OTT services. The majority of global OTT service providers are cloud-native, including Netflix, Twitter, Alibaba, Uber, Facebook, and many others. Why? Because cloud-native provides the opportunity for business growth based on the platform's ability to adopt, recover (from failures), deploy, and scale. On the other hand, the telecommunications industry has yet to fully exploit the potential of cloud technology (e.g. automated operations of telecommunications networks and services tailored to a better user experience and edge-native applications/services).

Clouds together with mobile technologies (e.g. 5G and 6G) and IoT are fundamental in developing new services, creating new ecosystems and associated business models. In that manner, for example, the 5G expansion in the 2020s further changes the way CSPs may do business. While 5G offers enhanced mobile broadband and improved features, it also provides the possibility for telecom operators to offer new services (e.g. edge-native) including various IoT services, XR (which includes AR and VR), and smart factories and cities. Also, 5G and beyond mobile networks as well as fixed access networks and future satellite networks will be cloud native (e.g. software based, with virtualized resources and virtualized network functions), which bring flexibility, scale, and agility of deployments required for new innovative business use cases, including consumer and enterprise segments. According to the predictions [28], public cloud services revenues as well as edge cloud services revenues are expected to exponentially increase in the 2020s (until 2030), and that trend will continue further in the 2030s, while telecom/ICT revenues from legacy services (which include broadband Internet access, mobile voice and IPTV services, and VPN business services) will remain unchanged as a whole in the near future. The main reason is due to saturation of broadband markets in most of the countries, so the possibility for revenue growth is seen in vertical markets (with digitalization of initially non-ICT sectors and industries), which is primarily fueled and supported by increasing use of various cloud services from both telecom operators (e.g. edge clouds) and OTT providers (e.g. global online clouds).

7.8.2 Regulation Aspects of Cloud Computing

Considering the cloud services dispersing characteristic, the question is how to govern and regulate it? The right answer, as in many regulation approaches, is that it depends.

There are several laws that are applicable to cloud computing, which is country-specific. They may include telecommunication law, competition law, consumer protection law, as well as environmental and jurisdictional concerns. Also, certain regions (e.g. European Union) have created digital strategies, which include cloud computing for providing the data processing capacities required to enable data-driven innovations [29].

Considering the categories of service providers, it is clear that telecom operators as providers of broadband communication services for access to cloud services will be regulated by the telecommunication law, including national, regional, and international. It can also include the CSP if the type of service includes network infrastructure (for example, for NaaS) or it is about edge cloud services that require access through edge federation (for example, for autonomous vehicles and other V2X services, for industrial IoT, etc).

A specific concern for regulators is the transnational nature of cloud computing, resulting in a multitude of jurisdictions potentially "competing" to govern the regulated activity of the services as well as to have jurisdiction over data stored in the cloud, including privacy policies that may be applied on them. The movement of data in and out of a cloud service often (as with many other online applications over the public Internet) results in scenarios where the data becomes subject to the rules and jurisdictions of both the cloud user and the CSP, as well as to all cloud infrastructure providers who could potentially be located in different countries (for example, CSC is located in country A, CSP is in country 2 and in federated cloud scenarios in multiple countries, and traffic on the way between CSC and CSP(s) passes through several other countries).

Data transfer to a cloud that is placed outside the user's jurisdiction (country or region) can be problematic to the cloud user (i.e. CSC) and can raise questions about the user's control over own data in the cloud, as well as raise concerns for the national regulator (in the given sector to which the data is related) regarding the effective supervision of data (when needed). For some regulated sectors, such as financial services, cloud-related transfers and storage outside the jurisdiction of the regulated entity may itself breach certain national rules.

Usually national regulators in each country are not willing to hand over jurisdiction to any foreign authority unless appropriate arrangements are in place for mutual recognition of certain rules and regulations. In that direction, the increasing involvement of cloud computing also by national telecom operators (subject to strict regulation at the national level) and their federation and roaming at the national, regional, and international levels, will require greater transparency and cooperation between national regulators in the direction of resolving potential conflicts that may arise from laws and regulations for cloud computing environments, as well as for data stored in the clouds.

Cloud services are not a single service, but they cover a range of categories of cloud services (in general, one may write X as a Service – XaaS, where X can be software, infrastructure, platform, and network). Thus, some cloud services are based only on software running on a server in a data center (e.g. SaaS services), while others also include a platform (e.g. PaaS) or infrastructure (e.g. IaaS) or even an entire network (e.g. NaaS).

The data governance in the clouds is also crucial, but there also should be made distinction between large global cloud providers (e.g. Amazon, Google, and Microsoft), governmental/institutional cloud service, and smaller ones. One may say that a light touch approach in the cloud regulation and governance by monitoring of the cloud services (e.g. offered on the single digital telecom market) will be a good approach in developed, i.e. competitive telecom markets. Anyway, cloud computing is an emerging technology that will also shape future networks and services. For example, developed regions such as EU have set a target to have 75% of businesses using cloud services by 2030 [29]. So, the long-term goal in digitization, which will be an infinite (always-on) process in the future, is to succeed in the future with almost all businesses (including small and medium-sized enterprises) using cloud services (for example, in the 2030s and beyond), which on the other hand should be regulated by certain rules (e.g. cloud rules) that will not threaten such a process with too much bureaucracy.

Cloud computing has emerged as an important and disruptive technology of the current decade (2020–2030), also called the digital decade [29]. For the cloud to develop further (as planned and needed), one key aspect is "trust", which generates many questions from the industry, users, governments, and regulators. Why? Well, because cloud computing as a technology is reshaping the boundaries of companies and institutions that are using clouds (and the long-term target all of them to use cloud services) as well as consumer habits.

Analysis of cloud adoption [27] shows that it requires a framework of trust that applies to CSPs, with the four pillars of security, transparency, control, and business continuity being the most important, where:

- Security refers to protecting the cloud infrastructure and data, which is crucial for all parties, including CSPs, CSCs, governments, and regulators (e.g. data regulation).
- Transparency is one of the main components of a trusted cloud ecosystem, which aims to create trust on the part of cloud customers (as consumers) to use cloud services.
- Control refers to the confident movement of data by business entities, public entities (e.g. government, local authorities, and institutions), and citizens into the cloud, or in other words, CSCs should be in control of their data at any time, including the ability to leave the CSP whenever they want and be able to fully recover their data (when switching one CSP with another [30]) and to trust the CSP not to use their data for any purpose without permission other than to provide the service they have agreed to.
- Business continuity is becoming a critical element for adopting cloud computing (including public and private institutions as well as critical infrastructure), which means that CSPs should provide clear and transparent SLAs so that CSCs are confident that the SLA covers client workloads running in the cloud and that client data are accessible.

Cloud computing provides the data processing capabilities that are needed to enable data-driven innovation. It is essential to enable AI/ML, new IoT services in different verticals and for usage of 5G/6G cloud capabilities in terms of cloud service (e.g. cloud-native design and edge-native applications and services). Lack of interoperability between cloud providers creates risks of vendor lock-in, undermining user confidence and cloud adoption. Countries at the regional and international level should agree to work together toward a functional cloud federation and cross-border roaming (e.g. required for mobile IoT services, such as V2X, UAV, and others) to shape the next generation of secure, energy-efficient, and interoperable cloud services, based on a set of common technical rules and norms. Cloud federation (e.g. defining a common approach for the federation of cloud entities [31]) aimed at developing synergies between national and cross-border cloud regulations toward improving and expanding their scope and coverage.

Finally, cloud infrastructures and services in various forms, provided by regional or global OTT service providers or by national telecom operators (e.g. edge clouds), will have increased use toward the future as an enabling environment for the digitalization of all verticals and society as a whole all over the world. It will also bring many new business opportunities for all parties in the cloud value chain (from designers and providers of cloud services to telecommunications operators and end users, including citizens and businesses), which will require continuous support, monitoring, and adaptation of policies and regulations by national regulators (not only for the telecom/digital sectors) and governments.

References

1 Janevski, T. (April 2019). *QoS for Fixed and Mobile Ultra-Broadband*. USA: John Wiley & Sons.

2 Janevski, T. (November 2015). *Internet Technologies for Fixed and Mobile Networks*. USA: Artech House.

3 Janevski, T. (April 2014). *NGN Architectures, Protocols and Services*. UK: John Wiley & Sons.

4 ITU-T Rec. Y.3500, "Cloud computing – Overview and vocabulary", August 2014.

5 ITU-T Rec. Y.3511, "Framework of inter-cloud computing", March 2014.

6 ITU-T Rec. Y.3515, "Functional architecture of Network as a Service", July 2017.

7 ITU-T Rec. Y.3531, "Functional requirements for machine learning as a service", September 2020.

8 ITU-T Rec. Y.3530, "Functional requirements for blockchain as a service", September 2020.

9 Docker documentation, "Docker overview", 2019.

10 Google Kubernetes Engine (GKE), https://cloud.google.com/kubernetes-engine, accessed in September 2023.

11 Amazon Elastic Container Service, https://aws.amazon.com/ecs/, accessed in September 2023.

12 Microsofot Azure Container services, https://learn.microsoft.com/en-us/azure/containers/, accessed in September 2023.

13 ITU-T Rec. Y.3532, "Functional requirements of Platform as a Service for cloud native applications", May 2023.

14 ITU-T Rec. Y.3508, "Overview and high-level requirements of distributed cloud", August 2019.

15 ETSI GS MEC 003 V3.1.1, "Multi-access Edge Computing (MEC); Framework and Reference Architecture", March 2023.

16 ITU-T Rec. Y.3526, "Functional requirements of edge cloud management", November 2021.

17 GSMA, "Operator Platform Telco Edge Requirements, Version 4.0", March 2023.

18 ITU-T Rec. Y.3537, "Functional requirements of a cloud service partner for multi-cloud", September 2022.

19 Microsoft Azure, "Azure Quantum cloud service", https://azure.microsoft.com/en-us/products/quantum, accessed in September 2023.

20 Google, "AI and machine learning products", https://cloud.google.com/products/ai, accessed in September 2023.

21 Fernando, N., Loke, S.W., and Rahayu, W. (2013). Mobile Cloud Computing: a Survey. *Future Generation Computer Systems* 29.

22 Lei, L. et al. (2013). *Challenges on Wireless Heterogeneous Networks for Mobile Cloud Computing*. IEEE Wireless Communication, June.

23 ETSI, "Multi-access Edge Computing (MEC)", https://www.etsi.org/technologies/multi-access-edge-computing, accessed in September 2023.

24 3GPP TS 23.548 V18.2.0, "5G System Enhancements for Edge Computing; Stage 2 (Release 18)", June 2023.

25 3GPP TS 23.558 V18.3.0, "Architecture for enabling Edge Applications; (Release 18)", June 2023.

26 3GPP TS 28.538 V18.3.0, "Edge Computing Management (ECM) (Release 18)", June 2023.

27 ITU-D Report, Study period 2018-2021, "Emerging technologies, including cloud computing, m-services and OTTs: Challenges and opportunities, economic and policy impact for developing countries", 2021.

28 NGMN, "Cloud Native Enabling Future Telco Platforms", May 2021.

29 European Commission, "Shaping Europe's digital future - Cloud Computing", https://digital-strategy.ec.europa.eu/en/policies/cloud-computing, accessed in September 2023.

30 Official Journal of the European Union, Regulation (EU) 2018/1807, "Framework for the free flow of non-personal data in the European Union", 14 November 2018.

31 European Union, "Declaration: Building the next generation cloud for businesses and the public sector in the EU", 15 October 2020.

8

Future Fixed and Mobile Services

The telecom world has converged on IP networks and services in this 21st century. This has had an impact on all networks, terrestrial (fixed and mobile access networks) and non-terrestrial. Legacy telecom services, such as voice and television, evolve over time, with the change of networking technologies (e.g. from circuit switching to packet switching, where the latter appeared to be Internet/IP networking) and the broadband access (it has an impact on TV and video services). Services are provided either as telecom services with guaranteed quality of service (QoS) [1] or as Over The Top (OTT) services over the open Internet, based on network neutral principles and best effort approach in transferring the IP traffic through the networks [2]. Also, new services emerged with the convergence of telecom networks and services (that is, a single telecom network is used for almost all services to end users, including residential users and businesses), such as Internet of Things (IoT) services which have a plethora of use cases in many sectors, including natively non-ICT verticals. This chapter covers the future development of legacy and emerging telecom and OTT services, and their impact on citizens, business, and society as a whole through the process of digitalization.

8.1 Future Telecom and OTT Voice Services

The voice services have transited to voice over IP (VoIP) in telecom networks with the application of Next Generation Network (NGN) framework on telecom voice services [3]. The transfer was initially realized in fixed telecom network by the introduction of IP Multimedia Subsystem (IMS) for signaling, based on protocols from IETF, Session Imitation Protocol (SIP) and Diameter (for communication with users' databases in telecom operator's core network) [2]. In fact, SIP replaced the Signaling System 7 (SS7) which was used in the 20th century for signaling in telecom networks based on circuit switching (one should note that SS7 was based on packet switching, although it was not based on Internet technologies at that time, however, it opened the door for packet switching in telecom operator's networks). The VoIP in 3GPP mobile networks was standardized for the first time (as the main voice service by mobile operators) in 4G LTE networks, which were defined as all-IP networks in ITU umbrella specification for 4G, called IMT-Advanced [2, 3]. The services were based on using IMS for signaling, where common IMS is also standardized by 3GPP in the sale release which introduced the LTE (that was 3GPP Release 8). So, the mobile VoIP was introduced in mobile networks as Voice over LTE (VoLTE). In a similar manner, although with different

Future Fixed and Mobile Broadband Internet, Clouds, and IoT/AI, First Edition. Toni Janevski.

protocol stack (in parts) was defined Voice over New Radio (VoNR) as the carrier grade voice service over 5G mobile networks.

Why do voice services continue to live on both landline and mobile networks? Well, that is because, unlike machines that can only use digits to communicate, we (as humans) use voice communication to talk.

When voice goes over IP networks (almost all telecommunication networks have become IP-based), then we call it VoIP. However, in terms of QoS, there are two main types of VoIP services, which include QoS-enabled VoIP services (also known as carrier grade VoIP) and OTT voice services.

QoS-enabled VoIP services are actually the replacement of legacy digital telephony (i.e. circuit-switched) by telecommunications operators (including landline and mobile) with VoIP with guaranteed end-to-end QoS. On the other hand, OTT VoIP services, such as Skype, Viber, WhatsApp, and others, are provided as online services over network-neutral Internet access (whether via fixed, mobile, or satellite access), without any QoS guarantees per flow or type of traffic, with the same treatment as all other Internet traffic.

What is fundamental about VoIP services? Well, to provide voice services in all-IP environments, the fundamental part is the signaling (belonging to the control plane). As has been the practice globally for legacy telephony for the past century, signaling should be standardized globally because voice services are global (in order to be vendor-independent and telecom interoperability). That is true for voice services from telecom operators, which are declining in fixed networks since 2005 [3], while such services are already saturated in mobile networks (as of the 2020s), however, they are not based on standards.

How is standardized VoIP signaling provided in the telecom operators' environment? This was achieved by the standardization of the NGN standards umbrella by the ITU. It defines end-to-end QoS support for voice using standardized architectures (from ITU-T) and protocols (from IETF). The standardized signaling system (mainly for VoIP) is IMS, standardized by 3GPP in its common form.

IMS is access-independent, which means that it can be used in various IP-based fixed and mobile networks. So, it is used in 4G mobile networks (where VoIP is used for the first time as the main carrier-grade voice service by default for mobile operators), and it is also used in 5G networks, for VoNR and other services that need signaling.

What was initially the problem in the 2010s to go after stronger VoLTE penetration? The problem was the user devices and the equipment of the mobile operator. In the initial rollout period, when most devices do not have VoLTE support on their mobile devices, when connected to a 4G network for voice calls, the network does a circuit switched fallback (CSFB). It is the approach used in many 4G mobile networks as a result of the lack of VoLTE network capabilities of smartphones.

8.1.1 Voice Over NR (VoNR) in 5G

The 3GPP has specified that 5G uses the 4G voice/video communication architecture and still provides voice/video communication services based on the IMS in 5G systems. When the 4G radio access technology is LTE and the voice/video over the LTE network is called VoLTE, in a similar manner, 5G radio access technology is NR (New Radio), so voice over the 5G network is called Voice over New Radio (VoNR) or Voice over 5G (Vo5G).

One may note that VoLTE and VoNR are different access modes for IMS voice/video communication services, which (will) exist in parallel in the late 2020s, while at the beginning of 2020s the focus was still on the VoLTE which had started its global penetration in the second half of the 2010s, close to the end of the 4G era. But that means that most of the 5G era (i.e. the 2020s) is, in fact, VoLTE era regarding the mobile operators' voice services.

In addition, with the deployment of 5G, 5G NR, and 4G LTE network coverage overlaps, EPS Fall-Back (EPS FB) can initially be selected for voice/video communication services. The backup EPS FB for 5G to 4G for voice is similar (as access) to CSFB from 4G (LTE) to 3G (circuit switched part).

In fact, in the first period (regarding the 5G), it is most likely that the dual registration approach will be the favorite one. Why? Because at the initial stage of 5G deployment, due to overlapping of 5G NR and 4G LTE network coverage, mobile operators can select the dual registration solution which means that a mobile device (e.g. a smartphone) can access the Evolved Packet System (EPS) in 4G) and 5G System (5GS) at the same time (i.e. simultaneously). Then, the voice/video communication service is carried using VoLTE in the EPS, and data services can be carried either over the NR/5GS (as 5G access/core networks) or LTE/EPC (as 4G access/core networks) based on the radio access coverage (is it 4G or 5G at a given location).

In 5G deployments in StandAlone (SA) mode, with 5G Core, 4G base stations will be connected to it besides the 5G base stations (gNodeBs). So, eLTE eNBs (i.e. 4G base stations connected to 5G Core) will continue to exist, and LTE and NR coverage will continue to overlap in the 2020s and later in the 2030s (the 6G era). In such cases, Radio Access Technology (RAT) fallback between LTE and NR can be used for conversational voice/video services.

Figure 8.1 shows the protocol stack for VoNR, including the transfer of voice data and signaling traffic. Let's compare VoLTE and VoNR protocol stacks. In both cases, SIP is used for signaling, although HTTP (HTTP 2.0, standardized in 2015, and also in future can be used HTTP 3.0, standardized in 2022) also is an option for signaling in 5G networks. So, signaling is either SIP/TCP/IP or HTTP 2.0 over TCP/IP (the latter for VoNR). In the future, it may be HTTP 3.0 over QUIC/UDP/

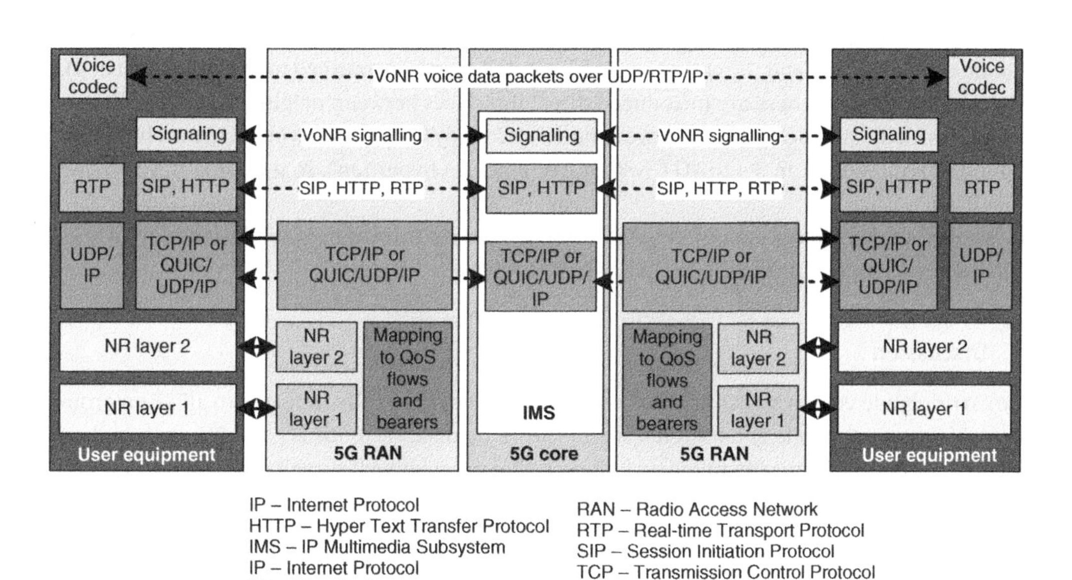

Figure 8.1 VoNR (Voice over New Radio) protocol stack.

IP, considering that HTTP 3.0 works over QUIC (it is a transport layer protocol standardized by the IETF in 2021) which in turn works over UDP/IP.

On the other side, as typical for carrier grade voice, the voice data is based on RTP/UDP/IP protocol stack (RTP stands for Real-time Transport Protocol, also an IETF standard) [2], due to UDP being less sensitive to delays than TCP, and the human ear being error prone up to several percents of error ratio [1]. And, of course, signaling flows and voice flows are logically separated end to end, although they may travel through the same access, core, and transport networks.

Similar to OTT voice services (e.g. Skype, Viber, and WhatsApp), mobile operators will be able to combine carrier grade VoIP services with other services (such as messaging and data services), all with guaranteed QoS. Also, with 5G network slicing, there will be the possibility of introducing new competitive mobile virtual network operators (MVNOs), which can provide various offers for voice services by "purchasing" network slices from the 5G operators. Thus, many future options exist for voice services in mobile networks.

The QoS requirements are important for voice services in all networks and at all times, today and in the future. Before 5G (and LTE-Advanced-Pro from 3GPP 4G family of standards) mobile voice was the service that had the most stringent performance requirements, especially in terms of delay (i.e. latency). It is well known in the telecommunication world that for excellent voice communication, one-way latency should be less than 150 ms but never more than 400 ms [3]. When the latency is above 400 ms (in one-way) people start interrupting each other and then it wouldn't really be voice communication. Based on this, we cannot have real-time communication from Earth to our friends (in the future) on the moon, because the distance between the Earth and the Moon is approximately 300000 km and signals travel at a maximum speed of 300000 km/s. That means from the Earth to the Moon we will have a one-way delay of over 1 second, that is, 1000 ms and that is not acceptable for real-time voice communication. So, with the current state of the art of technology science, the voice services can only be used on our planet.

Also, voice is intolerant of delays in handover of an ongoing voice call (handover occurs when the mobile user is moving and the call needs to be handed over from one cell to another target cell). For low handover delays, there are introduced direct interfaces between neighboring base stations in both 4G (initially) and 5G. Also, for lower end-to-end delays, the 3GPP evolved the architecture to a flat one on the way from 3G to 4G (System Architecture Evolution), so we could have VoLTE and VoNR.

But, in terms of performance requirements, voice has lost its 'throne' in 5G and beyond networks to critical services such as critical IoT (e.g. URLLC services).

8.1.2 Discussion

Mobile carrier grade voice services are the last legacy services to be fully transited to all-IP environments (4G is the first mobile generation which is all-IP by default). Both 2G and 3G are based on 20th century circuit-switching technology; however, that provides well-established roaming among operators from different countries (since the 1990s), and therefore it is not easy to switch it off. The first standardized mobile voice service in an all-IP environment is VoLTE which is provided over 4G LTE. However, the transition from 3G CS (Circuit-Switched) voice to VoLTE is still ongoing although it started years ago. So, VoLTE will be the major mobile carrier-grade voice service during the first part of the 5G era (e.g. until 2025), while VoNR will expand on the mobile "scene" toward 2030 and beyond.

8.2 Future TV/IPTV, Video, and XR/AR/VR Services

The most used services in both managed IP networks (from telecom operators) and OTT services is video; it is the "king" of the Internet and generally all IP networks. It is noted in many places that video contributes to over 70% in the total traffic on the Internet and all other IP networks (including here also managed/private IP networks that transfer IPTV). The video services can consume the bandwidth with more video content being offered (generated from many sides, including the end users as video creators, which generates traffic in upstream) and with higher video resolutions that are available on the Internet (e.g. with access speeds of 50 Mbit/s or more one can watch 4k video streaming over the public Internet access [1]). In the future, the resolutions will increase further at certain time intervals so that we will have 8K, then 16K (in the longer run), and so on over several decades. As one may expect, the Video on Demand (VoD) and IPTV services from telecom operators have a higher part of the total video traffic in fixed access network (including FWA and satellite broadband access), while online video sharing platforms and social media videos (e.g. YouTube, Facebook, Instagram, and TikTok) have a larger share in mobile access networks. With fixed-mobile convergence (FMC) in 5G and beyond mobile networks, the same services (including TV provisioning from telecom operators) can be provided through both fixed and mobile access networks.

Provisioning of TV services over managed IP networks of telecom operators is referred here also as IPTV (in that sense IPTV is not an online OTT service, while web-based or app-based online TV streaming is referred to as online TV services).

8.2.1 Scope of TV/IPTV Services

The scope of video services includes many different services, applications, and technologies, which can be split into two basic forms [4]:

1) Linear services: The television service provider distributes audiovisual content based on a given TV schedule, mainly continuous distribution each day in the week.
2) Non-linear services, also known as VoD: In this case, the end user determines the time to watch certain video content by selecting it from a library of structured video contents. These non-linear services also include playback of TV linear content after their initial broadcast.

Linear and VoD services can both be offered with different resolutions (e.g. Standard Definition – SD, High Definition – HD, and Ultra HD) and in two main forms:

1) Video services with managed QoS: This is typically a form initially used by fixed telecom operators, and with higher throughputs and capacity in mobile network it also becomes a service offer by mobile operators.
2) OTT video services over public Internet (no QoS guarantees): In this case, the video content providers provide services as OTT by using network neutrality on the public Internet. In this case, video picture quality and service availability are not guaranteed, because they are dependent upon the availability of broadband (fixed, mobile, or satellite) access to Internet. Such OTT video services can be linear (e.g. live streaming and web-based streaming of TV services as an additional offer by telecom operators) or nonlinear, either with free content (e.g. YouTube) or subscription-based (e.g. Netflix).

Regarding the underlying technologies, video services can be delivered over different technical platforms, which include legacy broadcast networks (e.g. Digital Video Broadcasting – DVB) and IP-based networks (e.g. managed IP networks).

The TV/IPTV platforms come with different end user equipment. In fixed, FWA, or satellite access networks, the TV/IPTV equipment is called Set-Top-Box (STB), which is typically connected to the TV set one side, and on the other side it is connected with Ethernet or Wi-Fi to the home gateway which is the terminating point of the access network (e.g. Fiber To The Home – FTTH and cable access network). When the TV content is provided online over the Internet, the range of end user equipment is much wider, including smartphones, tablets, laptops, desktop, and game consoles. In all IP-based networks, the TV delivery is multiplexed with other traffic. However, in cases where TV/IPTV is distributed with QoS guarantees, there is dedicated bandwidth per IPTV STB. Each STB is used for one TV channel streaming at a time in the case of IPTV, which is different from legacy cable systems for TV broadcast where all TV channels are broadcasted on different frequency bands on cable and the end user tunes to the desired one via the STB.

8.2.2 Future 5G and Beyond Broadcast and Multicast Services

Mobile TV offered as a service by operators began to be standardized by 3GPP through Multicast and Broadcast Service (MBS) back in 3G. But TV requires a higher bitrate for simultaneous transmission of multiple TV channels in the same network cell (to different users), which is why it did not really take off in 3G or in the first 3GPP releases for 4G, due to the really limited capacity in mobile networks and priority of mobile Internet access and voice services (in relation to mobile TV). Since 3GPP released 3GPP 13 and especially Release 14 (which address LTE Advanced Pro), major improvements have been introduced in mobile TV systems.

However, the 3GPP Release 13 evolved Multimedia Broadcast Multicast Service (eMBMS) system, when deployed in Single Frequency Network (SFN) mode, allows a maximum distance of only 10 km between adjacent transmitters which is much shorter than the transmitter distance of about 90 km in typical systems for broadcasting with Digital Terrestrial Television (DTT). This made these eMBMS systems relatively expensive for mobile TV services.

Furthermore, 3GPP Release 14 through the introduction of a longer so-called cyclic prefix of 200 μs on the radio physical level symbols (which is six times longer than the prefix in Release 13) enabled greater distance between transmitter sites for mobile TV (multicast and broadband) to increase to 60 km.

Further, in 5G standards, 3GPP is developing two main approaches for broadcast and multicast services in 5G mobile networks [5]:

1) 5G Broadcast, which is dedicated to standalone broadcasting (standardized in 3GPP Release 16), and
2) 5G Multicast and Broadcast Services (5G MBS), including both downlink and uplink, which is initially developed in 3GPP Release 17.

5G Broadcast is delivery of linear TV and radio services from one or multiple transmitters to an infinite number of receivers without the need for a subscriber identity module (SIM) card or even any uplink capabilities [6]. It permits to tune audiovisual services on the smartphone without depending on other standards and chipsets than the ones already integrated in all smartphones. In that manner, 5G Broadcast can be operated on a broadcast infrastructure and can be received without being linked to a particular operator (the signal is offered free-to-air, or better said, free-to-tune, without a contract and the corresponding SIM card). 5G Broadcast technologies rely on the LTE frequency bands on 700 MHz.

8.2.2.1 Delivery Methods of 5G Multicast Broadcast Services (5G MBS)

5G system continues the standardization of multicast and broadcast services with the general MBS, which is not limited only for mobile TV services, but also can be used for public safety, V2X application, group communications, and IoT applications. Such 5G MBS service also have two modes:

1) MBS Transport-Only Mode, in which the multicast and broadcast contents are transparent to the 3GPP network functions, and
2) MBS Full-Service Mode, in which the 3GPP network functions are aware of the contents.

From the viewpoint of 5G Core (5GC) network there are two traffic delivery methods for 5G MBS service (Figure 8.2):

1) Individual MBS traffic delivery method: In this case, the 5GC network receives a single copy of MBS data packets from a given content source and then multicasts and delivers separate copies of those MBS data packets to individual UEs via per UE Packet Data Unit (PDU) sessions.
2) Shared MBS traffic delivery method: In this case, 5GC network receives a single copy of MBS data packets and delivers a single copy of those MBS packets packet to a RAN node in the access network, which then delivers them to one or multiple UEs.

When MBS is used 5G Radio Access Network (RAN) delivers MBS data to UEs using either Point-to-Point (PTP) delivery or Point-to-Multipoint (PTM) delivery. With the PTP delivery method, 5G RAN node delivers separate copies of MBS data packet to individual UE. On the other side, in PTM delivery method, the 5G RAN node delivers a single copy of MBS data packets over radio to a given set of UEs. So, each 5G RAN node (e.g. gNodeB) may use a combination of PTP and PTM to deliver the MBS data to end users. Both methods can be used simultaneously for a multicast MBS session.

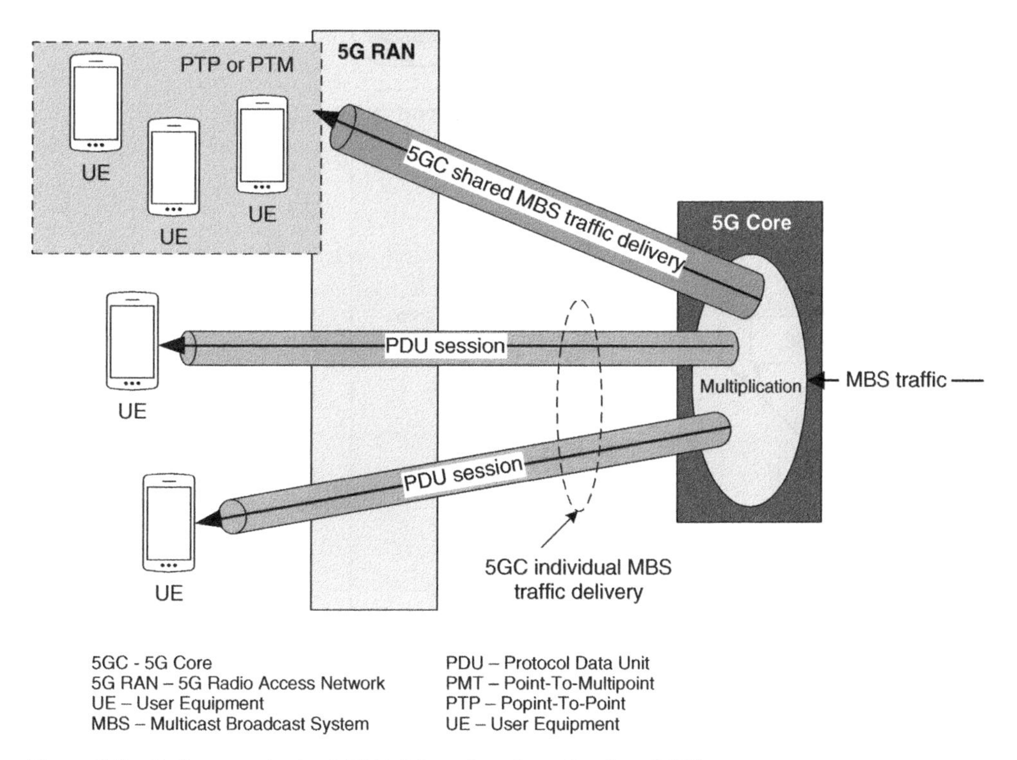

Figure 8.2 Delivery methods of 5G Multicast Broadcast Services (MBS).

The MBS in 5G is based on NR and also uses non-SFN deployments, so there are changes in the MBS architecture (when compared to MBMS in 4G). As shown in Figure 8.3, the MBS related entities are located in 5GC. There can be distinguished four entities for deployment of MBS in 5G network [7]; they are:

1) Multicast/Broadcast Session Management Function (MB-SMF): It is used for management of MBS sessions and interactions with the 5G RAN and Multicast and Broadcast User Plane Function (MB-UPF) for MBS data transport.
2) Multicast/Broadcast User Plane Function (MB-UPF): This MBS entity does QoS enforcement and delivery of multicast and broadcast toward UPF or 5G RAN nodes.
3) Multicast/Broadcast Service Function (MBSF): It provides service level functionality for MBS, that is, interaction with Application Function (AF) and MB-SMF for the MBS sessions.
4) Multicast/Broadcast Service Transport Function (MBSTF): This function is media anchor for MBS data traffic by provision of transport functionalities.

For the purpose of 5G NR MBS QoS enforcement, 5GC network functions, including the Access and Mobility Management Function (AMF), SMF, UPF, or NG-RAN, perform classification of UEs

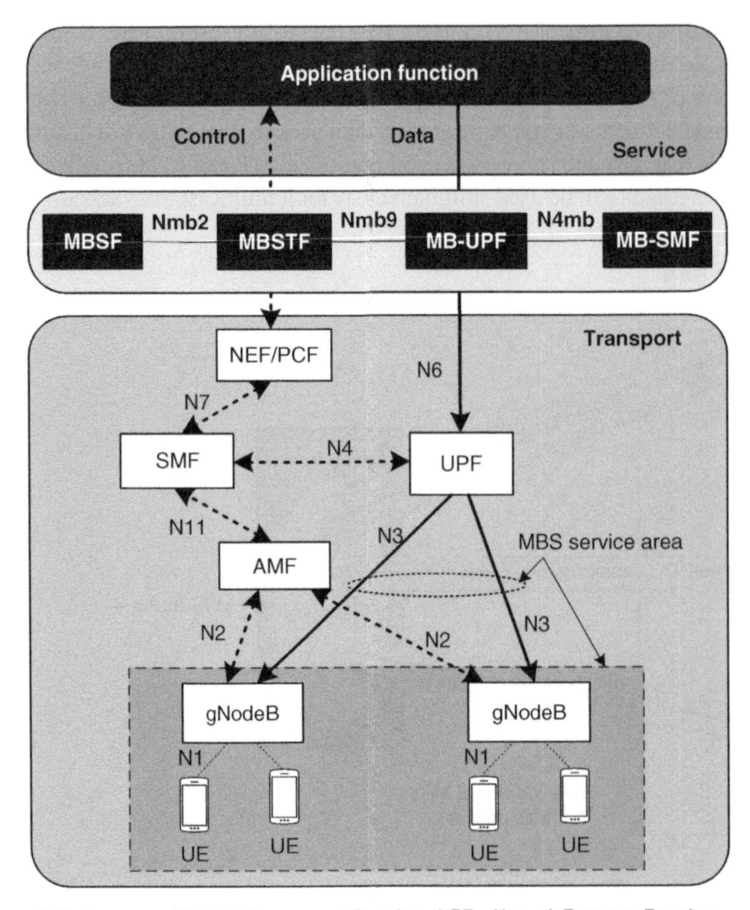

AMF - Access and Mobility Management Function
PCF - Policy Control Function
MBS – Multicast and Broadcast Service
NRF - Network Repository Function
NEF – Network Exposure Function
SMF - Session Management Function
UE – User Equipment
UPF – User Plane Function

Figure 8.3 5G MBS architecture.

within the MBS service area based on the association of UEs with MBS flows or QoS profiles. In the case of 5G MBS, the UPF performs QoS enforcement while the SMF does the mapping of Service Data Flows (SDFs) to MBS QoS flows. When QoS application is done by NG-RAN, it can associate UEs with the MBS QoS flows based on the QoS profiles provided by the UPF in the 5GC.

8.2.3 eXtended Reality (XR) Services in 5G and Beyond Mobile Networks

The main approach regarding the media delivery, such as media streaming services in 5G is the provision of flexibility (for different scenarios) and quality (in certain cases, the best effort approach used in open Internet simply is not good enough). In 5G mobile systems the so-called 5G media streaming (5GMS) is defined. That is a completely new standardized architecture by 3GPP for media streaming over 5G networks. The 5GMS provides several different scenarios for different types of collaborations.

What types of collaborations are possible for 5GMS? Well, as all other components in 5G network, the 5GMS is based on the use of 5G network functions. It uses 5GMS AF and 5GMS Application Server (AS) as well as other functions in the 5GC (e.g. policy management functions). Then, different collaboration scenarios are provided with the positioning of the application and server functions – they may reside in the mobile operator's data network (called trusted network in such a case) or in the external data network. So, with 5GMS, the mobile operator can provide full streaming service "in house" (i.e. own network), or can host the content of a third party, or can be only an enabler of streaming services which are hosted externally. Of course, 5GMS provides the possibility for the content to be hosted inside the operator's network which may be very important for some XR services sensitive to delays.

Are there any prerequisites required at the end user's side? Well, there is a requirement for the UE to have 5GMS client built in the device. So, it depends also on the devices, whether they have 5GMS client or not [8]. If not, they will not have the possibility to use standardized 5GMS services, and vice versa.

8.2.3.1 Different Realities of VR, AR, MR, and XR

Different types of streaming services can be provided through 5GMS, including VR, AR, mixed reality (MR), or all in one called XR. Let's define AR, VR, MR, and XR in short.

VR is a rendered version of a delivered visual and audio scene that mimics the visual and audio sensory stimuli of the real world in a way that should appear as natural as possible to the observer or user. VR content usually, though not always necessarily, requires the user to wear some form of head-mounted display (HMD) with headphones, which usually completely replaces the user's field of vision with a simulated audiovisual component. In doing so, VR applications track the movement of the user's head which results in updating the simulated visual and audio components from the user's perspective.

AR is the ability to mix real-time digital content that is appropriately positioned in the 4D image of the real world that surrounds the user. Thus, unlike VR, with AR we actually have an augmented reality that does not block our environment, but complements it with digital content, with the aim of improving presence in the real world, whether for work, entertainment, or any other purpose of a given service. AR technologies are those that are expected to play a significant role in the development of many new services in different verticals, such as the digitalization of industry and the successful deployment and adoption of smart city services by citizens. Other examples of AR are retail (AR shopping in real stores), healthcare (user health data while being physically examined by a doctor), education (enhancement of lectures with digital content added to the real environment in classrooms or laboratories), public safety, and tourism (AR guides through tourist places and attractions). In doing so, AR can be combined with various sensors and other IoT devices, laptops, IoT data, and AI.

MR, also known as hybrid reality, represents the merging of the real and virtual worlds in order to create environments where physical and digital objects coexist and can interact in real time. For example, it means placing new images in real space that can react to some degree with images in the real physical world. In this way, MR can be seen as a combination of VR and AR contents in the same service or application.

According to the 3GPP [9], XR refers to all real-and-virtual combined environments and human-machine interactions generated by computer technology and wearables. The XR content includes different forms of realities such as AR, MR, and VR as well as their various combinations. In XR, the levels of virtuality may range from partial inputs from various sensors to fully immersive VR. The main target of XR services is the extension of human experiences relating to the human senses of existence (represented by VR) and also the acquisition of cognition (represented by AR).

In XR, the user interacts with the digital content using movements, gestures, and body reactions. To define the movements in XR content, so called degrees of freedom (DoS) are defined in 3D space. Typically, there are several DoS, which include the following:

- 3DoF includes three rotational and unlimited movements around the *X*, *Y*, and *Z* axes (respectively pitch, yaw, and roll). Typical use case is 3D 360 VR content on an HMD.
- 3DoF+ includes additional limited translational movements such as head movement along *X*, *Y*, and *Z* axes. (e.g. sitting and looking at 3D 360 VR content on an HMD).
- 6DoF is DoF that is based on 3DoF experience with added moving up and down, moving left and right, and moving forward and backward (e.g. free walking in 3D 360 VR content).
- Constrained 6DoF is 6DoF with constrained translational movements along all three axes (e.g. freely walking in VR environment on an HMD within a defined walking area).

In general, various digital technologies continuously transform the media landscape. Looking into the future, it is not enough to just have very high resolution video content and, as the new type of content is coming with VR/AR (and MR, based on combinations of VR and AR content) or, in other words, XR – to mark all various realities with a single term.

Such advances in multimedia are changing the world, especially the entertainment industry which consumes much of the broadband traffic in every network. However, richer communication also brings improvements (for example) in medical care and education, besides the richer communication between friends and families.

8.2.3.2 XR Architecture

Many smartphone-based XR applications are possible. Cheaper end user equipment includes smartphone VR headsets which provide limited 3DoF tracking. Higher computation processing capabilities can be achieved with tethered XR headsets such as AR glasses tethered to a smartphone or a VR headset tethered to a desktop or laptop, where better user experiences regarding media quality can be achieved, but that is accompanied with tethering wires (for example, Universal Serial Bus – USB), which may be inconvenient. On the other side, wireless tethering can be done with Bluetooth, Wi-Fi and 5G Sidelink. The desired use cases in mobile networks are tailored to standalone XR headsets, which offer the convenience of no dependency on another device. So, one may say that it is a long-term future direction of the market for XR devices and wearables.

For XR services, there is a trade-off between convenience of using XR services and connection types for both standalone and tethered XR headsets. On one side, USB tethering offers high data throughputs and can power the tethered XR headset, but it has the inconvenience of the wires.

On the other side, Bluetooth offers a lower power wireless connection, but data throughputs are limited. Then, Wi-Fi and 5G Sidelink are technologies that are expected to provide trade-offs between the data throughput and the power consumption.

For XR service provisioning, there are required XR engines as a middleware that abstract hardware and software functionalities from the XR application, which is needed for developers. However, such engines are mainly proprietary solutions. The future goal is the creation of standardized abstraction layers and Application Programming Interfaces (APIs) for XR (e.g. OpenXR and W3C's WebXR) [9].

The integration of XR applications within the 5G system is approached following the model of 5GMS [8]. A given 5G-XR application provider will make use of 5G functionalities for XR services. On the side of UE, there is required a 5G-XR client for the application to be able to use network interfaces and APIs.

The 5G XR architecture given in Figure 8.4 includes the following three main functions:

1) 5G-XR AF: It is the application function defined in 5GMS, which is dedicated to 5G-XR services.
2) 5G-XR AS: It is dedicated to 5G-XR services.
3) 5G-XR Client: It is a UE internal function dedicated to 5G-XR services.

The XR application provider may be a different business entity than the telecom operator that has the mobile network which is used for provision of 5G XR services. The external Data Network (DN) used by 5G-XR application provider is the DN trusted by the telecom operator. In such a case, AFs in trusted DNs may directly communicate with all 5GC functions. But, the main point for communication with such external DNs with 5GC functions is the Network Exposure Function (NEF) via the interface N33.

Normally, for provision of XR services, there is required certain QoS support by the 5G system. Values of 5QI potentially relevant for XR applications include 1, 2, 3, 4, 7, 8, and 80 (of course, different 5QIs for different purposes in 5G XR applications). One may note that 5QI = 80 is in fact introduced for XR applications to go over extended Mobile Broadband (eMBB, in particular for the AR part of the XR), considering the defined delay budget (for 5QI = 80) of up to 10 ms and the

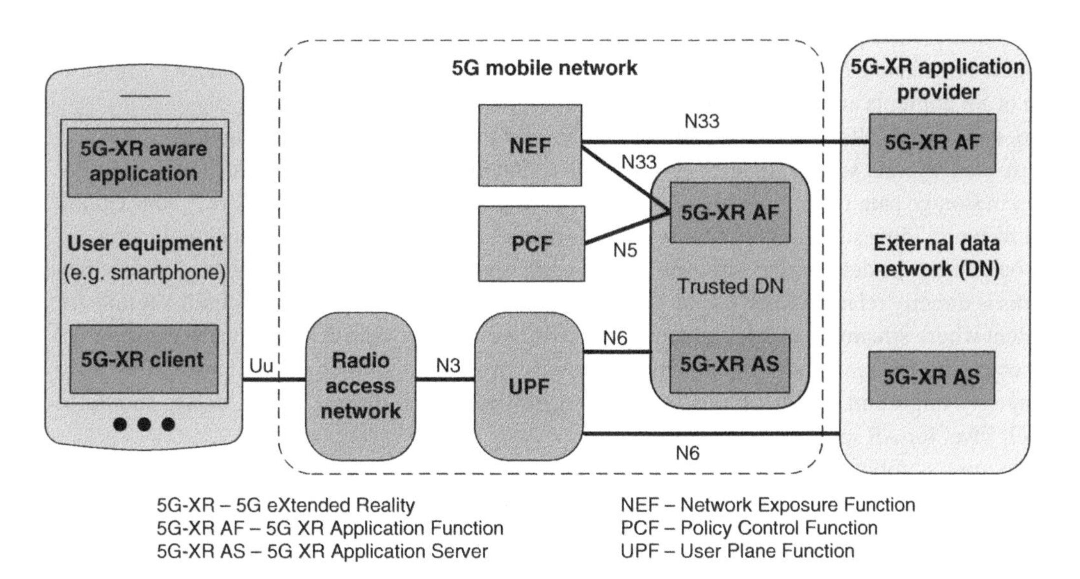

Figure 8.4 5G-XR architecture.

requirement on packet loss rate to be less than 10^{-6}. So, in this way, XR can also be provided via eMBB network slices in a 5G network.

Additionally, XR applications may use (or need for their operations) edge computing. Such edge resources or platforms should be able to be discovered by using legacy mechanisms (e.g. DNS) or something new. However, for using the edge resources, the edge computing should be first integrated in the 5G system of the telecom operator.

One use case for mobile users in the future will be XR content sharing. Thus, to share real-time XR between two mobile devices A and B, where one device is capable of XR rendering (for example, device B) using AR glasses or with a mobile phone sending a real-time video stream of the XR experience to the other user (user A). The rendering experience includes real-time placement of 3D objects in a 2D scene and avatars, which can be sent by user A to user B or downloaded from the cloud (which is used for XR). Between the two users A and B, a two-way or one-way audio-space channel can be used, depending on the use case.

In general, XR mobile services in the future will be highly dependent on the availability XR-capable devices. For XR applications over 5G/5G-Advanced networks, they would be connected via a mobile device (a smartphone) via a local radio connection with a very high capacity such as 802.11ay Wi-Fi (which operates at 60 GHz and has the required capacity of tens of Gbit/s). Of course, the battery of the smartphone should be able to support the intensity of XR communication.

8.2.3.3 The Future of XR Services

What are the possible services with XR? Well, they are different in different environments. For example, enterprises may use them to increase the efficiency and productivity. That can be accomplished via XR applications such as virtual trainings, then augmentation of manufacturing (e.g. augmented workers) or richer and enhanced collaboration on a distance. Additionally, XR can be used also for professional applications including real estate, tourist guides, and even psychological counseling.

What about XR in the consumer domain? Well, XR in the consumer domain has more focus on entertainment, where the main adoption can be given to gaming and 360° panoramic video content. But one should not forget the social aspects with use of XR, which can potentially help people to lighten the feeling of isolation by using the VR. But, for XR to be used, there is a need for production of XR contents and development of XR ecosystems.

So, end-to-end XR ecosystems and delivery to end users are crucial for establishing XR as a future mass market service. In such process, one of the most important moves is changing the XR content storage pattern by moving it to a cloud environment close to the network (e.g. edge cloud), and then providing streaming to the end user devices (from the cloud), thus reducing the processing burden on the devices (e.g. smartphones). So, the success of XR in 5G and beyond mobile networks is directly related to the use of clouds. Regarding the delay, XR includes both VR and AR content where streaming of VR content is less sensitive to delay than AR systems, because in AR the objects in the real world are supplemented with artificial digital objects, and that makes the delay very important and AR typically has stringent latency requirements. So, delays should be much fewer for AR services than for VR.

The most significant factor for XR success is the use of clouds (e.g. mobile edge clouds). With clouds near the XR clients, there is a transfer of the processing capability from devices to the cloud. And with high-speed and low-latency ultra-broadband network such as in 5G and beyond, it is possible to have interactive feedback and real-time delivery of the display content from the XR cloud.

Who has higher promises in the 5G era, VR or AR? Well, the overall VR ecosystem is more mature (by the 2020s). But, in the long run, the AR ecosystem is the one that may provide more opportunities, especially in later phases of 5G development and in the 6G era (the 2030s), e.g. with included URLLC network slices in most of the networks.

The XR services can be provided either via the network neutral broadband Internet access in 5G or via dedicated (managed) network slices in 5G which will not be part of the public Internet. For example, XR gaming can use eMBB, while an augmented worker (based on AR) may use private URLLC network slice in 5G/6G (e.g. provided to the given factory).

8.3 Telecom and OTT Massive IoT Services

Massive IoT refers to IoT services that use low-cost devices and low-cost Internet connectivity. What is massive IoT in device density? Well, one can refer to the massive IoT requirements set for 5G by the IMT-2020 umbrella specification from the ITU, in which massive machine type communication (mMTC) actually refers to massive IoT services [1]. We have already classified IoT services based on their performance requirements, the most important of which are latency and reliability requirements. So, in short, IoT services that do not have strict QoS requirements and have massive deployments (many low-power IoT devices deployed) are called massive IoT services. Then, in 5G terminology, massive IoT services are called mMTC. However, massive IoT is also provided through broadband or narrowband access (when low bandwidths are required for communication to/from IoT devices), which is not only limited to mobile networks, but they are also provided in fixed broadband and satellite networks.

What is the difference in terms of MTC/IoT in 5G compared to previous mobile generations? Well, MTC/IoT solutions in 5G are aimed at many IoT devices being served in a single cell. The requirement for 5G specified in IMT-2020 (by ITU) was support for 1 million devices per square kilometer (although the same density requirement in IMT-Advanced was 100000 IoT devices per square kilometer, which is also a large number, but 5G is increasing it about 10 times. That is why we say it is massive MTC (i.e. massive IoT), not just MTC (or IoT) like before. As covered in Chapter 5 of this book, the upcoming 6G is expected to increase the density of massive IoT devices up to 100 million per square kilometer.

So, massive MTC includes IoT services based on a huge (massive) number of deployed devices with low cost, low power consumption, small volume of data generated or transmitted, and massive deployments. This group includes capillary sensor networks, smart agriculture, smart metering, fleet monitoring and management, smart buildings, and smart cities. However, massive IoT does not have strict QoS requirements, i.e. they are more tolerant of packet losses, delays, and require lower bitrates. These services may also be provided over the Internet on a best-effort basis, including landline, mobile, and satellite access networks. Considering the network slicing approach in 5G and beyond cellular networks, massive IoT services can be provided through eMBB or mMTC network parts.

In countries that have a wide variety of marine life, environmental monitoring is considered a very important use case for massive IoT (mIoT, interchangeably used here with mMTC), including agriculture (in terms of environmental monitoring). In all these areas, the 5G network with massive MTC/IoT is expected to boost things going forward, supported by the ubiquitous coverage of mobile networks, their standardization and deployment globally (the same 5G standards are implemented in all countries of the world, so roaming of IoT becomes possible in the future) and extremely high density for connected devices supported by 5G.

Furthermore, mMTC with 5G and wider mobile networks (including terrestrial networks – TNs and non-terrestrial networks – NTNs) are intended to connect many new "things" on the network, taking into account the global wide coverage of 5G and the standardized approach for IoT, which was one of the segments that were missing in the past for mass deployments.

Massive IoT services can be based on their implementation by telecom operators (e.g. fleet tracking) and OTT service providers (e.g. home surveillance cameras). Telecom mIoT services can be provided over private network segments or over the public Internet. Massive OTT IoT services are typically provided over open Internet access.

We call everything connected to the Internet smart. So, any appliance, device, vehicle, or machine, which is connected to the open internet (for any purpose, including configuration or measurement) is called a smart device. However, not only is the device smart, if we have a lot of smart devices in our home, it is called a smart home (although there is no strict requirement for how many smart devices make the home smart). We can go to bigger things, like a smart building or a smart city.

What are the new mMTC services for telecom operators in 5G going toward 5G-Advanced? Well, that is RedCap (Reduced Capabilities), as explained in Chapter 6 of this book. Why is it referred to as RedCap?

The standardized cellular IoT approach in 3GPP standards started with LTE Advanced Pro standardization, when NB-IoT (NarrowBand) and enhanced MTC (eMTC) were introduced (we covered this in Chapter 6). These services continue to be standardized and used also in 5G mobile networks. However, they are not 5G-native, which means they do not use 5G NR interface. So, 5G needed massive IoT technology which will be NR-native, and can be convenient for low-end IoT devices (i.e. massive IoT). So, RedCap was introduced in 3GPP release 17, to expand in the 5G-Advanced era, to fill up the empty area between very light IoT on one side (provided with NB-IoT or LTE Cat-1 and LTE Cat-4) and very high demanding IoT on the other side, such as critical IoT, provided via URLLC services over 5G. So, some kind of 5G mid-tier services were needed, and so RedCap was created. It is expected that this will expand the 5G-native services in this mid-tier IoT arena, which can be used for many consumers as well as non-critical industrial IoT services (e.g. sensing and metering).

8.3.1 Massive IoT Ecosystem and Interoperability

The primary element for IoT services is connectivity. It consists of infrastructure, spectrum (for wireless connectivity) and devices [10]. For massive IoT technology expansion, it is required to have a certain infrastructure for coverage (e.g. for mobile IoT services) and connectivity.

Massive IoT services are intended for different verticals, such as smart transportation, traffic management, logistics, and smart vehicles, and it is necessary that the infrastructure supports different IoT environments composed of devices, sensors, and user terminals with different requirements, which will be mutually connected to provide massive IoT services. Therefore, it is essential for the real implementation of massive IoT services (which can be of different heterogeneous purposes) to take into account the necessary expansion of the telecommunication infrastructure, including coverage, connectivity, capacity, and high density of connection of devices. For example, a plan for massive IoT at the national level can or should focus on fixed optical access networks, 5G and further mobile networks as well as satellite networks in countries that cannot ensure the coverage of the entire territory with telecommunication terrestrial infrastructure. For massive IoT services that require connectivity, of particular importance is spectrum and its availability in a given country for a given IoT connectivity, which may include licensed spectrum (e.g. 5G spectrum used for mobile IoT services) or unlicensed spectrum.

There can be defined different unlicensed spectrum portions which are needed for massive IoT technologies and applications, or vice versa – massive IoT services should be deployed in existing and regionally or globally harmonized frequency spectrum, for both terrestrial and non-terrestrial networks [11]. Identifying the frequency range in a given country for a specific IoT application should take into account several aspects that naturally include licensed or unlicensed use of the above bands, according to their intended use such as IoT equipment cost, network coverage, reliability, and resilience. Table 8.1 presents several examples for spectrum allocations for terrestrial access networks for massive IoT services [11].

The huge number of devices required for massive IoT services can bring certain risks to the network, because the devices are of low cost and performance and accordingly cannot have the same level of protection and security as more powerful user devices (for example, computers). Therefore, IoT devices should be based on internationally recognized technical standards or best known practices for IoT implementation. For this reason, regulators and legislators in each country need to prescribe conformity assessment approaches to ensure that IoT device manufacturers implement the necessary technical standards (and harmonization of such regulations is also needed in order for IoT device manufacturers to have a larger market such as regional or global market). In doing so, the conformity assessment should not impose additional delays for the deployment of new IoT services that normally require IoT devices (e.g. self-assessment of devices by the manufacturer that may be accepted as satisfactory by regulators and policy makers for IoT services, if they meet the set requirements for the same).

Due to the high heterogeneous nature of IoT since the beginning of the 21st century, including technologies, devices, and services, the interoperability is a continuous IoT issue and challenge. That refers to both massive and critical IoT. According to Figure 8.5, interoperability in the IoT ecosystem can be classified as device, networking, syntactic, semantic, and platform interoperability. Devices by their resources and capabilities can be classified as constrained (with limitations) or non-constrained (without strict limitations). Further, network interoperability is required for exchange of messages related to IoT services for end-to-end communication (e.g. between IoT devices and the IoT platform). For the understanding of such messages there is also needed

Table 8.1 Selected examples for spectrum allocations for terrestrial IoT services.

Frequency range	Applications and/or standards
169.4–169.8125 MHz	Meter reading, data acquisition
433.05–434.79 MHz	Short Range Devices (SRD)
410 MHz 450 MHz	Mobile IoT (MTC/eMTC. NB-IoT), LPWAN
700 MHz	EC-GSM-IoT, LTE MTC, NB-IoT
800 MHz	EC-GSM-IoT, LTE MTC, NB-IoT
900 MHz	EC-GSM-IoT, LTE MTC, NB-IoT
862–870 MHz	Wideband SRD, Smart Grids, RFID
1800 MHz	Mobile IoT (EC-GSM-IoT. LTE MTC, NB-IoT)
2.1 GHz	Mobile IoT (EC-GSM-IoT, LTE MTC, NB-IoT)
2.4 GHz	SRD, Wide Band Data Transmission Systems
2.6 GHz	Mobile IoT (EC-GSM-IoT, LTE MTC, NB-IoT)

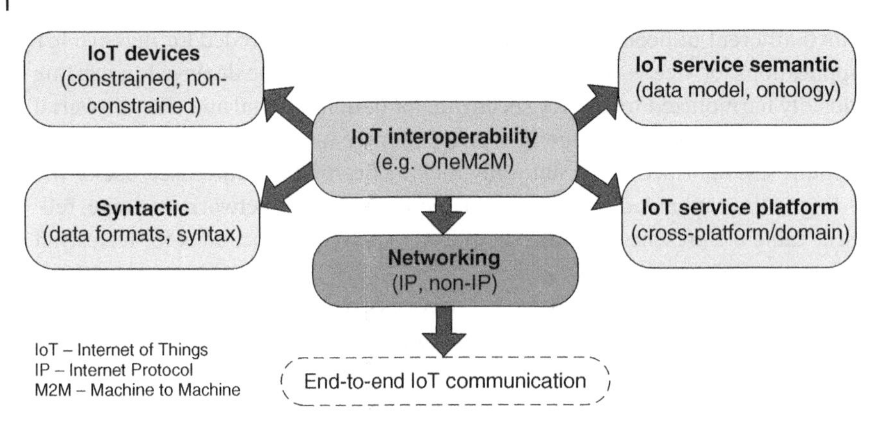

Figure 8.5 Massive IoT ecosystem and interoperability.

syntactical interoperability which refers to IoT data formats. The syntax rules of data messages between the sender and recipient of IoT messages should be based on compatible rules. Then, semantic interoperability (which includes data model, ontology, and information model) is needed for exposure of data and metadata in IoT services via APIs, which enable end-to-end IoT (i.e. M2M) communication. Finally, interoperability of IoT platforms is targeted to adapt IoT applications which are deployed on one IoT platform to IoT applications that work on another platform. Such platform interoperability can belong to one of the following two types:

1) Cross-platform interoperability: It enables the interoperability between different and separate IoT platforms for the same vertical (e.g. smart home and smart vehicles).
2) Cross-domain interoperability: It enables the interoperability between different IoT platforms that are used in different vertical domains.

Figure 8.6 shows the IoT reference model for interoperability. The main segment in this model is IoT application/service support layer. There is a standardization work done for the IoT interoperability, and one of the main such examples is oneM2M standardization [12]. OneM2M standard is a cross-domain hub model that uses interworking standards to support data exchange between IoT systems.

When deploying the massive IoT services (including IoT devices and IoT platforms), there is needed work on data privacy and security, based on the user-centric approach. There are required regulatory frameworks on IoT services that will cover the approaches risk assessments and data protection and privacy impact assessments in different phases of IoT services design and deployment process for different verticals [10].

8.3.2 OTT Massive IoT Services

OTT services use the network neutrality policy for open Internet access for all services and to/from all devices, based on Internet technologies from network layer up to the application layer, which includes IP addressing (on network layer, Layer 3) and the mapping of IP addresses to domain names (via DNS) [2]. On the other side, massive IoT typically includes constrained IoT devices such as sensors and meters, which typically run on battery (which should last for many years) and hence have very limited storage and processing capabilities (when compared, for example, with smartphones or computers as end user devices). In such cases, some of the IoT devices may have limited possibility to use IP protocol stack, so they typically may use other non-IP protocols for

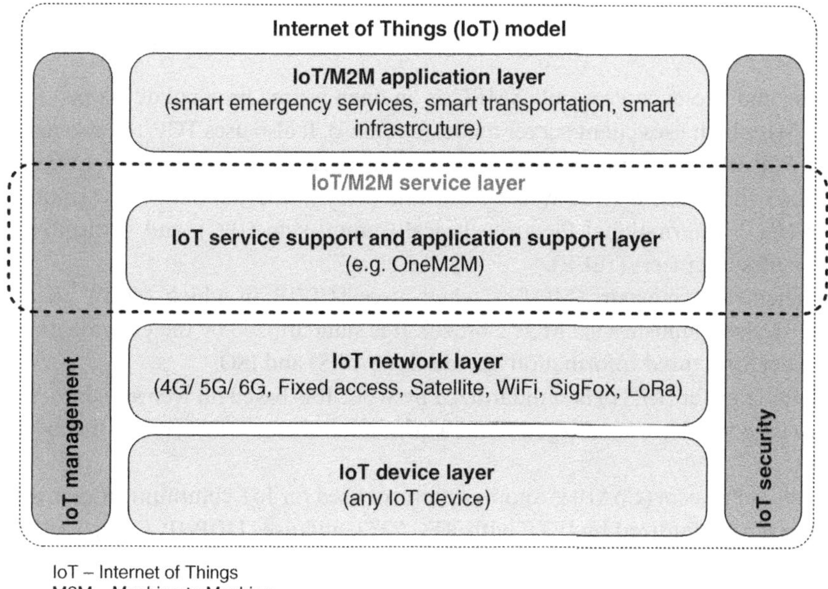

Figure 8.6 IoT reference model for interoperability.

communication with other such IoT devices in a given massive IoT network domain, and then to map such non-IP addressing to IP addressing via an IoT gateway, which should be placed at their proximity (due to limitations of massive IoT devices).

When using OTT IoT, the main interface that is possible for use is Web, considering that it is based on HTTP protocol. The Web technology provides the possibility for users to interact with heterogeneous IoT devices by using the standardized (by IETF) Web interfaces. Such an approach is also referred to as Web of Things (WoT).

In WoT for OTT IoT services, the Web technology is used for the exposure of physical devices as resources on the Web using. Then users can interact with the devices using Web interfaces. Interaction with devices (i.e. the things) in the IoT by using Web interface is referred to as WoT. In WoT, there are also two general types of devices (as common in the IoT segment):

- Constrained devices, which require so-called WoT broker, because they cannot connect directly to the Internet by using Web technology due to limited capabilities that do not allow running HTTP protocol stack which is needed for Web applications.
- Fully-fledged devices, which have all necessary web functionalities such as running HTTP client process, so they do not need WoT broker, however, it is possible to interact with it.

In a WoT environment, there can be used several different service types, which include WoT service: a service (it has one-to-one mapping with services and/or functions in a physical device through the appropriate adaptor), mash-up service (combines WoT services in a WoT broker with Web services outside of the WoT broker, as well as Web service (they can be accessed on the Web).

The main standardized protocol for the Web is HTTP. There are three versions of HTTP from which HTTP 1.0 is standardized in 1996, HTTP 2.0 is initially standardized in 2015, and HTTP 3.0 is published in 2022 (as given in Chapter 2 of this book). With HTTP 3.0, the HTTP became lighter and better than before, and can be used for various IoT services.

Besides HTTP, there are several other protocols that can be used for OTT IoT services [13], which include the following:

- eXtensible Messaging and Presence Protocol (XMPP) is an open source protocol developed for instant messaging (IM), which uses client-server model (as email). It also uses TCP/IP (also may go over HTTP over TCP/IP). Standards organizations which cover XMPP include XSF (XMPP Standards Foundation), IETF, World Wide Web Consortium (W3C), International Organization for Standardization (ISO), International Electrotechnical Commission (IEC), and Institute of Electrical and Electronics Engineers (IEEE).
- Message Queuing Telemetry Transport (MQTT), which uses TCP/IP, in which MQTT client (deployed in the IoT device) publishes via MQTT broker. It is standardized by the Organization for the Advancement of Structured Information Standards (OASIS) and ISO.
- Representational State Transfer (REST) is standardized by W3C. It is based on web standard, so uses HTTP. It is also used in telecom managed IP networks, not only for OTT services (i.e. over the Internet).
- Constrained Application Protocol (CoAP) is another protocol used for IoT communication over the Internet, which was standardized by IETF with RFC 7252, and uses UDP/IP. Compared to HTTP (i.e. Web) it has low overhead (the UDP header is 8 bytes long, which is many times smaller than the TCP header, and also UDP does not have state diagram as TCP, which requires more resources on the IoT devices' side). On the other side, CoAP is simple and easily translatable to HTTP, so it is a convenient protocol for constrained IoT devices which connect to the Web via a broker (that translates CoAP to HTTP and vice versa).

Either from telecom operators or from OTT service providers, massive IoT is a never-ending evolution of the ICTs that is targeted to connect every physical or virtual object to the Internet that can provide any form of benefit to the people.

8.4 Future Critical IoT/AI Services

We have classified IoT services into massive IoT and critical IoT. Critical IoT services have stricter requirements on performances which require application of AI techniques for their operations. In this part, we refer to critical IoT/AI services that include (but are not limited to) 5G URLLC, industrial IoT, and smart city services.

8.4.1 5G URLLC Services

As already noted in previous chapters 4 and 5, Ultra Reliable and Low Latency Communications (URLLC) services require very low delays and ultrahigh reliability, as the name denotes. The reliability in telecom networks in general is provided by using different mechanisms. For example, one approach is to have at least two separate paths between the end-systems, because the probability that two or more paths will be broken is much lower than the probability that a single path will be broken (due to any reason, such as hardware, software, or human failures, or disaster reasons). Of course, more paths with the same quality increase the reliability in the network (and the cost) and vice versa. So, URLLC services require reliability better than 1 fault in 100000 events, such as 1 fault in a million or 1 fault in 10 millions (that is 10exp-7) or better (means lower value) of events for the URLLC service. Example QoS KPIs for different URLLC use cases [14], is given in Table 8.2.

Table 8.2 QoS requirements for critical IoT (URLLC) services.

Use case	Use case example	End-to-end latency	Jitter	Round trip time	End-to-end reliability	User experienced throughput
AR/VR	Augmented worker	10 ms	—		99.9999%	—
	VR broadcast	—	—	< 20ms	99.999%	40–700 Mbit/s
Energy, smart grids	Differential protection	<15 ms	<160 us	—	99.999%	2.4 Mbit/s
	Fault location and identification	140 ms	2 ms	—	99.9999%	100 Mbit/s
	Management in distr. power generation	<30 ms	—	—	—	1 Mbit/s
	Advanced industrial robotics	<2 ms	—	<1–5 ms robot ctrl	99.9999% to 99.999999%	—
Factory of the future	AGV (Automated Guided Vehicle) control	5 ms	—	—	99.999%	100 kbit/s (downlink) 3–8 Mbit/s (uplink)
	Robot control	~1 ms robotic motion; 1–10 ms machine control	<50%	—	99.9999%	—

The other quality requirement for URLLC is the delay. For such services, the delay should be around 1 ms. This is possible in 5G. In fact, that became possible even in 3GPP Release 14 (in LTE Advanced Pro) by adding the possibility to transmit different OFDM (orthogonal frequency division multiplexing) symbols in downlink and uplink direction within the same time slot (that is called a mini-slot, the reader may refer to Chapter 4 on this). However, as a main target, it is specified in 5G mobile systems for URLLC service types. But how is it possible to achieve this very low latency?

Well, let's calculate this. The radio signals, as every other signal in the existing and near future telecommunication systems and network, travel at a speed which is close to the speed of light, and with the speed of light signals can reach only 300 km distance for 1 ms, and less for less than 1 ms. We can also note that this is only the theoretical maximum distance, because in practice there are different delays added to the delay budget on the path of IP packets. Also, longer IP packets require longer time (longer delay) to be delivered to the destination. Then, due to such reasons it is straightforward to conclude that many URLLC applications will be using smaller IP packets, with the aim to "push down" the delay as much as possible, of course, in cases (services) where it is needed. The conclusion is that the other end for the URLLC services (e.g. server side or control side) should be close to deployed devices that execute the tasks (e.g. for controlling machines in a factory, the processing server unit that manages the manufacturing process should be located in the same factory at the same location, a short distance away).

Let's give some examples of critical IoT services which can be provided over URLLC network slices. One example is the control of industrial manufacturing via 5G mobile network. Then, automation in a smart grid, or automotive safety, driverless cars, communicating with other vehicles and infrastructure, e.g. safety purposes, and so on. One of the foreseen frontrunners for URLLC services with 5G and beyond mobile networks can be the automotive industry. Further, URLLC includes remote control of machines, mission critical applications for e-health, then public protection and disaster relief communications, and other services.

For many URLLC services, there is required automation based on use of AI/ML, which requires data analytics (regarding the data collected from the mobile network and its functions). For such purpose, 3GPP has standardized NWDAF [15], which provides network data analytics (i.e. load level information). Typical services provided by NWDAF are [16]:

- Subscribe to/unsubscribe from notifications for different analytics information from the NWDAF, or when a ML model matching the subscription parameters becomes available, or notifications for the ML model training.
- Request and get specific analytics from NWDAF.

In both cases, typical network functions that use are Policy Control Function (PCF) and Network Slice Selection Function (NSSF).

NWDAF uses the interfaces of the 5G service-based architecture to deliver network analytics functions aimed at automation or reporting, solving more complex challenges such as network optimization to maximize QoS/QoE for different types of services expected to exist in future mobile networks, such as XR services, URLLC services, and V2X services (Figure 8.7). Most of such

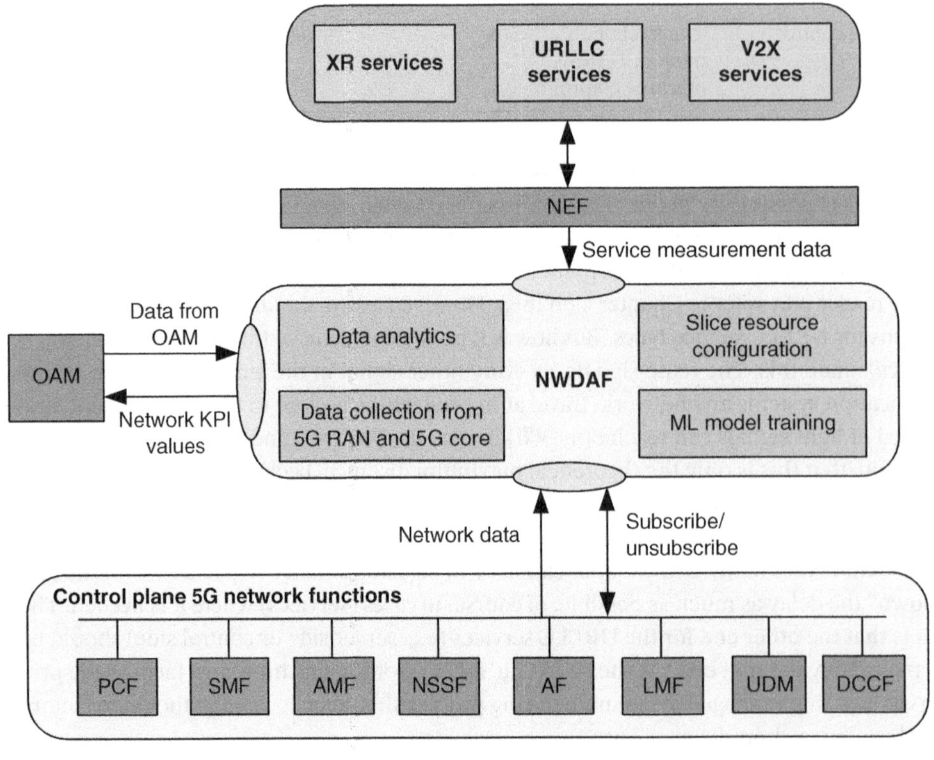

Figure 8.7 Network data analytics function (NWDAF).

critical services will be delivered through heterogeneous networks, including public and private network slacks, which increases the complexity and number of parameters that need to be analyzed for efficient network operation (from the Operation and Maintenance – OAM – point of view) and optimized use of network resources in the radio access network and in the core network of the mobile operator. For analytics of the data collected from the 5G (and, tomorrow, 6G) network and services, NWDAF uses ML algorithms, which must be trained with valid data collected from the network and services. In addition, NWDAF can have a distributed architecture that provides real-time analytics at the edge of the network (e.g. edge NWDAF) and a centralized NWDAF (in the 5G/6G core network) for data analytics at the network level.

Regarding the deployments, the URLLC services may need extra effort and business needs to reach some significant market share on the global scale.

8.4.2 Industrial IoT

The use of telecom/ICT technologies in industry verticals contribute to their further automation in all aspects, from logistics to delivery of the product to the customers. What is the driving force for use of ICTs in factories?

Well, advanced industrial automation is targeted to maximize productivity, and also economies of scale and quality of the production process. So, in the current widely connected world, we also want to connect the factories and for that purpose we need human-machine interfaces, then programmable logic controllers, as well as control of motors and sensors in a scalable and efficient way. What are the roots of ICTs in the industry?

Initially, ICT in industry was developed on serial-based interfaces. However, such interfaces were designed and offered by different companies, and some of them later have become standards in the industry. But, that has resulted in many different heterogeneous standards on the market (for the use of ICT in industry). One of the main communication technologies (on physical layer and MAC layer, considering the protocol layering model in telecommunications) used nowadays in factories is the already mentioned industrial Ethernet. The industrial Ethernet is also a LAN, but with different requirements regarding the speeds, precision, and synchronization, as well as its adaptation to function in harsh working conditions in factories (high temperature or high humidity, or a lot of motion by different machines and so on). Due to such conditions, the Ethernet switches and cables (used in factories) feature a more robust construction than, for example, Ethernet equipment used in data centers or offices (for LANs). Regarding the speeds, the industry automation needs less bitrates because typically there is exchange of smaller control messages in master-slave manner between different components on the factory floor. However, even with Ethernet, there is needed deterministic work (with time precision communication), hence, there is a standard called Time-Sensitive Networking (TSN) as an extension of the Ethernet standard by IEEE. Typically, changes are implemented in the MAC layer (i.e. protocol layer 2) in such cases. The TSN Ethernet LAN can guarantee timeliness and limit the maximum latency, which is essential for industry automation [17]. So, industrial Ethernet has become important in the Industry 4.0, which is based on M2M communication for industrial and robotic automation.

The success of industrial automation depends also on the reliability of the networks and their efficiency regarding the end-to-end delays and time precision. Looking into the future, with introduction of AR-based management by using "digital twins" of the factories, the speed of the telecommunication infrastructure in factories will also matter more than before.

What is needed with the aim to provide equal grounds for automation of different industrial sectors and to provide equal opportunities for different players in the telecom/ICT industry?

Well, that is standardization of the solutions. Such umbrella standardization is carried with the Industrial IoT (IIoT) reference model for smart manufacturing, which refers to three main dimensions: system hierarchy, product life cycle, and intelligence. So, the newer dimension in this framework is the "intelligence".

What is the relation between the Industry 4.0 and 5G and beyond mobile networks? Well, one of the main targets of 5G URLLC is, in fact, the industry automation with building smart factories. And the crucial requirements for URLLC are high reliability (up to many nines) and very low delays (in range of around 1 ms), and also very important is the requirement for time synchronicity. But, why will 5G and future 6G be good for smart factories for IIoT?

There are several reasons. One is that 5G is a unified global mobile standard (similar to Ethernet for fixed LANs). Then, 5G provides radio communications with high reliability in its URLLC network slices (which is, however, a matter of network design), which is more convenient for connecting different machines in the factory in a more flexible way. Also, 5G and beyond mobile networks provide the possibility (in mmWave bands) for very small delays, which is crucial for smart factories.

What are the possible deployment models of smart factories with 5G and beyond? The deployment typically will be with a private 5G network in a given factory, also called Non-Public Network (NPN) deployment solution. However, there are several "flavors" of deployment of 5G in smart factories. In that way, there is a standalone mode (private 5G network for the factory has completely separate infrastructure than public 5G mobile network), hosted mode (factory network is not public, however, it is hosted partly or completely by the 5G mobile operator), and integrated mode (in this case, 5G is used together with non-3GPP networks such as the already mentioned industrial Ethernet).

Also, the smart factories can be considered as part of the smart cities, which is another emerging trend/technology, targeted to building smarter cities with the use of ICTs.

8.4.3 Smart Cities

With the increasing use of the ICTs in all segments of living, one of the most remarkable use cases of smart services is through the development of Smart Sustainable Cities (SSC) services. According to the ITU's definition [18], smart cities and communities refer to effective integration of physical, digital, and human systems in building an environment that can deliver a sustainable, prosperous, and inclusive future for its citizens.

So, the ICT architecture for the SSC aims to provide certain services for the citizens, which include intelligent transportation services, e-government, e-business, smart health services, smart energy services, smart water supply services, and so on. For provision of such services, SSC is based on a multi-tier architecture from the communication point of view.

The ICT architecture for smart cities (Figure 8.8) consists of several layers; going bottom up, they include sensing layer (it connects various terminals, such as sensors, RFIDs, and cameras), network layer (it consists of different fixed and mobile networks), data and support layer (it stores the collected data from the sensing layer into clouds), and application layer (it includes applications for the SSC services, such as smart healthcare, smart governance, public safety, and environmental protection).

Overall, SSC architecture is integrating many emerging ICT technologies based on IoT services including massive and critical ones. The main target is to provide a sustainable quality living environment for the citizens based on use of the ICTs for provision of "digitalized" city services. So, with the digitalization of different city services and layering human and nonhuman objects into

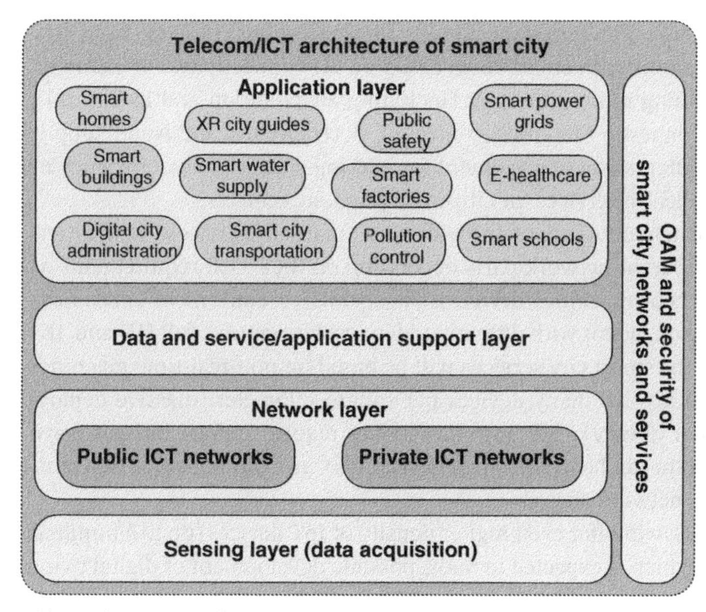

Figure 8.8 ICT architecture for smart cities.

layers (e.g. based on whether they are static, such as buildings and streets, or mobile, such as humans and vehicles), cities are being mapped into their digital twins [19].

Similar to the use case of 5G and beyond in Industry 4.0, one may note that it is also convenient for building smart cities, because it provides all the needed capabilities for such purpose. Considering the focus of 5G and future mobile networks to network slicing, one may expect wider deployment of smart city services with 5G-Advanced developments in the 2020s and beyond in the 2030s (with 6G) by using different network slices for different smart city services (of course, multiple services can be also provided over the same network slice). So, different smart city services may use different network slices which are dependent upon their QoS requirements as well as end-points (e.g. human users or machines/devices/vehicles). In that way, for example, mobile broadband Internet access is convenient for using tourist guide smart city service based on video services with different video resolutions (e.g. HD, Full HD, 4K, and 8K) and Augmented or Virtual Reality (AR/VR). On the other side, vehicle to infrastructure assisted driving requires strict QoS such as guaranteed bitrate, very low delay, and ultrahigh reliability. This service belongs to critical IoT services, for which URLLC network slices in 5G and 6G mobile networks are the most convenient.

The citizens and companies can also contribute to smart cities by providing open data, i.e. allowing the collected data to be used for improvement of certain smart city services, such as traffic control, environment control, and energy control.

In this process of making cities smart, 5G and more broadly mobile networks, with their integration with AI, big data analytics, cloud computing, and IoT, form the basis of the digitization of every aspect of the city, including manufacturing, daily life, as well as city management. However, smart city services result in the generation of huge amounts of data on an hourly and daily basis, so their analysis requires AI/ML and cloud computing technologies (e.g. edge clouds).

Also, different people usually have different expectations of smart cities, ranging from low expectations to super high expectations with smart cities based on 5G connectivity. For example, environmental protection is becoming more important (including air pollution and its control), then smart transport and traffic congestion in cities, control of electricity use and water supply, improved safety in urban areas, better waste management for making cities cleaner, efficiency of urban functioning and services, e-health services for citizens, and smart vehicles.

How can 5G and beyond mobile networks support/provide such smart city services? Well, different smart city services may use different network parts depending on their QoS requirements as well as endpoints (e.g. human users or machines/devices). Thus, eMBB is convenient to use for a smart city tour guide service based on videos with different video resolutions (e.g. full HD and 4K) and AR or VR. However, most of the smart city services will be based on non-real-time machine-type communications, which will involve many devices per square kilometer (massive deployments) connected over the cellular network. The mMTC services require smaller bitrates, have higher tolerance to delays (e.g. delays can be in the range of seconds), and can be served in a best-effort manner through the mMTC network slices.

The future mobile generation, 6G, will offer even higher density of IoT devices (up to a hundred times higher than 5G standards), which is expected to make possible development of digital twins of cities in 2030s and beyond.

8.4.4 Vehicle to Everything – V2X

Work on vehicular communication began in 1999 when the US Federal Communications Commission (FCC) allocated 75 MHz of spectrum in the 5.9 GHz band for intelligent transport systems (ITS). At the time, the allocation had sparked significant research activity worldwide for the development and deployment of V2X communications over the past decades. The first set of V2X standards was completed by IEEE in 2010, based on IEEE 802.11p technology [20]. In general, V2X covers the following four specific cases:

1) V2V (vehicle-to-vehicle) refers to communication between two vehicles (e.g. for collision avoidance and other safety features).
2) V2I (vehicle-to-infrastructure) refers to communication between vehicle and infrastructure such as traffic lights and Roadside Units (RSU). For example, used to understand traffic light conditions and priority.
3) V2P (Vehicle-to-Pedestrian) refers to communication between vehicles and people, bicycles, animals, etc, for provision of safety alerts to pedestrians and bicyclists, and similar services.
4) V2N (Vehicle-to-Network) refers to communication between vehicle and cloud services, typically edge clouds, and for provision of traffic management and rerouting.

3GPP Release 12 was the first standard to introduce direct device-to-device (D2D) communications for proximity services (ProSe) based on the use of mobile (i.e. cellular) technologies. This work was used by 3GPP to develop LTE V2X, which was the first Cellular V2X (C-V2X) standard from 3GPP (based on the 4G LTE air interface). LTE V2X was further developed in Release 14 and was further improved in Release 15 (the release in which 5G NR was also standardized). However, the initial V2X standard for 5G NR was introduced in 3GPP Release 16.

The 5G NR standard in its initial release 15, however, did not include so-called sidelinks (SLs). Why is SL important? Well, that is because in some use cases, it may not be necessary, possible, or desirable for data to pass through the cellular network. So, SL refers to direct communication between mobile nodes (e.g. UE and infrastructure terminals) without the data passing through the mobile network.

What are UEs in 5G V2X? In NR V2X UEs include vehicles, RSUs, or mobile devices that are carried by pedestrians. They were included for 5G V2X in Release 16. However, 5G NR V2X has been developed to complement and not replace LTE V2X communications (including SL).

The V2X application layer can be divided primarily into the following:

- V2X application-specific layer consisting of V2X specific applications (e.g. platooning and safety); and
- V2X application support layer consisting of V2X enabling services (e.g. V2X service discovery, message delivery, and service continuity) and common enabling services (e.g. group management, configuration, and location management).

8.4.4.1 V2X Architecture in 5G

From Release 16, NR RIT includes support of V2X, mainly by means of NR sidelink communication over the PC5 interface, partly leveraging what was defined for E-UTRA V2X sidelink communication.

In 5G NR, there are also two modes of operation for V2X communication, which are:

1) V2X communication over PC5 reference point and
2) V2X communication over Uu reference point.

The UE can obtain the V2X parameters via PC5 and Uu from various sources, which include PCF via AMF using non-access layer signaling (NAS) in the 5GC, or from the V2X AS. It is also possible for V2X parameters to be pre-configured in the Universal Integrated Circuit Card (UICC) or UE.

In 5GC, the PCF is the network function that can provide the UE with authorization and policy parameters for V2X over both reference points (radio interfaces), PC5 and Uu. The PCF further provides to the AMF parameters that are part of the UE context, which is needed for configuration and management of V2X communication (e.g. parameters related to PC5 QoS flows and profiles). As usual, for 5G mobile networks, such configuration parameters are taken from the Unified Data Repository (UDR). On the other hand, Network Function Repository (NRF) is used for UEs in V2X to discover other Network Functions (NFs).

Figure 8.9 shows the path of data packets from V2X applications to the radio bearers [21]. The V2X layer does the mapping of V2X data packets to PC5 QoS flows and applying the PC5 QoS Flow ID. All V2X data packets belonging to the same PC5 QoS flow are tagged with the same PFI (PC5 QoS Flow Identifier). Then, the V2X packets in the protocol stack are passed through the lowest two layers, Layer 2 and Layer 1, which belong (in the V2X model) to the access stratum (layer. The access stratum layer does the mapping of the incoming packets from the V2X layer onto the access stratum layer resources. Such access stratum resources consist of Layer 2 (L2) links, where each L2 link may contain one or more radio bearers used to transmit V2X packets. For a given pair of source and destination Layer 2 IDs, there can be multiple radio bearers, each corresponding to a different PC5 QoS level. Multiple PC5 QoS flows can be mapped the same radio bearer (the transmission mode can be broadcast, groupcast or unicast).

For NR based PC5, the QoS model is based on 5QIs (5G QoS Identifiers) which are relevant to V2X use cases. Overall, for NR based communication over PC5 the 5G per-flow QoS model for PC5 QoS management is applied.

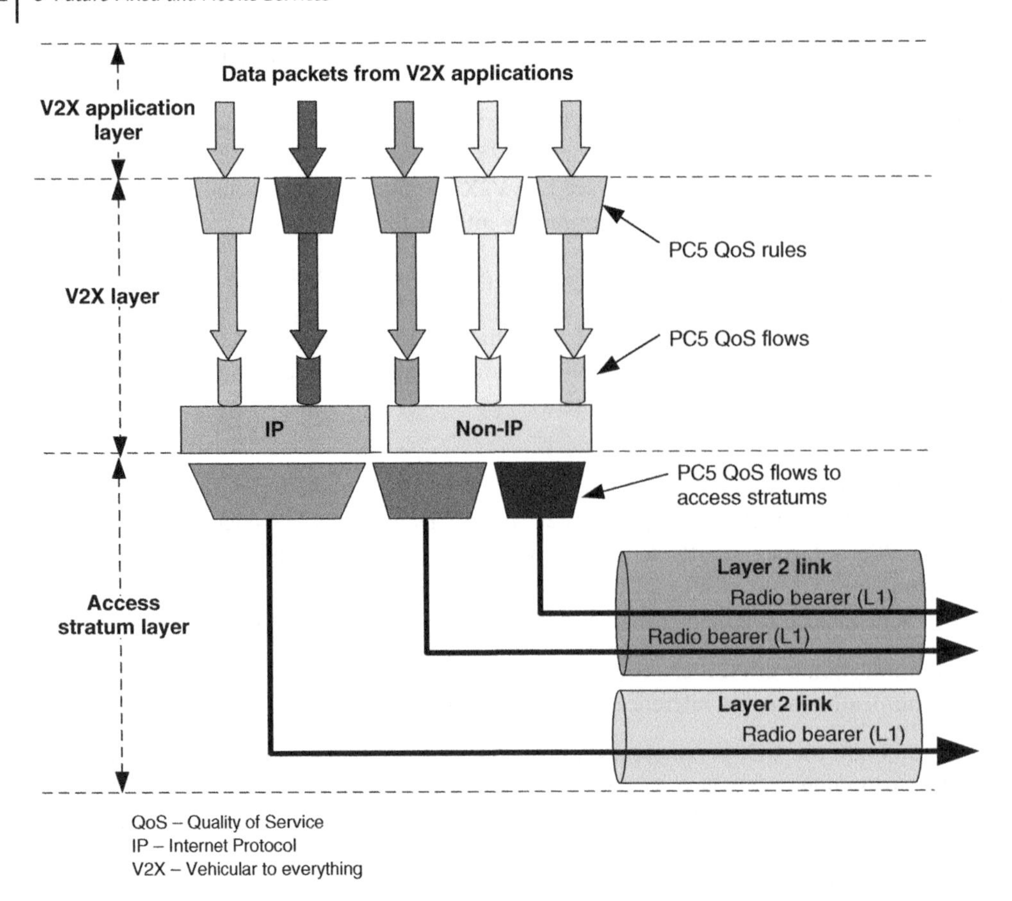

Figure 8.9 QoS for V2X in 5G networks.

Due to the V2X specifics in all aspects, there is standardized separate Slice Service Type (SST) for V2X, with SST = 4, as shown in Table 8.3.

The V2X communication will not be possible without participation in the process of car producers, besides vendors (for V2X equipment) and normally the mobile operators that implement the mobile network infrastructure and provide functions for V2X services. Thus, the 5G Automotive Association (5GAA) [22] was established as the largest association created to connect the telecommunications industry and vehicle manufacturers to develop end-to-end solutions for future mobility and transportation services. The 5GAA works tightly with 3GPP, considering that its V2X

Table 8.3 Standardized 5G slice/service types.

Slice Service Type (SST) name	SST value
enhanced Mobile Broadband (eMBB)	1
Ultra Reliable Low Latency Communications (URLLC)	2
Massive IoT (mIoT)	3
Vehicular to everything (V2X)	4
High-Performance Machine-Type Communications (HMTC)	5

solutions are based on 3GPP mobile technologies, in particular LTE (Releases 8 to 14, especially Release 14) and 5G (Releases 15, 16, 17, 18, and beyond). But, V2X services are more about regulation and its regional and global harmonization than upon the technology behind it.

8.5 Future OTT Services

OTT services are services provided over the open Internet based on the principles of network neutrality and a best-effort approach in serving application traffic. Most broadband access networks, including fixed, mobile, or satellite networks, are developed and deployed primarily for broadband access to the open Internet, also known as cyberspace. There are millions of OTT applications (one can only refer to many online application stores), some are standardized (for example, web-based and email), and others are proprietary applications that work with both client and server applications developed by a company or developers. However, the fact is that the main driver for the need for higher speeds in all telecommunications networks is the Internet access service as a whole, which on the other hand is used for OTT applications. And, traffic demands in telecom networks can be expected to largely rise for future OTT services, at least in the near future (e.g. 2030s).

When referring to digitalization, cyber world, online applications, email and Web, social networks, and online clouds, we are, in fact, talking about the OTT services. In short (although there can be found slightly different definitions in other literature or online), the OTT space includes all services offered over the open Internet, so simply said the OTT services space = cyberspace = open Internet.

8.5.1 Future Web

Looking into the future, Web services can be expected to continue to be the dominant online graphical user interface for accessing websites and other online content, primarily because they are based on standardized HTTP stack. For example, if one wants to access metaverse (or any other online service) through an application, then he/she needs to download it through a website first. Thus, the Internet is actually an interface to the digital world.

The Future Web is called Web 3.0, but the name (Web 3.0) is used more as jargon. Web 1.0 thus referred to the 1990s and early 2000s when users were primarily consumers of web content (e.g. before the advent of broadband penetration, smartphones, and video sharing websites such as YouTube), while in Web 2.0 users have also become content generators (in addition to being content consumers) by using primarily their smartphones (which provide easy access to end user photo and video creation) and then uploading them in the clouds (for example, through social networking applications and client cloud applications).

In Web 3.0, the domain of what can become smart, contextual, and consequently "semantic" will not be limited to text, but expanded into the physical world itself, dominated by spatial objects, environments, and interactions. Web 3.0 will be a semantic web, but not because we have embedded intelligence in text. It will be semantic because it will theoretically embed 3D spatial intelligence into everything. In practice, we will have better versions of existing web services, as well as new bandwidth-hungry web services, such as XR web services (which include AR and/or VR).

The future web (i.e. Web 3.0) is not defined by any single technology, but by an integrated "stack" of technologies, consisting of an interaction layer, a computation layer, and a data layer (according to Figure 8.10). Thus, the future web will use spatial (AR, VR, XR), physical (IoT, wearables, robotics), cognitive (AI/ML) and distributed computing technologies (blockchain, edge).

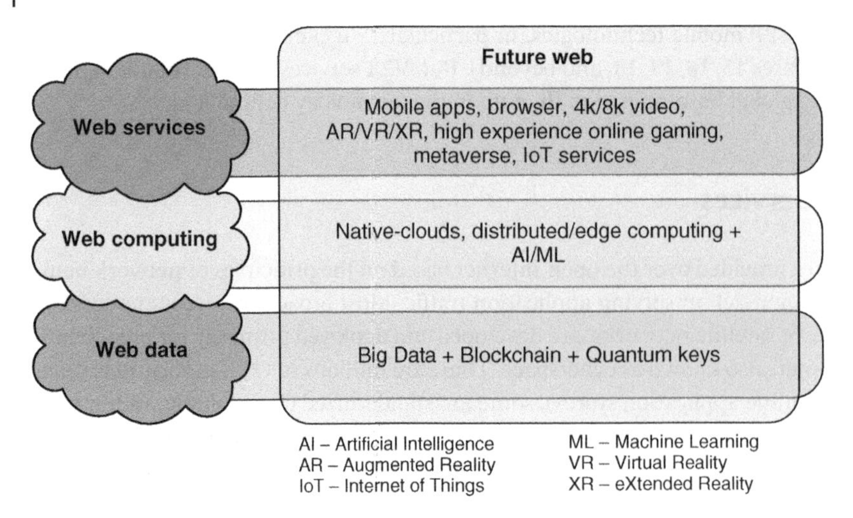

Figure 8.10 Future web.

Future web services are expected to use computing that takes place in a spatial environment, typically with special peripherals such as VR/AR/XR headsets or glasses, and haptic devices used to view, say, gesture, and touch digital content and subjects. Spatial computing allows us to naturally connect with computers, in the most intuitive ways, best aligned with our biology and physiology. However, most of the people will not walk around or work with AR/VR headsets – for example, initially, they may be used more for online games, virtual meetings, or some key tasks of the mission (e.g. mining and rescue). So, in addition to AR/VR/XR technology, the future web is also adapted to further improve existing online services (demanded by end users, e.g. online video services) toward a high user experience (most people would like to continue to have contact with the real physical world while consuming future web services). Thus, given that video services now account for over 70% of all Internet/IP traffic, the future web is also geared toward improved video streaming, video on demand, and online TV services (online TV was emerging in cyberspace due to broadband Internet access accompanied by unlimited data caps in developed telecommunications markets), with higher video resolutions than before. Thus, by the end of the 2020s, 4K and 8K online videos can be expected to become the "mainstream" video quality (around 2030), and then 16K and higher resolution video content can be expected to start appearing (which, however, also depends on the availability of consumer electronics for consuming such video contents). The demands placed on customer equipment are even higher for VR/AR/XR content.

If one looks into the Web computing approaches in future Web, then the native cloud approaches (already practiced by large global OTT service providers) will further emerge, with the possibility of using edge computing where it will be possible (because the edge infrastructure is owned by national telecom operators, while service providers for Web services are thousands of entities, locally and globally, some big ones and many smaller). With such large number of Web applications and users and IoT services also embedded in Web services, the Big Data from the network will be possible to be efficiently processed only by use of AI/ML techniques. The security of transactions will be possible to be improved by use of efficient blockchain approaches. And, if one implements Quantum Key Distribution (QKD) network as an overlay (or underlay) network to open Internet (between end nodes that will demand it), then certain services that could be provided over specialized/managed IP networks (in the past) could be provided over the open Internet in the future (e.g. by using quantum keys for securing data in some transactional Web services).

8.5.2 Cloud Gaming

Online gaming is constantly growing in popularity on the Internet since the beginning of the 21st century. The overall gaming includes casual online gaming, networked console gaming, and multiplayer virtual world gaming.

Cloud gaming technology is different from traditional online games because the games are displayed and processed in the cloud server and the video is streamed to the end users and the player requests the thin client to play the games. Users receive the video frames from the cloud to watch game video and send input data to a cloud server. The user sends commands to play the game and receives a display from the cloud data center.

New AR/VR/XR services are expected to emerge in 5G and beyond mobile networks due to the expected high bandwidth and very low latency required for cloud gaming, because with cloud processing the mobile terminal (e.g. smartphones) will consume less power of the battery and need less processing power (since the main processing, such as the Graphics Processing Unit (GPU) in this case resides in the cloud).

However, in order to have very low latencies, the use of edge computing is required. In such a case, mobile telecom operators can act as a service enabler (providing the necessary infrastructure) and provide cloud gaming with AR/VR/XR as a specialized service, where the game will redevelop the online gaming industry.

However, it is up to sustainable business cases whether these possible applications (such as XR gaming) will become a reality in the future with the involvement of telecom operators or will they remain completely in the OTT space where they currently exist.

Overall, cloud OTT servers offer platform-independent gaming so that a user can play the same game using any device, anywhere in the world. Furthermore, cloud games can be updated more easily by developers considering that the game exists in the cloud servers. Changes/updates to cloud games are automatically reflected to users. Also, cloud servers may capture user experience and use such data to improve their services, as well as adapt them to the user's current environment and speed of movement in the case of mobile networks (e.g. use of AI/ML techniques).

When games are running on the cloud side, the game images are compressed and transmitted to users through the network after rendering in the cloud. Cloud game user uses cloud game apps to play games, so the user equipment can be without expensive graphic cards, while higher processing is executed in the cloud which, in such cases, needs more resources. In such case, the costs of GPUs will be transferred to the cloud and to game providers which is a different business logic than today when the costs of graphic cards are covered by the end users themselves. So, future cloud gaming success depends also on the business logic at a given time. Overall, gaming is a multi-billion dollar global business in all segments of the value chain, from software to hardware components, and typically provided via the open Internet (as OTT service). Also, one may expect it to expand in the digital (cyber) world by offering new services or merging with other online service providers, such as future social networks (e.g. metaverse), as one of the main possible developments.

8.5.3 Future of the Social Media – Metaverse

Social media are defined as Internet-based applications that enable the creation and exchange of user-generated content. The broader definition of social media refers to the many relatively inexpensive and widely available electronic tools that allow anyone to post and access information, collaborate, or build relationships. Furthermore, social media depend on mobile and web-based technologies to create highly interactive platforms through which individuals and communities share, co-create, discuss, and modify user-generated content.

Having an avatar in the digital space is similar to online games. The difference between social network and the online game is the target of engagement, playing a competitive game or having a digital life.

On the other side, the AR/VR equipment is initially used for online gaming. Game makers are those that are developing 3D virtual worlds because they know how (from developing 3D games). If the context is changed we get a social network (from a game environment), and vice versa. Also, the emerging metaverse is in fact similar to 3D game, with avatars, digital currencies (or tokens), and various digital objects and environments, only the context is socialization of individuals, groups, and businesses. While in the 2010s the future of social media was considered to be social business, in the 2020s the future is called metaverse.

Also AI techniques can be applied in such metaverse, but what is the gain in such cases? Well, metaverse is the web of virtual reality spaces where online users interact with each other and within an entirely computer-generated 3D environment, based on video/VR/AR/XR contents/services. Then, in metaverse one is more likely to be able to apply AI-driven predictive models with their potential for continually accelerating knowledge acquisition. For example, the metaverse offers the potential to function as a decentralized decision-making platform for different kinds of users. However, for metaverse it may be necessary to define technical frameworks through standards so everyone can have a dialogue on the same "channel". In that direction, for example, AI-supported metaverse applications could become a vital tool in the pursuit of UN's Sustainable Development Goals (SDGs) for 2030.

Overall, many see the metaverse as the next evolution of OTT social services based on the integration of physical and digital experiences. Also, there are different foreseen technologies for the metaverse, including gaming engines, VR/AR/XR, digital twins, and blockchain and cryptocurrencies.

8.6 Open Internet vs. QoS, QoE, and Network Neutrality

The Open Internet is based on network (net) neutrality, which means that all Internet traffic should be treated equally, without prioritization of certain OTT services over the others [1]. So, simply speaking, telecom operators that provide Internet access should treat equally the traffic to/from Google search engine or YouTube as the traffic to/from local news portal.

Internet communication is used to refer to the general end-to-end communication provided over the public Internet. Typically, the provider of Internet Access Service (IAS) is referred to as Internet Service Provider (ISP). In the 2020s, the ISPs are typically national telecom operators who provide fixed, mobile, and/or satellite broadband services in a given country.

Also, network neutrality and open Internet typically refer to the same, that is, Internet access for which network neutrality policies are in force. However, network neutrality refers to treating all legitimate Internet traffic equally but, of course, viruses, worms, and other security threats are not under the umbrella of Internet network neutrality.

Network neutrality is in the basic philosophy of the Internet so that it does not disappear, but usually changes its definition or positioning in the telecommunication market (or one may also say "digital market") from time to time. Also, it has significantly sped up the innovation in the telecom/ICT sector with the appearance of the open Internet available to everyone since the middle of the 1990s. Why?

Because OTT service providers (who provide services on the basis of Internet network neutrality) do not need to oblige to strict national regulations unlike national telecom operators. With network neutrality, the responsibility of actions in the cyberworld (that is open Internet) is transferred to end users (not to telecom operators as ISPs). Then, what is illegal offline is also illegal online and the same laws can be applied in the cyberspace (or digital world) as they are applied in the physical world.

Also, with the aim to prohibit censorship on contents and applications or their differentiated treatment, the network neutrality (as a native and inseparable feature of the Internet) was accepted on the global scene by most of the countries worldwide. That also allows Internet traffic to traverse over networks deployed in different countries (in most of them) without blocking or filtering. Why?

Well, it is related to the individual's freedom to choose what she/he wants to use as a service and what content that individual wants to consume. Such approach can be named as "freedom" in the telecom/ICT space and in the so-called cyberworld (or cyberspace).

The cyberworld by itself is mimicking the real world and all of its functions, so freedom in the real world (to move freely and to act freely) is translated in the Internet via the principles of network neutrality (it can be named also an Internet freedom or Internet equality). On the other side, as in the real world, there are legislations and laws in different fields (e.g. financial or banking sector laws, criminal law, and human rights laws) which are obligatory for the Internet users (when they are using OTT services/applications) as they are obligatory for citizens in the real world. However, laws can differ from country to country (e.g. online betting is allowed in some countries while it is forbidden in others) and hence such rules are mapped into the Internet cyberspace although it is network neutral (it does not mean that it is neutral to law enforcement).

8.6.1 Regulatory Aspects of Network Neutrality

The need for intervention from the NRA will depend on the quality of the Internet access service and its monitoring and evaluation is necessary to detect degradation. Degradation of the IAS as a whole includes the following possible cases:

- Telecom operators prioritize specialized services at the expense of the IAS as a whole, which is typically a case of not enough capacity to provide guaranteed bitrates to specialized services as well advertised bitrates for IAS service during the peak traffic period (with the highest traffic load during the day). If the bottleneck is the access network, then such degradations are typically local (e.g. in city area, in limited rural area, or a given part of a highway), while in cases when degradation occurs due to lack of capacity (at peak periods) at interconnection points, then the IAS in the whole network is affected.
- Internet traffic load grows faster than the increase in available capacity. This is typically the case when the data cap per user is increased (e.g. by increasing weekly or monthly traffic allowance) or number of subscribers of IAS (and other specialized services) without proper traffic and capacity planning in advance.
- IAS of sufficient quality is accessible to only a limited number of users. In such cases, typically, the reason is capacity of access network (e.g. a mobile network) and/or fronthaul and backhaul networks (for that access network) which have less capacity than needed to serve the traffic from users at peak traffic periods. In mobile networks, such case may happen due to unexpected circumstances (e.g. natural disaster, accidents, and mass gathering of people), while in all other cases that is a result of inappropriate network design or capacity planning.

Given the possible cases of IAS degradation, the question arises: when is regulatory intervention required? The Internet access service market situation in a given country should be assessed to determine whether the problem requires intervention by taking into account [1]:

- Availability of IAS offers by telecom operators in the country (more offers with desired QoS, such as sustainable bitrates, with bitrates that correspond to the one advertised by provider);
- Ease of switching between ISPs (i.e. telecom operators that provide Internet access service) where the term "switching" refers to a change of ISP by the customer.

When degradation of services such as IAS by an ISP is noticed in the country, the NRA in that country may intervene and consider imposing minimum QoS/QoE requirements (which are different at different times, considering the continuous development of telecom/ICT technologies, including expectations on speeds and latencies from the end users' perspective). The legacy approach to the noticed degradations would be to require the telecom operators (as ISPs) to improve service quality until degradation is eliminated in a reasonable time period.

There is no need for intervention when there is a good availability of Internet access service offers of satisfactory quality (i.e. without degradation) at a reasonable cost, and the possibility and ease of switching are sufficient (see Figure 8.11).

Regarding the NRA's role in imposing minimum QoS requirements on IASs, first, the NRA will typically identify situations that may need additional attention. Then, an evaluation of the quality indicators will need to be performed to confirm the clues and analyze the results to determine if intervention is needed. If regulatory intervention is required (due to violations of net neutrality and QoS assessment analysis), the telecom regulator (the NRA) will have to choose which regulatory tool to use, such as encouragement or enforcement.

8.6.2 QoS/QoE vs. Network Neutrality vs. Traffic Management

Network neutrality is often seen as a key driver of innovation in the applications and services space, the OTT space. However, with the convergence of telecommunications networks and

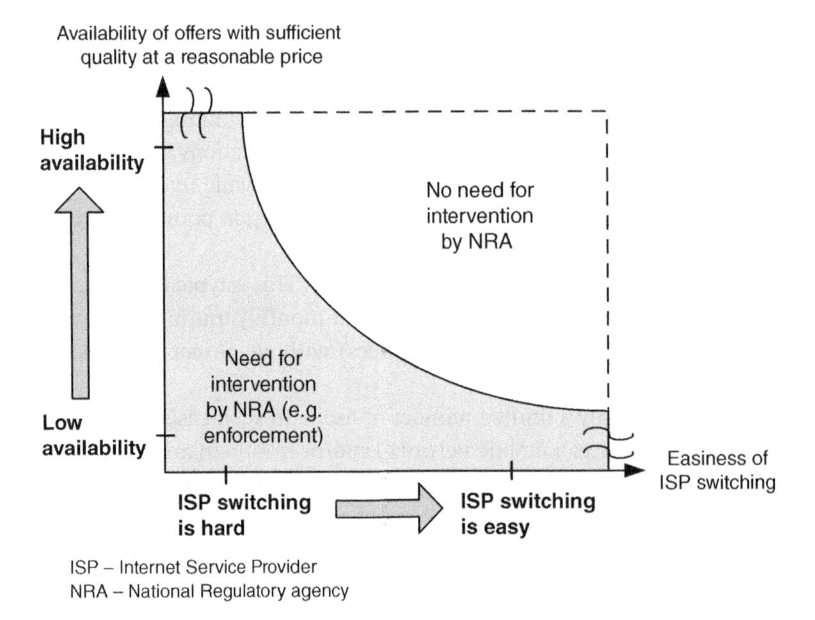

Figure 8.11 Regulatory interventions in regard to network neutrality.

services onto IP-based networks and services, there is a need for end-to-end QoS in such all-IP environments. In all-IP environments, all services are (or will be) carried over IP-based networks, but not all IP networks belong to the open (i.e. public) Internet space. On the other side, all parts of the open Internet are IP-based networks (either access, core, or transport/transit ones).

Network neutrality only applies to IAS, that is, to the open Internet, not for every IP-based network and service. For example, telephony provided over all-IP networks by the telecom operator, with QoS guarantees, i.e. operator-type VoIP (as a replacement for Public Telephone Networks (PSTN) [3], is transmitted over managed (i.e. specialized) IP networks, hence, it is not part of the public Internet and net neutrality does not apply to it. On the other hand, OTT voice services such as Skype, Viber, WhatsApp, and others are provided over the open Internet (i.e. via IAS) and hence the principle of network neutrality applies to them.

Figure 8.12 shows the relationship between network neutrality and managed IP networks versus OTT services and specialized services.

Is traffic management (as part of QoS/QoE management) by ISPs/telecom operators against network neutrality? The clear answer is, it is not. Traffic management is a tool for providing a better IAS that directly affects all OTT services that are actually provided on a network neutrality basis. In practice, traffic management potentially offers benefits not only to network operators (i.e. telecom operators), but also provides better QoS and QoE to OTT content and application/service providers (either on their customer side or on the side of their servers in data centers connected to the Internet), and also to end users.

What about QoS/QoE for OTT services? Well, it is possible (and typical) for broadband Internet traffic to be delivered on a network neutrality basis and the same traffic to be exchanged between different data centers of OTT service providers over managed IP networks (which can provide strict QoS guarantees). It enables service providers to place Content Delivery Network (CDN) data centers (as well as caches) as close as possible to end users in different regions and to apply differentiation and aggregation by traffic type (e.g. VoIP, video, IoT, and other data) and transfer such different types of traffic between their own data centers with different QoS (e.g. using different virtual private networks – VPNs over the transport network). In this way, network neutrality does not prevent OTT service providers from improving the QoS/QoE of their services. Thus, to recall, as noted in Chapter 3 of this book, the largest global OTT service providers since the 2010s have also been investing in transport submarine cables globally in order to have better control over QoS/QoE for their OTT services.

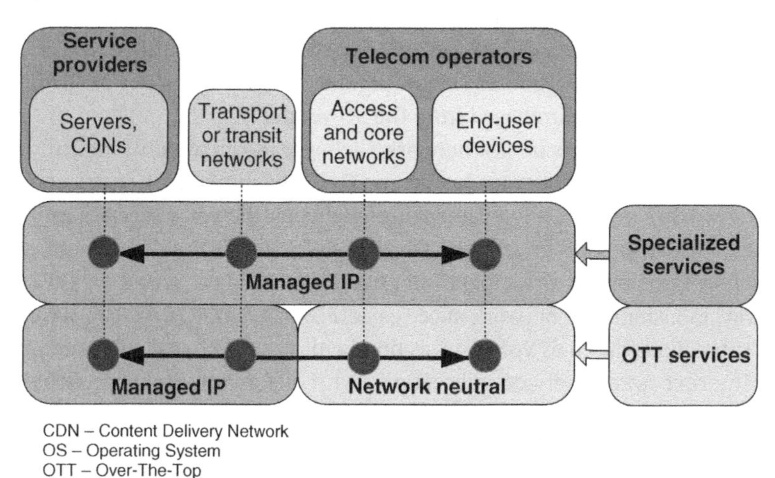

Figure 8.12 Relations between Internet/managed IP networks and OTT/specialized services.

On the other side, the telecom operators directly influence the QoS of the OTT services with the provided bitrates and delays (i.e. latencies) over their own broadband access networks as well as data caps offered to the end users via Service Level Agreements (SLAs).

Overall, sustainable broadband speeds for IAS benefit all parties, including OTT service providers (better QoS on average for their services), telecom network operators (satisfied customers with IAS from the telecom operator) and, finally, the most important, the end users who are customers of both the telecom operators (from whom they buy Internet access service) and the OTT service providers (from whom they use the services, either free or for a fee).

Regarding the business side of services in the 2020s, we have digital single market (e.g. in European Union), which refers to all services, including telecom services as well as OTT services (provided over the network neutral Internet access service) [23]. This positioning of services also makes more transparent the regulatory work, especially considering the emerging technologies and shifting of the telecom networks to software defined and virtual networks that can be shared, used for different types of services provided by different parties to different parties, and so on.

However, all services that are not provided via the open Internet access are referred to as specialized (i.e. managed) services.

For example, is machine to machine communication part of such specialized service? Yes, it is when it requires strict QoS from the networks (e.g. critical IoT services, such as motion control in smart factories). And no, it is not when it is provided through the open Internet access (e.g. sensors connected via the open Internet).

Let us apply this analysis further to the future network framework such as 5G that will be used in the 2020s and 6G that will be deployed in the 2030s. For example, in 5G, open access to the Internet is provided through eMBB network parts. Also, mmTCs can be provided over public Internet access, but can also be provided as specialized (e.g. private) services (depending on the use case from a business point of view). However, services that require ultra-reliability and very low latency communications will likely be provided as specialized (i.e. managed/private) services over separate network segments (logically separate from open Internet access, although they may share the physical network infrastructure, which is however dependent on the use case scenario).

8.6.3 Future Services vs. Network Neutrality

Network slicing is a core capability that enables telecom operators to create different service offerings over the same (e.g. 5G) infrastructure, including slice types for eMBB, mIoT (i.e. mMTC), URLLC, V2X. In that manner, a given 5G operator can have multiple eMBB slices, multiple mMTC slices, multiple URLLC slices, and multiple V2X slices (assuming that the given number of slices per slice type is supported by the equipment purchased from the 5G vendors).

Just to recall (from previous chapters of this book), network slicing is the creation of logically isolated network partitions, but actually it is not entirely new in 5G. For example, even in 4G and earlier (in IP-based mobile networks) we have a logical isolation between the voice services provided by the mobile operator with strict QoS guarantees (for example, VoLTE) and the mobile Internet access service (which is based on the principle of net neutrality and best access for OTT applications). In the same way, in a single 5G network slice (e.g. eMBB slice in 5G), we can have multiple logically separated "subsections", such as VoNR subsection (with strict QoS), AR subsection/VR/XR (with strict QoS), Internet access service subsection (with net neutrality approach), and so on.

Given that different verticals are entering 5G and beyond the telecom world, there may be confusion among them about network neutrality and the open Internet (which are connected to each other) and various other services with QoS guarantees that are not provided over the open Internet (as already noted, such other services are referred to in this book as managed or specialized services).

Some issues related to the principle of the open internet (i.e. net neutrality for OTT services) may be seen as concerns that in the future regulation related to the open internet and net neutrality may prevent verticals from taking full advantage of opportunities such as cutting the net to offer tailored services to vertical sectors (e.g. instant critical software updates for driverless cars versus streaming entertainment videos on a smartphone or desktop computer). However, V2X critical services and Internet access should not be provided over the same network sub-parts (or slices); they should be logically isolated. But, if the software update is for the infotainment system in the car (which is not a critical use), then it can be also done over the open Internet.

Further, there are possibilities, for example, with 5GMS, for provision of OTT media streaming services by using functions in the 5GC of the mobile operator. In such cases, the 5G operators will not be only a "pipeline" to open Internet, but also the enabler of certain OTT services. Of course, that is also not a completely new case – for example, mobile telecom operators also host in their network own Web and email servers which both belong to the OTT service/application space (i.e. to open Internet). Also, telecom operators provide TV streaming via IAS (as an OTT service) in parallel with IPTV provision as a managed IP service.

Without jeopardizing the network neutrality principles, telecom operators can implement appropriate and reasonable traffic management to different traffic types within the aggregate Internet traffic (e.g. video, audio, Web, messaging, VR/AR/XR, and IoT), with a target to provide better user experience for all different OTT services and managed IP services. It may be accompanied by network slicing, but network slices configured by a given telecommunications operator typically share the same network infrastructure of that operator, including access, core, and transport network resources and functions. Given the complexity of aggregated traffic, in the future, it will be necessary to use AI/ML techniques to obtain the best KPI values for different services provided over the same telecom network infrastructure, including OTT services over the Internet and managed IP services (which include critical services, such as URLLC services in 5G and beyond mobile networks).

In general, regulators should interpret the open Internet principle in a way that encourages flexible and efficient networks and results in the most enjoyable user experience when using OTT applications over mobile internet access (e.g. via eMBB in 5G and ultra MBB in 6G). However, user experience is directly related to bitrates (with 5G/6G they increase) and delays (with 5G/6G they decrease), so a simple conclusion is that 5G/6G mobile networks increase QoE for all OTT applications on the open Internet (taking into account that mobile communications are mostly personal and all people carry their smartphones with them almost all the time), which on the other hand being the basis for digitalization all over the world.

8.7 Future Digital Economy and Markets

New broadband services based on cloud computing, IoT, VR/AR, and AI/ML, including telecom services and OTT services, have transformed the economies of both developed and developing countries, and will further lead the digital transformation. As the process of digitization accelerates, and as more and more people worldwide are connected to the Internet and generally to telecom/ICT networks and services, these benefits increase to all countries.

The Internet has enabled the free flow of data that has become a significant driver of the global digital economy [24]. In the digital world, such free flow of data between different countries has proven to be a force for economic growth and increased globalization that has an impact on all regions and countries of the world. Which services primarily contribute to digital economy?

Well, those are OTT services provided over the open Internet, that is, online services. Advances in online services and platforms made easier for businesses to operate through on-demand or pay-per-use access to applications and technology, rather than making more complex ICT infrastructure investments. In theory, with just a laptop or smartphone and broadband Internet access, it has become possible to create an online business or work via Internet for other companies around the world. This approach has dramatically reduced start-up costs for small and medium-sized enterprises (SMEs) as well as provided the possibility for many skilled workforces to remain in one country (e.g. home country) and work for companies in other countries.

The basis for development and growth of the digital ecosystem is the flow of data [25]. Data has become the new currency for businesses, and a significant subset of this is personal data. Personal data is the primary driver for the digital revolution, including direct marketing as well as sale of goods and services to individuals and businesses. Global OTT service providers (e.g. Google, Amazon, Facebook, Apple, and Netflix) obtain a portion of their revenue from user profiling and using the personal data they collect, including data about user interests online, which can be collected using combinations of HTTP cookies on visited Web pages, IP address allocated to the network interface, social network activity, personal data when registering online, credit card data when purchasing online services and goods, and so on. That allows OTT service providers to directly target their users with different products and services, considering that (unlike Telecom operators) many online services are offered free to use (e.g. free search engines, free Web content, free video content, free social networks, free voice services, and free email services). But, each service has its own costs for development and operation and maintenance, which need to be obtained via different revenue channels so such free OTT services can continue to exist. Some online platforms provide the options to both leverage data and collect it.

8.7.1 Digital Transformation

By the decade 2020s, digital technologies have become an integral part of all aspects of work, learning, entertainment, socializing, shopping, and accessing information, as well as services that were previously obtained through physical access to institutions, such as health services, culture, and government services. In this direction, digital policy strategies have been established in various countries and regions of the world that empower people and businesses to seize a human-centered, sustainable, and more prosperous digital future [23].

The digitalization can connect people for private or business reasons regardless of where they are physically located. At the same time, digitalization can become a decisive enabler of individual rights and freedoms, as it allows people to have information and contacts outside certain territories, social positions, or community groups. In that way, many new opportunities can become available in the future for work, study, and entertainment, considering that the digital infrastructure (broadband Internet) is widely deployed in all regions in the world. Considering the importance of the digitalization as a process, many countries have created strategies for digital transformation [23]. Such strategies tackle different aspects of the digital transformation, with the aim to avoid appearance of digital gaps between urban and rural areas in a given country, as well as to build the environment for digital services, including the human skills for their usage and trust in them.

The digital transformation in the near future (toward 2030) is targeted at different ecosystems, which include the following as the main ones:

- Manufacturing: With the use of 5G and 6G mobile networks, as well as the increased use of AI/ML in the ICT area, factories will be able to be even more automated and able to collect industrial data, improving workers' jobs, safety, productivity, and well-being.
- Smart health: Introduction of more online interactions, paperless medical services, digital medical records, and automation in patient monitoring. The long-term goal in digitalization is for all citizens to have digital medical records.
- Agriculture: Digital technologies can help better production results in the agriculture sector via monitoring and automatic irrigation, protection from pests, and harvesting, as well as better communication between the farmers and the ecosystem for selling the agricultural products.
- Smart transportation: Digital technologies may help toward better connected and automated transportation for reducing traffic accidents and improving the efficiency of transportation systems. This includes various transportation means on Earth, sea, and air, which include cars, buses, trains, ships, drones, and planes.

There are also specific digital transformation goals set for businesses that relate to enterprises' use of cloud computing services, Big Data, and AI. For example, in the Digital Compass, the European Union set as a goal that 75% of enterprises use advanced technologies (clouds, Big Data, and AI) by 2030 [23]. In the longer term, after 2030, it can be expected that the goal will be for all enterprises to use advanced digital technologies. However, technologies that are advanced at a given time will not be advanced in the long-term future. However, the approach of using advanced digital technologies by businesses can be considered as a long-term future goal. As usual, such developments will appear first in developed digital markets, followed by emerging digital markets (which have greater growth potential). To avoid the emergence of digital gaps (such as the difference between the digital transformation of different regions and countries), it is necessary to bridge the gaps for broadband connectivity around the world [26]. Of course, it should be known that the term broadband is relative to a given time, that what was broadband a decade or two ago is narrowband today, and this can be mapped to today's broadband when looking into the future (i.e. what is broadband today will surely be considered narrowband in a decade or two).

The same as for the development of broadband Internet can be said for AI mechanisms, which will also change over time, and some will be abandoned while new ones appear (for example, future mechanisms will be trained with new and greater volumes of data due to the constant increase in the speed of broadband Internet). When it comes to the application of AI in digital services, a key aspect is trust. Trust is about transparency in building an AI algorithm, but it comes hand in hand with the ability for a larger audience to understand what it does and how it works by using Big Data, which includes a lot of personal data (e.g. user activity in the digital online world).

Another important future target is digitalization of public services, including governmental services as well as local administration. That is a process started in the 2010s in many countries, which continued in the 2020s, and will continue in the 2030s and beyond.

For use of digital services by citizens and businesses are needed skills which can range from basic digital/ICT skills (e.g. using browsers, email, social networks), medium level skills (e.g. using digital tools for business goals), and advanced skills (being able to develop complex ICT solutions, executing complex tasks with the help of digital tools and services).

The mid- and long-term goal in digitalization is obtaining digital ID, which will uniquely represent the citizens and business in the digital ecosystems by preserving their security and privacy. Overall, the so-called digital citizenships [23] are expected to be based on the following principles:

- People at the center: Digital transformation based on digital (i.e. ICT) technologies should protect the human rights of all people, including the disabled.

- Inclusion: The prerequisite for every digital transformation is availability (i.e. broadband coverage) and affordability (which is related to costs) of broadband Internet access to all citizens and business, with the aim for all players to be able to participate in the digital economy and digital society.
- Freedom of choice: The digital (i.e. online) environment should be safe for everyone regardless of the level of digital skills and every citizen should be able to choose which user equipment and which access technology to use to access digital services. Such freedom of choice is native to the Internet ecosystem, based on numerous OTT applications/services that users can choose to use based on their own decisions.
- Control of own data: Most of the digital services are OTT services that store data in the clouds, in which cases the citizens should be able to have control over their own data, and for some critical services (for example, online voting, government services, local administration services, and bank customer data) the data mandatorily needs to be stored in data centers located in the given country.
- Trust, security, and privacy: The future digital environment should be safe and secure for all users during the whole life (from childhood to old age), regardless of the technologies being used (e.g. clouds, Big Data, AI, and 5G/6G mobile).

For the digital future, it is not enough to deploy infrastructures, build digital skills and capacities, and digitize businesses and the economy and public services (national, local), but it is also necessary to enable all citizens to fully utilize the opportunities that arise with the availability of digital technologies and services [27]. In the digital space, it should be possible to exercise the same rights that apply in the real physical world. To this end, people should first have access to affordable, secure, and high-quality broadband networks, built basic digital skills as a right for all, which are needed to enable them to fully participate in economic and social activities, including the present and the future.

Finally, such human-centered, safe, trusted, and open digital environment should be in accordance with the law but, on the other hand, it should enable people to exercise their rights, including the rights to privacy and their data protection, freedom of choice, and freedom of expression (in the digital world), rights to the children online, as well as consumer rights.

8.7.2 Business Aspects for Future Telecom and OTT Services

New services based on cloud computing, IoT, VR/AR/XR, or AI/ML, including telecom services and OTT services, are transforming the economies of both developed and developing countries and will continue to lead the digital transformation. As the digitalization process accelerates and more people around the world are connected to the Internet and generally to telecom/ICT networks and services, these benefits increase for all countries. But also affordable access to broadband as a human right is a prerequisite for digitization in all sectors in both the public and private spheres.

OTT services enable accelerated digitization in many countries of the world that do not have their own ICT industry, either hardware or software. On the other hand, there are stakeholders (also present in developed countries) that global OTT services (which have hundreds of millions of users globally and from almost all countries in the world) have an impact on the revenues and profits of traditional telecommunications operators, and an impact on the investments needed in the infrastructure, primarily the fixed fiber-optic infrastructure that is necessary for new access technologies such as 5G and the future 6G mobile networks, satellite broadband, as well as optical access networks. Also, global OTT service providers have an impact on trends in different sectors, including entertainment, politics, education, and many others.

On the other hand, OTT services affect lower effective prices for consumers, and this not only transfers the profit to consumers (for example, the use of free services such as websites and social networks, through the same Internet access that they pay to telecom operators), but also motivates them to consume more services, which will benefit not only consumers but also generate new revenues for network (telecom) operators due to the need for higher bitrates and higher data cap. Also, at the global level, the number of users of broadband networks is continuously increasing, partly due to the improvement of the price/performance ratio (different end users have different willingness to pay for telecom/ICT services), thus generating new revenues. At the same time, the continuous improvement of the price/performance ratio of telecom networks (for example, by using AI/ML techniques and cloud native approach in their design, operation, and management) and equipment reduces the unit costs for telecom operators.

Telecom operators used to provide legacy telecom services (voice, IPTV, Internet access service, and business VPN services), so they need to rethink their businesses in order to capitalize on new technologies and developments with the digital transformation of verticals and industries. For end users and the society as a whole, the potential benefits of online OTT services are significant and clear. In fact, the OTT services that came with network neutral access to the Internet through the broadband networks of telecom operators provided an unprecedented development in the telecom/ICT sector via innovation of many applications and services for different purposes. In that direction, OTT services are certainly good for users (to have the option of choosing online services), but also for telecom operators since OTT services are the main reason why telecom customers (including residential and business customers) around the world need access to the Internet, for which they pay fees to telecom operators.

In both basic types of services, telecom and OTT services, AI technology is increasingly being used in both networks and services. The economic impact of AI is likely to be greater compared to other general purpose technologies that have emerged in the past. However, the profits from the application of AI in the telecom/ICT sector are unlikely to appear immediately. The impact of AI is likely to increase over time in the future, so the benefits of the initial investment are unlikely to be seen in the short term. At the same time, there is a risk of opening an AI gap between those countries and societies that move faster to adopt AI technologies and those that do not, even more so between workers who have skills that match the demand in the era of AI and those who do not [27].

The benefits of AI are likely to be unevenly distributed, much like digital transformation in different countries and regions of the world, and over time such inequality may deepen. In that direction, long-term strategic thinking is needed by governments and regulators, from the aspect of applying AI technologies and training people to use them. For example, critical V2X or smart cities cannot be provided locally without governance by local or national administrations.

Most services (both telecom and OTT) used cloud computing, where the already mentioned (in Chapter 7 of this book) problems appear in terms of data localization requirements, data security, and their protection according to the applicable laws in that country. Moreover, such data localization requirements can have a major impact on the adoption of cloud-based services, since the requirement itself limits the choice of available offers for such services, which can affect the quality of services, especially in countries with digital markets in development, where there is not enough human capacity or legislation to set up reliable and secure cloud solutions in the country (which may not be the case in developed regions). Also, more stringent requirements may result in higher costs of providing cloud computing services.

Despite all the potential barriers to using clouds, the cloud computing services are a key driver of growth in the digital world and they are required for almost all future services. Why? Because all offline data is actually stored in clouds, or at the network edge, centralized or distributed (only local disk drives can remain under the formal management of ordinary end users). Furthermore, edge and distributed computing will also influence the type of telecom/ICT equipment that will be deployed at the network edge and at end user premises.

8.8 Regulatory Challenges for Future Telecom and OTT Services

Future OTT services will be alongside telecommunications services in the digital single market. But future services in various verticals aim to transform the entire society and the entire economy by adding more security, more automation, more efficiency, and a higher quality of life.

On the other hand, given the heterogeneous nature of services and entry into different sectors (e.g. different industrial sectors through Industry 4.0, different utility sectors through smart cities, intelligent public transport and so on) there are also certain challenges [28].

Businesses must re-skill employees to face the future world of the ever-changing ICT. Telecom/ICT is also constantly changing, and now it is driving the changes in society in the future.

It is also possible that certain national and regional regulations may need to be revised in relation to future services [29]. Privacy and competition are continuously important topics. Privacy is increasing in its importance in future services due to the use of different data in different AI/ML algorithms. Given that all such AI/ML algorithms are created by humans, their use of personal data must be transparent. So, this means that "transparency" will be very important in the regulation of future services as well.

There are also challenges in terms of billing and taxation models for new services that are split across multiple domains, because different parties may want to add charges (e.g. local taxation of digital services that are offered by some global service provider). In the same way, the responsibility matrix for new services is also divided into multiple domains. Typically, each such domain will want to lock in the customer, while on the other hand, it is difficult to have a single business entity that is responsible for the whole end-to-end delivery of digital services. The multi-tenant nature of future services may lead to blame-shifting when the service fails (e.g. for V2X services, where car vendor and communication service provider are different business entities). So, it will be a challenge to regulate who is responsible for what in the value chain, especially for future mission critical services. And, it can vary from one critical service to another. For example, factory traffic control has a limited impact on the factory floor and factory workers, while driverless vehicles and drones may have a broader impact (e.g. in dense urban areas).

Another aspect is the jurisdiction regarding customer protection regulations, data control, privacy aspects, as well as AI ethics. It can become more complicated in a distributed cloud environment, which includes distributed application software, hosting hardware and software, and various networks and network operators. A possible example is the case of autonomous vehicles or drones – who to blame when things go wrong. Also, the answer (who to blame) may be different in different countries depending on their culture and legislation.

Finally, it can be said that the biggest challenge for both business and regulation will be to find the right balance and model that allows innovation to continue to develop on the telecom and OTT side in the future, while ensuring that telecom network operators continue to receive sufficient economic return on their investment to build and maintain the next-generation infrastructure needed to support future services.

What about the path to an all-digital future in the 2030s and beyond? Well, as already noted earlier, the "main" goal in the 2020s is universal broadband access for all people living on the globe. As usual, this is not enough, so it is necessary to develop digital skills in order for people to be aware of digital services and use them daily in their lives and work.

Digitization touches/will touch all sectors, including finance, justice systems, local and national administration, and businesses in many different verticals. But that is not enough. All citizens in the future society should be able to use digital services on an equal basis, to realize their basic human rights, as well as to enjoy their "digital life" in "digital twin" worlds.

References

1 Janevski, T. (April 2019). *QoS for Fixed and Mobile Ultra-Broadband*. USA: John Wiley & Sons.

2 Janevski, T. (November 2015). *Internet Technologies for Fixed and Mobile Networks*. USA: Artech House.

3 Janevski, T. (April 2014). *NGN Architectures, Protocols and Services*. UK: John Wiley & Sons.

4 ITU, "The future of cable TV: Trends and implications", 2018.

5 3GPP TS 23.501 V.18.3.0, "System architecture for the 5G System; Stage 2 (Release 18)", September 2023.

6 EBU TR054, "5G for the Distribution of Audiovisual Media Content and Services", Geneva, May 2020.

7 3GPP TS 23.247 V18.3.0, "5G multicast-broadcast services; Stage 2 (Release 18)", September 2023.

8 3GPP TS 26.501 V18.0.0, "5G Media Streaming (5GMS); General description and architecture (Release 18)", 2023.

9 3GPP TR 26.928 V18.0.0, "Extended Reality (XR) in 5G (Release 18)", March 2023.

10 ITU-T Series Y Supplement 68, "Framework for Internet of things ecosystem master plan", May 2021.

11 ECC Report 305, "M2M/IoT Operation via Satellite", February 2020.

12 OneM2M the IoT standard, https://www.onem2m.org/, accessed in September 2023.

13 Amazon, "AWS IoT Core", https://aws.amazon.com/iot-core/, accessed in September 2023.

14 NGMN Alliance, "Verticals URLLC Use Cases and Requirements", February 2020.

15 3GPP TR 28.864 V18.0.1, "Study on enhancement of the management aspects related to NetWork Data Analytics Functions (NWDAF) (Release 18)", June 2023.

16 3GPP TS 29.520 V18.3.0, "5G System; Network Data Analytics Services; Stage 3 (Release 18)", September 2023.

17 ITU-T Rec. Y.2623, "Requirements and framework of industrial Internet networking based on future packet based network evolution", April 2021.

18 ITU-T Rec. Y.4602, "Data processing and management framework for Internet of things and smart cities and communities", March 2023.

19 ITU-T Rec. Y.4600, "Requirements and capabilities of a digital twin system for smart cities", August 2022.

20 IEEE 802.11p-2010, "Wireless LAN Medium Access Control (MAC) and Physical Layer (PHY) Specifications Amendment 6: Wireless Access in Vehicular Environments", 2010.

21 3GPP TS 23.287 V18.1.0, "Architecture enhancements for 5G System (5GS) to support Vehicle-to-Everything (V2X) services (Release 18)", September 2023.

22 5G Automotive Association (5GAA), https://5gaa.org/, accessed in September 2023.

23 European Commission, “2030 Digital Compass: the European way for the Digital Decade”, 9 March 2021.
24 United Nations, “Digital Economy Report 2021”, August 2021.
25 ITU, “Powering the digital economy: Regulatory approaches to securing consumer privacy, trust and security”, 2018.
26 ITU, Broadband Commission, “21st Century Financing Models for Bridging Broadband Connectivity Gaps”, October 2021.
27 ITU-D, “How broadband, digitization, and ICT regulation impact the global economy”, November 2020.
28 ITU, “Regulatory challenges and opportunities in the new ICT ecosystem”, 2018.
29 ITU, “Digital Regulation Handbook”, 2020.

9

Conclusions

The future is always interesting and intriguing. Why? Well, simply because every sector wants to see growth, and the same is true for the telecom/ICT sector, which is fueled by possible future growth. However, the world of telecommunications/ICT will not change drastically (although it may seem that way at times), but it typically goes through evolution – a continuous evolution of technologies with backward compatibility of newer technologies with earlier ones (where possible). Of course, this is accompanied by the setting of performance targets for various services (especially by telecommunications operators) [1], and the use of Internet technologies such as packet switching technologies [2], which have replaced legacy circuit switching technologies that were used throughout most of the 20th century.

Let's take a look back at new technologies. In this century, the focus was first placed on the noted development of Next Generation Networks (NGNs) [3], created for the transition of legacy voice and TV to an all-IP telecommunications infrastructure such as carrier-grade VoIP and IPTV. However, NGN was designed based on network functions, which provided the basis for the development of future networks, aimed at the use of Software Defined Networking (SDN) and Network Functions Virtualization (NFV). Furthermore, that image has been accompanied by cloud computing technologies (developing further to the edges of the network, to be closer to end users, for lower end-to-end latency) and the Internet of Things (IoT), and more recently Artificial Intelligence (AI) applications and services using Machine Learning (ML) approaches.

If one looks at mobile networks, one can see NGN approaches even in 4G mobile networks (e.g. LTE/LTE Advanced/LTE Advanced Pro), while Future Networks approaches SDN and NFV (via network slicing) together with the critical IoT and anticipated application of AI/ML algorithms can be seen in 5G/5G-Advanced standards, and is expected to further evolve in 6G and beyond mobile networks.

Also, each new technology provides higher bitrates, has more capacity than the previous ones (including fixed and mobile technologies), which increases aggregate speeds as well as individual end-to-end speeds. With more capacity, for example, in the same spectrum (when talking about radio technologies such as mobile), the average price per byte or GB falls, and this is due to the development of the technology over time and its greater efficiency.

So what do we have if we look to the future? Well, we have all-IP networks, including the public Internet, as well as managed IP networks (which can be called private IP networks, private Internets, and so on, depending on the wording used). After that, the capacity is continuously increased. If one compares individual user bitrates of today, let's take as an example 100 Mbit/s in the early 2020s (of course, one can have fixed access of several hundreds, Mbit/s, 1 Gbit/s, but also less of 100 Mbit/s on average, depending on the region – e.g. in underserved or unserved areas) and

Future Fixed and Mobile Broadband Internet, Clouds, and IoT/AI, First Edition. Toni Janevski.

compare this to an average bitrate of about 50 kbit/s about three decades ago (e.g. typical residential Internet access via dial-up modems with 56 kbit/s in the 1990s). It seems that in two decades we have increased the bitrates in telecommunications access networks by about 1000 times. What does that mean?

Well, if we apply a simple calculation, this means that on average access bitrates double every two to three years (this is similar to Moore's Law of computer processing power, which is not independent on data rates because all sent and received bits must be processed). Then, if we apply this, let's name it a "law" (which we have now derived), to say 2030, then by doubling bitrates on average every two years we will reach approximately 1 Gbit/s average access speed (by 2030), at least in developed telecommunications markets (it should be noted that this is only an approximate calculation). Of course, it may be slowed down by the telecommunications business side, because there is less potential for horizontal growth (in the number of subscribers for voice, TV, Internet access, including fixed and mobile) in most countries of the world, so the focus is targeting new verticals, especially from telecom operators although such offers for some verticals may also come from OTT service providers.

So, we calculated that the speed (bitrates) should grow to Gbit/s by 2030 and beyond, if they continue at the same pace as in the past two decades. But why would anyone need such high bitrates?

Well, high bitrates are needed for bandwidth-hungry applications and services, such as very high-resolution video (e.g. 4K, 8K, and above), augmented reality (AR) and Virtual Reality (VR), and so on. And if we go even further, to higher bitrates after 2030, then we can also talk about holograms as telecom or OTT services.

In terms of bitrates, the question that always come up is, who is driving who – are applications causing higher bitrates to be introduced or vice versa. This seems like a typical dilemma – which came first, the chicken or the egg.

However, the ultra-broadband speeds in the access network, supported by increased speeds in transport networks by using SDN, NFV, and network slicing approaches provide the possibility of using the same infrastructure for delivery of existing services (e.g. carrier-grade voice and TV services), OTT services over Internet access service (provided by telecom operators), and new emerging services (e.g. eXtended Reality – XR services, critical IoT services, and so on) in various verticals.

But, is it all about the bitrates in the telecom world? Well, it is not only about the bitrates. For delivery of mission critical services, there are requirements on ultra-reliability and very low latency. However, that is not enough for some services. Why?

Because even latency in range of ms may be not enough for high precision services (e.g. for high precision control of certain automation). What is needed for having high precision?

Well, that is high synchronization, in range of microseconds (that is a thousand times smaller than a millisecond). But, the Internet and IP networks are created in an asynchronous manner, packets are sent when they are created and when there is the possibility to send them (e.g. empty queue ahead of the IP packet). Also, different IP packets can have different lengths (e.g. up to 1500 bytes, considering the limits of payload in Ethernet networks, which are typically used on the access side). So, we want high synchronization on originally asynchronous IP networks which can have variable IP packet lengths?

Yes, that is correct, but only for certain services that will demand such high precision in synchronization. Someone who is a longer time in the telecom world may say – well, we had such networks, called SDH/SONET, which were centrally synchronized with Primary Reference Clocks (PRCs) that provided synchronization in the range of nanoseconds (that is a million times smaller than a millisecond). However, those technologies built in the 1990s were based on 64 kbit/s time slots, as the main bitrate for voice (in one direction) in digital networks at the end of the 20th century. And, the point here is that the wheel is rolling, and we need the same high synchronization for some critical future services.

Of course, such synchronization will be provided (and it is being provided over the industrial Ethernet now) over IP and Ethernet networks, which dominate the networking infrastructures of today and tomorrow. But, can we have strict synchronization with very large IP packets, such as 1000 bytes or 1500 bytes?

Well, that is really hard to do. Why? Because larger IP packets require more time to be transmitted than smaller IP packets. So, in cases where delay and synchronization are very important for a given (critical) service then very small IP packets need to be used (e.g. smaller than 100 bytes). Of course, some future services may look futuristic from today's perspective, but if we go back three decades and talk about YouTube and Facebook and similar services in the future, they would look very futuristic to us. And today, they are part of our everyday lives.

But, not every future service will be a success. It depends upon the business side of the story, who will have an interest to invest in it; is it a legacy telecom operator, is it some service provider, or is it a city or government, or someone else. And an even more important question – are there customers for such service in a given country.

Then, going toward 2030 and beyond, we also have services with complex constraints which refer to time and data varying granularities. Also, there is the coexistence of heterogeneous network infrastructures, which is a present-day scenario and it will also be part of all future scenarios. There will be more private transits (e.g. submarine cables owned by global service providers), integrated space-earth communication (e.g. with LEO satellites, to have the smallest possible round-trip delays in satellite communication), as well as distributed edge clouds.

When horizontal markets are being saturated, the only way forward for telecoms is to open new markets, called new verticals. That also includes smart factories (e.g. Industry 4.0), smart cities, smart transportation, and so on. For many of the new verticals, crucial is high bandwidth, low latency, localization, reliability, precision, and so on. Also, new verticals are expected in the long run to strengthen the telecom sector, while in the short term, gradual and incremental changes can be expected.

But, what will happen with OTT services in the future? Well, all services and applications provided over the public open Internet access (via fixed or mobile access) can be referred to as OTT. They include legacy standardized services such as Web and email (standardized by IETF), and many proprietary applications in many OTT ecosystems (e.g. Google's Play Store, Apple's iStore, and others). The beauty with OTT is in their high speed of innovation. How?

Well, few programmers can develop an application and provide (share) it on some website. And, if the application will gather crowds and will find investors (for investing in equipment, networks, and people) then it can become a global success. However, there are thousands of applications (e.g. online games, online applications and tools, and so on) which have a smaller impact on the cyber community, but they are valuable for certain users. So, OTT is not only about the big global service providers, but also about smaller ones, which are perhaps invisible to many.

What will be the future of OTT services over the Internet? Well, public Internet is network neutral and best effort based. But, the speed matters and delay (i.e. latency) matters a lot. So, higher speeds and lower delays are also required for OTT services. Higher throughputs are ensured with higher bitrates in the access networks (e.g. fiber access networks, mobile Internet access via 5G and future 6G), where average access speeds range from tens and hundreds of Mbit/s to multiple Gbit/s, looking toward 2030 and beyond. Then, with higher speed and lower delays (which can be accomplished with content delivery networks by OTT providers, or with edge computing via caching of certain contents) we can have a variety of services. Which of them are the most emerging ones?

Well, "the number one" service is video; it is the "king" of the Internet and generally all IP networks, at least in the 2020s. It is noted in many places that video contributes over 70% of the total traffic on the Internet and all other IP networks (including here also managed/private IP networks that transfer IPTV). The video services can consume the bandwidth with more video content being offered (generated from many sides, including the end users as video creators) and with higher video resolutions that are available on the Internet (e.g. with access speeds of 50 Mbit/s or more one can watch 4k video streaming over the public Internet access). In the future, the resolutions will increase further at certain time intervals so that we will have 8k, then 16k (in the long run) and so on. Of course, that development may take decades to become a main use scenario.

And, when we talk about video contents, the legacy video content is TV. However, TV appeared almost a century ago, and the first ITU standards on it are more than 70 years old. The first TV services were analog in black and white, then appeared color TV, and after that digital TV; in this century appeared IPTV (as a telecom managed service) and online TV. Today, TV service is delivered across various fixed broadband networks (e.g. passive and active optical networks, xDSL, cable networks), satellite networks and, lately, TV is also becoming a very interesting service offer for 4G and especially 5G mobile operators via the Fixed Wireless Access (FWA), thanks to much larger mobile capacity today (which is needed for many TV channels to be delivered over the mobile radio access network).

The TV penetration statistics over the past decade show that TV is still a promising service, given its vast global market (with nearly 2 billion TV subscriptions globally) and the habits of people (of all cultures) who have grown up with TV, watching their favorite shows and series from childhood onward. And, with higher video resolutions, the traffic share of TV also increases. But what is the future of TV?

Well, one future option is further development to higher resolutions for TV channels, which drives buying new TV sets by end users, new offers by telecom operators who are today also TV providers, and so on. The innovation wheel is rolling even faster than before for the TV services. The future also aims at using AI/ML techniques in different aspects of TV provisioning, from encoding of the content to automatic TV program creation.

What is the main difference between TV content and other video content (e.g. content on video sharing sites, including user-generated content)? Well, the main difference is the quality of the content and its guaranteed delivery when TV is delivered as a specialized service (for example, DVB and IPTV). But in the decade of the 2010s, major OTT providers launched online TV as an OTT service. So we have Amazon Prime Video, as part of the Amazon Prime "bundle" of various services from Amazon (not just online over the Internet). Then there is Google TV (since 2012), YouTube TV (since 2017), Apple TV+ (since 2019), and so on. So, what can we say about TV, about how does it survive today? Well, TV is still "alive" in both the managed telecom world and the Internet OTT space.

But are there services that are more demanding than video streaming or TV, and what are they? Well, they are AR, VR, or generally XR. They may require higher bitrates than video streaming and TV. They can also be used in a variety of services, from entertainment and online games to smart manufacturing (not just over the public Internet in such cases). While VR is the next step from video streaming, AR is more demanding (and a more complex service, targeting different verticals) because it combines a physical world with spatially mixed digital content. So, AR requires very small delays, which in turn requires implementing edge computing to support such delay requirements.

So there are many future services, some as enhancements or evolutions of existing services, and some completely new. Many of them are expected to be used in initially non-ICT sectors called verticals (e.g. industry and transportation), such as massive IoT as well as critical IoT services.

Also, the open internet and digital OTT services will continue to be drivers for the digitization of all segments of society, including private and public sectors.

What about the 2040s and 2050s as the more distant future? Well, we can look into the past and project the development of technology in the coming decades with some accuracy. However, digitization will certainly continue to cover all countries around the world and introduce new digital "layers" in human lives, society, and the environment. In any future, the work of future networks and services is not for machines or elsewhere, but for the people of this planet to have a better and more pleasant "digital" future.

References

1 Janevski, T. (April 2019). *QoS for Fixed and Mobile Ultra-Broadband*. USA: John Wiley & Sons.

2 Janevski, T. (November 2015). *Internet Technologies for Fixed and Mobile Networks*. USA: Artech House.

3 Janevski, T. (April 2014). *NGN Architectures, Protocols and Services*. UK: John Wiley & Sons.

Index

Future Fixed and Mobile Broadband Internet, Clouds, and IoT/AI, First Edition. Toni Janevski.

j

k

y

z

Printed and bound by CPI Group (UK) Ltd, Croydon, CR0 4YY

22/04/2024

14487226-0001